T0281651

BIRD SONG
SECOND EDITION

Bird song is one of the most remarkable and impressive sounds in the natural world, and has inspired not only students of natural history, but also great writers, poets and composers. Extensively updated from the first edition, the main thrust of this book is to suggest that the two main functions of song are attracting a mate and defending territory. It shows how this evolutionary pressure has led to the amazing variety and complexity we see in the songs of different species throughout the world. Writing primarily for students and researchers in animal behaviour, the authors review over 1000 scientific papers and reveal how scientists are beginning to unravel and understand how and why birds communicate with the elaborate vocalizations we call song. Highly illustrated throughout and written in straightforward language, *Bird song* also holds appeal for amateur ornithologists with some knowledge of biology.

CLIVE CATCHPOLE is currently Professor of Animal Behaviour at Royal Holloway, University of London. He has written, broadcast and researched on many aspects of bird ecology and behaviour for more than 30 years and published over 100 books, articles and scientific papers. He has studied birds in many parts of the world and has been a visiting researcher at the Max-Planck-Institute for Ornithology in Germany and the California Academy of Sciences in the USA.

PETER SLATER is Kennedy Professor of Natural History at St Andrews University in Scotland. He is a former Editor of *Animal Behaviour* and of *Advances in the Study of Behavior* and is a Past President of the Association for the Study of Animal Behaviour, which association awarded him its medal in 1999. In 1991 he was elected to Fellowship of the Royal Society of Edinburgh. He is the author of around 150 scientific papers and several books, and has been studying acoustic communication, largely in birds, for 30 years.

BIRD SONG

BIOLOGICAL THEMES AND VARIATIONS

Second Edition

C. K. Catchpole
Royal Holloway, University of London

P. J. B. Slater
University of St Andrews, Scotland

CAMBRIDGE
UNIVERSITY PRESS

CAMBRIDGE
UNIVERSITY PRESS

University Printing House, Cambridge CB2 8BS, United Kingdom

One Liberty Plaza, 20th Floor, New York, NY 10006, USA

477 Williamstown Road, Port Melbourne, VIC 3207, Australia

314-321, 3rd Floor, Plot 3, Splendor Forum, Jasola District Centre, New Delhi - 110025, India

79 Anson Road, #06-04/06, Singapore 079906

Cambridge University Press is part of the University of Cambridge.

It furthers the University's mission by disseminating knowledge in the pursuit of
education, learning and research at the highest international levels of excellence.

www.cambridge.org
Information on this title: www.cambridge.org/9781108469227

First published 2008
First paperback edition 2018

A catalogue record for this publication is available from the British Library

ISBN 978-0-521-87242-3 Hardback
ISBN 978-1-108-46922-7 Paperback

CONTENTS

INTRODUCTION

Bird songs are among the most beautiful, complex sounds produced in the natural world and have inspired some of our greatest poets and composers. Whilst biologists are equally impressed, their curiosity is also aroused. How and why has such an elaborate form of communication developed among birds? Charles Darwin was one of many who struggled to attempt an answer, and the elaborate songs of male birds such as nightingales clearly influenced his thinking as he developed the theory of sexual selection. Since then, biologists from many different disciplines, ranging from molecular biology to ecology, have found bird song to be a fascinating and productive area for research. The scientific study of bird song has made important contributions to such areas as neurobiology, ethology and evolutionary biology. In doing so, it has generated a large and diverse literature, which can be frustrating to those attempting to enter or survey the field. At the moment, the choice is largely between wrestling with the original literature or tackling advanced, multi-author volumes. Although our book is aimed particularly at students of biology, we hope that our colleagues in different branches of biology and psychology will find it a useful introduction. We have also tried to make it accessible to the growing numbers of ornithologists and naturalists who increasingly want to know more about the animals they watch and study.

The literature on bird song has seen rapid growth in the sheer number of publications, at least 1000 more in the 12 years since our first edition. The subject has also had its share of controversies. We have tried to do justice to the field without being too partisan and without burdening the reader with too many citations. We hope that our many friends and colleagues will excuse our omissions, especially where the examples we have chosen might have been from their favourite bird, but were not.

In our efforts to achieve a balanced view and widen our expertise, each chapter has been read by one or more experts in each particular area, and we are extremely grateful to them for their considerable help. We have not always followed their advice, and we cannot guarantee that they will be pleased with the result, but we do know that the book has gained immensely from their efforts. We remain grateful to those who read chapters in the first edition for us: Patrice Adret (2 and 3), Luis Baptista (3), Paul Handford (9), John Krebs (2), Bob Lemon (8), Peter McGregor (6), Anders Møller (5) and Bill Searcy (7) and Haven Wiley (4). We also thank those who have read the revised versions in this edition: Henrik Brumm (1, 4 and the reference list), Diego Gil (5), Michelle Hall (8), Rob Lachlan (9), Stefan Leitner (2), Peter McGregor (6), Katharina Riebel (3) and Bill Searcy (7). In addition, many of our immediate colleagues and respective research groups have answered questions and helped with discussion and clarification of numerous issues. Finally, we thank Karen Johnstone, who carefully redrew the figures for the first edition, Nigel Mann for his delightful vignettes, and Martin Griffiths of Cambridge University Press for seeing the book into print for us.

We are both ethologists, and so it is no coincidence that the book is structured around the four questions that Tinbergen prescribed. We start with causation, continue with development, and then move on to function and finally to evolution. Therefore, although the book has many 'themes and variations', it is very much about the biology of bird song. We hope that it illustrates what the scientific study of bird song has contributed to biology in the past, and what exciting developments it may hold for the future. A comparison of this edition with the last one will illustrate just how rapid progress has been in the past decade: in several places it has even been necessary to introduce new sections to take account of this.

Chapter 1 is an introduction to some basic theory, terminology and methodology. In Chapter 2 we attempt to summarise the dramatic and exciting recent advances made by neurobiologists, perhaps the biggest growth area in the whole field. This chapter centres upon the complex neural circuits concerned with song but also deals with sound production, hearing and perception. Chapter 3 deals with the development of song in the individual. Most birds learn their songs during a sensitive period early in life. The intricate interplay between the genetic and environmental factors involved has made the study of bird song a classic example of

behavioural development. Chapter 4 investigates the problems of sound transmission through the environment and illustrates how different habitats may have shaped the evolution of songs. Song structure can also give important information regarding the location and distance of the singing bird.

Chapter 5 considers the context in which song occurs: who does the singing and when? Is there really a dawn chorus, and why is this the best time to sing? Chapter 6 concentrates upon males and emphasises recognition of species, mates, offspring and territorial neighbours. Territorial defence appears to be an important function of bird song. Chapter 7 switches to the female and emphasises the role of song in sexual selection and female choice. Female attraction appears to be another main function of male song. In Chapter 8 we survey the extraordinary richness and variety of bird song and deal with complexities such as repertoires and duets. It searches for patterns and trends and offers some ideas concerning the evolution of such complexity. Finally, Chapter 9 ventures further along the evolutionary path and considers variation in both time and space. How do songs vary from place to place, do dialects exist, how do songs change as they are transmitted across generations?

To some of these questions this book may provide the answers, but to answer others we will have to wait for another generation of biologists. We hope that they will find investigating the biology of bird songs to be as fascinating, challenging and rewarding as we have done.

THE STUDY OF BIRD SONG

Marsh warbler

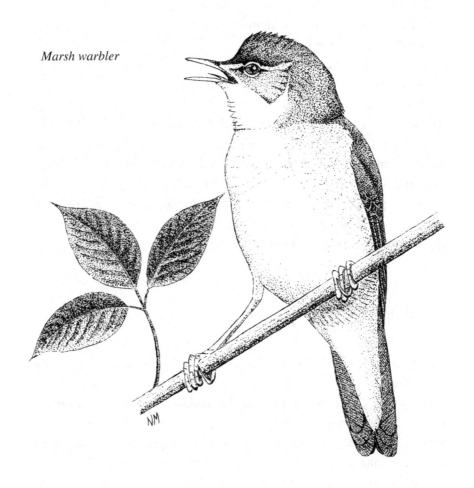

And your bird can sing
John Lennon *Popular song*

1.1 Introduction

This chapter is a brief introduction to the theory, terminology and techniques used in the scientific study of bird songs. Although everyone may assume that they do know what a bird song is, how does it differ from the other sounds that birds make? There are calls, notes, syllables and phrases to consider – and what are repertoires? Before we start using these words, it is just as well to define them and become acquainted with a terminology which can be confusing. Only then can we move on to consider the role of song in the lives of birds and to review the many studies that have attempted to shed some light upon it. Animal communication is a rapidly expanding field, and at this early stage it is also useful to consider some of the recent theoretical background. For example, what is 'communication' and how do we know it has occurred? What is 'information' and who benefits from sending and receiving it? What are 'signals' and why is sound transmission particularly effective? Finally, recording, analysing and experimenting with sounds is a highly technical field, currently being revolutionised by the use of computers at various points. We will also attempt to give the reader just a brief introduction to some of the more important techniques and any recent developments. But first, let us set the current study of bird song in its historical context.

1.2 History

Clearly, from references to it in music and in literature, particularly poetry, people have been interested in bird song for a very long time. But its detailed scientific study is a comparatively recent phenomenon. One reason for this is that making a permanent record of it, as we now do routinely on tapes or discs, was not easy until half a century or so ago. Song can be so rapid and complicated that only with such a permanent record that could be slowed down, repeated and analysed in various ways, is it possible to make a serious study of many aspects of it.

Some interesting work was done before that time. The Hon. Daines Barrington wrote a letter to the President of the Royal Society in 1773 recounting a variety of observations he had made. He established the existence of song learning, for example, because he heard the song of

a wren emanating from a house he was passing and, knowing how difficult such birds were to keep in captivity, knocked on the door out of curiosity, only to discover that the singer was a captive goldfinch. Presumably this bird had been exposed to wren song at some stage and had picked it up. At around the same time, in 1789, the great English parson and naturalist Gilbert White described how birds previously known as willow wrens could be separated by their songs into three separate species. These we now call the willow warbler, the wood warbler and the chiff-chaff. Those with a good ear were also able to detect that birds had repertoires of songs and study the way these were strung together into sequences, as Craig (1943) did with eastern wood pewee song, or that song could vary from place to place, as found by Marler (1952) for the chaffinches singing in different glens in the Scottish highlands.

The depth of such studies was severely limited, not just by lack of the possibility of recording, except latterly on wax drums, but most importantly by the lack of analytical equipment. The real revolution came with the invention of the sound spectrograph, first used to provide a visual representation of song by Thorpe in 1954. Such equipment was not cheap, and therefore its use was somewhat restricted, but it still led to a huge growth in studies of song. Today, equivalent visualisations of song, together with many other forms of analysis, can be carried out using a variety of computer packages at a fraction of the cost. The detailed study of bird song is within the scope and budget of many laboratories and even amateurs: as a result the subject is advancing with great strides. Thanks to these very powerful techniques, there are now few areas of animal behaviour research that have not been illuminated by studies of bird song.

1.3 Some basic theory

1.3.1 Signals and communication

When a bird sings, it produces a sound which serves to communicate with other members of the same species. Because the song is a special structure used solely in communication, we call it a signal. But how do we know whether communication has occurred? It is generally held that if the signal modifies the behaviour of the receiving animal then we can infer that communication has taken place (Slater 1983c). For example, if we play

back a tape recording of a male great tit song to another male, we may cause the second male to respond by approaching the speaker and displaying aggressively. As the song appears to have modified his behaviour, we are entitled to conclude that communication has occurred. This is a somewhat restricted definition of communication, as it relies upon a behavioural response and thus excludes passive signal detection by the receiver. For example, if we repeated the experiment on another great tit and obtained no response, it may be that the great tit had heard the song but decided for some reason not to respond. Such behaviour may often occur, so we must temper our definition with caution and also be sure to carry out a number of experiments before we draw any firm conclusions.

Although it may be relatively easy to demonstrate that communication takes place, it is much more difficult to suggest why it has evolved. Early theories emphasised the benefits that might accrue to both the sender and the receiver (e.g. Smith 1977) and saw communication as a sharing of information between individuals to their mutual advantage. Modern ethologists are much more inclined to view communication as the outcome of conflict rather than cooperation between sender and receiver.

Krebs & Dawkins (1984) examined how different kinds of signals may result from the benefits that accrue to senders and receivers. Sometimes, cooperation rather than conflict is involved, and they suggest that a system which benefits both receiver and sender would give rise to the evolution of relatively quiet, inconspicuous signals. For example, a great tit may give an alarm call to warn its fledglings that a sparrowhawk is approaching. The call should be loud enough to reach the fledglings but not loud enough to reach the hawk and give away the position of the caller. To be able to hear the call, the fledglings should develop sensitive hearing, and so a coevolutionary process will lead to the production of calls that are 'cost-minimizing conspiratorial whispers'. The fascinating story of the evolution of alarm calls, and their possible detection by predators, will be discussed in detail later in the book.

But where the interests of sender and receiver conflict, as in a territorial dispute, there will be a different kind of coevolution. Here, Krebs & Dawkins (1984) suggest that there is an evolutionary arms race between 'manipulation' by the sender and 'sales resistance' by the receiver. The male great tit this time sings as long and loud as he can manage, simply to force his message across. As we will see in later chapters, loudness and repetition are a particular feature of the songs of males when defending

their territorial boundaries. Although sales resistance may occur among rival males, there are even more compelling reasons why females should be wary of male signals. If a listening male makes a mistake, he may just waste energy in a display or a fight, but if a listening female chooses a male of the wrong species, or one of inferior quality, she may pay a severe penalty in reduced breeding success. We will also see in later chapters that there is now considerable evidence that females have been selected for fine discrimination of both quantity and quality of male songs.

So far, we have assumed that the signals transmitted give reliable information from sender to receiver. At this stage, we should mention that the word 'information' is used in two different ways. Technically, information theory considers it to be 'a reduction in uncertainty' about the sender's future behaviour on the part of the receiver. In other words, information is said to have passed from sender to receiver when the sender's behaviour becomes more predictable to the receiver (Halliday 1983). When information is transmitted between birds, it is generally about something quite precise, such as species, sex, identity, likely next action, and so on. There are some grounds for expecting that receivers will be selected to detect unreliable or false information. Zahavi (1979, 1987) has suggested that only signals which are honest indicators of size, strength or motivation should evolve. One reason for this is that many signals are costly to produce, and so it is difficult, for example, for a smaller, weaker animal to cheat or bluff the receiver into accepting it as a larger, stronger rival or mate.

The view that, because of costs incurred by the sender, evolution has generally favoured 'honest advertising' in communication has now become widely accepted. However, Dawkins & Guilford (1991) have pointed out that receivers also pay costs when assessing signals. If the costs of long, detailed assessment are high in relation to the value of the extra information gained, then receivers might settle for cheaper, less reliable signals. If so, the receiver may be open to being bluffed, cheated and manipulated to the sender's advantage. In their review, Krebs & Davies (1993) suggest that such coevolutionary arms races between sender and receiver may have two different end-points. In one, the outcome is the evolution of honest signalling, but in the other we may find unreliable signals have evolved as one animal attempts to manipulate the behaviour of the other. Recent developments in this field have also been reviewed by Maynard Smith & Harper (2003), and by Searcy & Nowicki (2005), who focus particularly on the evolution of reliability and deception, using bird

Table 1.1. A comparison of difference sensory channels of communication

	Acoustic	Visual	Chemical	Tactile
Nocturnal use	Good	Poor	Good	Good
Around objects	Good	Poor	Good	Poor
Range	Long	Medium	Long	Short
Rate of change	Fast	Fast	Slow	Fast
Locatability	Medium	Good	Poor	Good
Energetic cost	Low	Low	Low	Low

Modified from Alcock 1989.

song as a key example. The latter point out that reliability requires there to be a correlation between some aspect of the signal and some attribute of the signaler that the receiver benefits from knowing about. Hence the receiver benefits from assessing the signal rather than ignoring it. Deceit requires not only that the correlation between signal and attribute be broken, but that the signaler benefits from that breakdown. Searcy & Nowicki (2005) present a detailed review of this complex area and discuss how opinions about the prevalence of reliability and deception in animal communication have changed over the years.

The apparent coevolutionary arms race between senders and receivers involves many different aspects of communication, which we will be considering throughout this book. It is important to emphasise that we will not just consider the signal (song) itself, but how it is transmitted through the environment, how it is perceived by receivers and, in particular, how males and females react to both natural and experimental signals.

1.3.2 Why sound?

Sound is only one of several channels of communication that are open to birds, and the advantages and disadvantages of the different channels have been summarised in Table 1.1. In general, birds have rather a poorly developed olfactory system, and so this method is less important than the main channels of sound or vision. This contrasts with mammals, where olfaction is a very important method of communication. Olfaction is rather less important to humans, as their tiny noses indicate, and, like birds, humans rely particularly upon sound and vision. There is no doubt that

visual signalling is of great importance to birds, as indicated by their elaborate plumage and coloration and as seen in their eye-catching visual displays. What then are the particular advantages of sounds, especially when compared to visual signals?

Visual signals have several disadvantages, for example in darkness or poor light. But bad conditions for visual signalling can occur at any time in dense habitats such as forest or reeds and when animals move out of view behind objects. Try looking for a small bird as it moves through the canopy. Now you see it – now you don't! But if it calls or sings you can always hear it, long after it moves out of sight. Sound travels in all directions, it can penetrate 'through' or 'round' objects, and it travels over long distances. Sound is an ideal method for communicating over long distances, and although birds also call softly to each other, their songs are often loud and can carry for several kilometres. How natural selection may have acted to 'shape' song structures for optimal transmission through different habitats is one of many topics we will discuss later in this book.

Other advantages stem mainly from the rapid and transient nature of sound communication. A song or call is only produced when needed, and large amounts of information can be transmitted rapidly and efficiently through the sound channel. There might perhaps be one disadvantage if, as has sometimes been suggested, singing is very costly in terms of energetics. A song has to be 'made' each time it is produced, and some birds sing thousands of songs per day. However, recent evidence suggests that this is not the case and that song is comparatively cheap to produce compared, for example, with the cost to a bird of hopping or flying around its cage (e.g. Oberweger & Goller 2001, Ward et al. 2004). It does therefore seem that the many advantages of sound communication rather easily outweigh its costs. Birds, like humans, are intensely vocal creatures, and communication by sound has come to play a central role in their lives.

1.3.3 Songs, calls and terminology

Bird vocalisations can be divided into songs and calls. The distinction is both traditional and arbitrary, but as these terms are still retained in the literature we must attempt some clarification. There is also a taxonomic reason for the distinction. One particular group, the oscines, were originally separated from the rest of the order Passeriformes, primarily on the number and complexity of their syringeal muscles. As these birds

generally produced more complicated sounds or 'songs', the oscines became known as 'the true songbirds'. But, as we will see later, the differences between oscines and sub-oscines may have more to do with how they learn their songs (and the underlying brain structure) rather than with the actual complexity of their vocalisations.

As this book is largely devoted to the study of songs, it is only fair that we should attempt a definition. In general, 'songs' tend to be long, complex, vocalisations produced by males in the breeding season. Song also appears to occur spontaneously and is often produced in long spells with a characteristic diurnal rhythm. But to these features there are innumerable exceptions. Especially in the tropics, it is common for females to sing as well as males, and both sexes may do so throughout the year even though breeding only occurs during a restricted period (e.g. Langmore 1998). Even in temperate regions, song may occur well before egg-laying and there is also often a bit of a resurgence in the autumn. In the European robin, for example, song may be heard in every month of the year and in the winter it is produced by both males and females singing on separate territories (Lack 1946). However as far as complexity is concerned, it is not easy to generalise and, as we shall see in Chapter 8, species differ enormously in how varied their songs are. There are even songbirds that appear not to 'sing' at all, but the simple 'cheeping' of a male house sparrow on a rooftop may fulfil the same function so that it is, in effect, a very simple song.

What then are calls? 'Calls' tend to be shorter, simpler and produced by both sexes throughout the year. Unlike songs, calls are less spontaneous and usually occur in particular contexts which can be related to specific functions such as flight, threat, alarm and so on. As with the house sparrow example, there are obviously areas of overlap between simple song and complex calls, and plenty of exceptions to the criteria we have presented. But in general, ornithologists and ethologists recognise these distinctions and continue to find them useful. Why the oscines have evolved such complex songs, and a special brain pathway to learn them, is one of the central themes of this book.

Having stated that one of the main characteristics of most songs is their complexity, we have a number of other categories and units to define. Most birds have more than one version of their species song, and some have many. For example, most male chaffinches have more than one version of their song and will repeat one several times in a bout of singing, and then

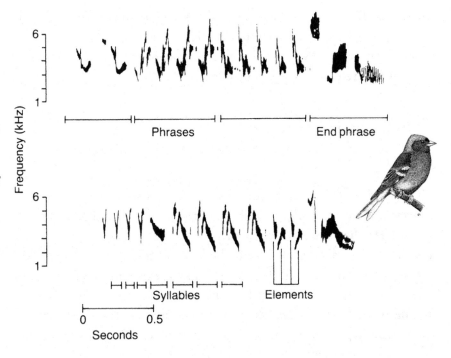

Fig. 1.1. Sonagrams of two different song types from the repertoire of a male chaffinch, also illustrating divisions into phrases, syllables and elements (from Slater & Ince 1979). As explained in Section 1.3.3, sonagrams are plots of frequency against time with the trace being dark where there is energy at that particular point, so giving a visual representation of the pattern of sound.

switch to another. Each version is called a song type, and the male chaffinch is said to have a repertoire of song types (see Fig. 1.1, and the more detailed discussion of repertoires in Chapter 8).

Moving our analysis to a more detailed level, we can also see that each chaffinch song consists of a number of distinct sections. These are called phrases, and each phrase consists of a series of units which occur together in a particular pattern. Sometimes, the units in a phrase are all different, as in the end phrase shown in Fig. 1.1. The units themselves are usually referred to as syllables. Syllables can be very simple or quite complex in their structure. When complex, they are constructed from several of the smallest building blocks of all, called elements or notes (but the latter is usually avoided because of its musical connotations). One definition of an element is simply a continuous line on a sonagram, as illustrated in Fig. 1.1.

Songs, syllables and elements can also be defined by the time intervals which separate them, intersong intervals are the longest, and so on downwards. Because of the great variety of form and structure in songs, individual workers often use and define their own terms, which may be slightly different from those given above. These can only serve as a general guide for this book, for there can be linguistic as well as technical problems with

definitions. For example, in German there are two different words for 'song': 'gesang' means the song of a particular species, whereas 'strophe' means a particular delivery of a song. This last word is now often adopted in English to refer to a single rendition.

1.4 Some basic techniques

1.4.1 Observing

The very idea of 'observing sounds' seems like a contradiction in terms. However, if the main objective is to determine what possible functions a sound has, then this is where to start. Currently, it has become fashionable in many branches of modern biology to construct a hypothesis, perhaps even a model, and then test selected predictions by experiment. Naturally, this book is full of such examples, as experiments have played a leading role in the scientific study of bird sounds. But to formulate an appropriate hypothesis or model, a period of observation should first be undertaken. This should preferably be a thorough field study which relates the singing bird to its habitat, to its other behaviour and to its general life history.

The experimenter may have rather less enthusiasm for this phase, regarding the necessary field work as difficult, dull and somewhat old-fashioned. However, it is vitally important for several reasons. For example, it will provide an accurate source of basic information from which a proper hypothesis can be constructed. It should reveal such essential information as when, where and to whom the bird sings. The first clues as to the probable functions of song invariably come from simple, contextual observations in the field. Does a male sing only in his territory? Does he countersing against rival males? Does he stop singing when he has paired with a female? These are very basic questions, and their answers will help to give initial clues to function and will allow appropriate hypotheses to be formulated. Does the male sing only at dawn? Does he stop when his mate appears? Does he sing more in her fertile period? More precise questions such as these can also be answered by careful observations and may lead to the eventual design of suitable playback experiments to test more detailed functional hypotheses. Nor need the modern field worker feel too old-fashioned. The traditional note-book can be replaced by an electronic one, and a number of software packages will allow a full,

integrated record of singing and associated behaviour patterns to be tapped into a portable computer in the field.

Apart from rather straightforward observation and later quantitative analysis, there are two more important techniques that should also be mentioned. These are correlation studies and the comparative approach. Although these are not simply observations, they rely upon the observational rather than the experimental approach. For example, we may wish to test the prediction that males in a population that sing at a faster rate attract females earlier than their rivals who sing more slowly. To do this, we need to have good observational data on song rate and pairing date, and we can then test the relationship by appropriate correlation analysis.

The comparative approach can also be used to test hypotheses, this time about how differences between species may have arisen during evolution. For example, are observed differences in song complexity related to the different mating systems seen in birds? Again, we need good, reliable observational data as well as a correlation analysis. But comparative studies may involve examining many variables and many species from different taxonomic groups. There are problems not only in selecting variables and controlling for confounding ones, but also in selecting the appropriate level of taxonomic comparison (species, genus, subfamily or family?). Although the most obvious method might be to use species as independent data points, this would bias the results towards genera containing large numbers of closely related species. What we should really do is reconstruct a phylogenetic tree and only make comparisons between data points that appear to be independent in terms of evolutionary events. This is a complex issue: the various ways of making truly independent comparisons were set out in detail by Harvey & Pagel (1991) and discussed more concisely by Krebs & Davies (1993). A more up to date treatment, in what is a very fast moving field, is provided by Felsenstein (2004). We shall describe some applications of the comparative method to song when discussing its evolution in Chapter 9.

It should be clear by now that the observational stage is of great importance for many reasons. It provides a great deal of valuable insight, as well as initial information which can be used for testing hypotheses by context, correlation and the comparative approach. It also prevents the creation and testing of hypotheses which may well be ingenious but may also be inappropriate or even irrelevant to the biology of the bird in its natural environment.

1.4.2 Recording

Having observed and listened to birds singing, the next step is to make a permanent recording of their songs. There are several reasons for doing this, and the first leads back to observation. We may need very accurate answers to the questions we are asking, such as 'how much' does a particular individual sing. Although some songs can be counted or timed, it is best to do this from a permanent record which can be analysed and reanalysed at leisure in the laboratory. As an alternative to the portable computer, a two-track tape recorder can provide parallel records of recorded song and associated behaviour patterns. There are also other reasons for recording songs. We may wish to investigate song structure and compare different males to see whether they share song types or have more song types than their neighbours or males in other populations. Finally, we may wish to conduct playback experiments, and for this purpose we will also need an adequate sample of songs from the birds in our population.

The recording equipment that we use should satisfy the following criteria. It should be of high enough quality to permit later analysis of recording in an acoustics laboratory and yet be light, portable and robust enough for use under field conditions. Fortunately, as we will see in Chapter 2, birds and humans tend to share much the same frequency range and so most commercial equipment is adequate for recording birds. There is now a considerable variety of formats and recorders available, some remarkably small but of very high quality. In the past few decades, there has been a gradual move from open-reel to cassette and, more recently, from analogue to digital recorders. Hard disc recorders are now becoming increasingly popular, especially given the ease with which the data can be downloaded from them onto a computer. Suitable equipment for recording birds is regularly reviewed in the journal *Bioacoustics*, as are the special microphone systems required. Here, the choice is between the long 'gun' type, or a standard microphone mounted in a parabolic reflector. Both of these are designed to meet the problems of recording a bird singing some distance away. The gun microphone is highly directional and tends to cut out noise from either side of the singing bird. Alternative, shorter, 'pistol' microphones also now give pretty good directionality without being quite so cumbersome. The parabola is also directional, but works by collecting a wider sound spread round the singing bird and then reflecting the sound waves onto

a microphone located at the focal point of a fibre-glass dish. The sound is thus amplified considerably. The only problem with a parabola is its size. Low frequencies have long wave-lengths which are not reflected by small objects, and so low, deep sounds, such as the hoots of owls, need a very large dish. Ingeniously, some recent makes can be rolled up for ease of travel.

1.4.3 Analysing

Having obtained our recordings, we may wish to ask very detailed questions about the precise structure of a song. Indeed, this is now a very early step in any serious research into bird song. Yet, as mentioned earlier, up until the 1950s bird song research literally stopped at this point, and there was really no way to examine, measure or compare permanent records of sounds. The sonagraph or sound spectrograph, developed at that time by Bell Telephone, was basically a frequency spectrum analyser which broke sounds down into their constituent frequencies. The main output from a sonagraph is a plot of sound frequency in kilohertz (kHz) against time in seconds. This has become the standard, conventional way to illustrate a song, and it is called a sonagram. As sonagrams will be used throughout the book, at this stage we will present a quick guide to their interpretation, using the sonagrams of different sedge warbler syllables shown in Fig. 1.2. With a little practice, it becomes quite simple to 'read' sonagrams and gain some impression of the original sound from the structure displayed. The main point to remember is that high-pitched sounds (with a higher frequency) appear higher on the y axis.

Perhaps the most common sound people associate with birds is a whistle. A short whistle of constant pitch will appear as a pure, unmodulated frequency trace on the sonagram (a). A whistle which starts at a higher frequency and drops to a lower one is said to be frequency modulated and appears on the sonagram as a slope from left to right (b). If more rapid modulations appear, as in a slow (c) or fast (d) vibrato, they are also easily recognised. But not all bird sounds are pure tones like these. A completely different sound is the harsh noise produced when a wide frequency spectrum is used. A short burst of such 'white noise' sounds like a click (e), and if several occur close together a buzzing sound is produced (f). Frequency modulations sometimes occur in more complex forms, and (g), for example, sounds like a chirp. Another complication is when a sound has higher frequencies occurring as multiples of the first or fundamental frequency.

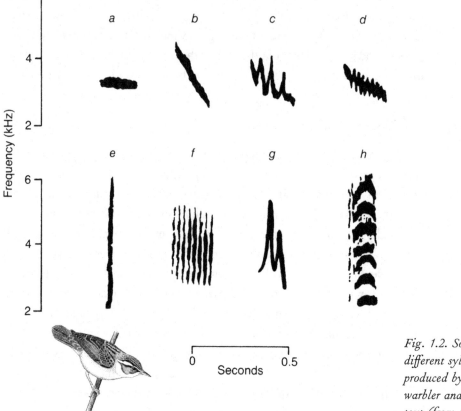

Fig. 1.2. Sonagrams of different syllable types produced by a male sedge warbler and described in the text (from Catchpole 1979).

These are called harmonics and in the example shown here (h) produce rather a gruff, barking sound. With the sonagraph, it became possible to analyse, measure, classify and recognise the different sounds birds make, and this is also true with the numerous computer packages that have superseded it in the past decade or so. It has become possible to discriminate between different species, populations, individuals, song types within individuals, and even different renditions of the same song type from an individual bird. Such visualisation has totally revolutionised the scientific study of bird sounds.

Computer based sound analysis systems are more powerful and faster, and songs and their analyses can now be stored and filed on disc as well as manipulated and even synthesised for experimental purposes. With such methods it is sometimes possible to relegate the still subjective and labour-intensive chore of analysing, recognising and classifying songs to the computer, and quantitative comparisons may even be made between them

using cross-correlational methods (e.g. Tchernichovski *et al.* 2000, Cortopassi & Bradbury 2000) or neural network analysis (e.g. Deecke *et al.* 1999). However, computer methods themselves have a variety of different assumptions and are likely to vary in their applicability depending on the particular task in hand. The human eye and brain are a sophisticated pattern recognition system, and for some tasks using it to scan sonagrams may still be the best means of splitting them into categories (Janik 1999) albeit necessitating care in assessing inter-observer reliability (Jones *et al.* 2001).

1.4.4 Experimenting

In spite of the importance of other methods, there is no doubt that the experiment remains the most powerful and often the final step in testing scientific hypotheses. This applies particularly to the study of bird song, where, as we shall see throughout this book, experiments have been used extensively in studies of causation, development, function and even evolution. Although many different types of experiment have been used, there is one technique above all others which has been highly developed and refined by those who study bird songs – the playback experiment. Playback, as the name suggests, is the technique of playing sounds to animals and observing their response. The sounds are usually recordings of natural signals, such as songs, but synthetic sounds can also be used. Playback of songs may occur in the field most often within the territory of a male bird, but playback can also be used in laboratory experiments, for example to captive females. In the past the sounds usually originated from audiotape, but are now more usually stored in the memory of a computer.

Playback techniques with birds originated in the 1950s, and the most important pioneer was J. Bruce Falls: his classic early experiments feature in later chapters of this book. Playback experiments have many advantages, not the least being that they are effective in both laboratory and field conditions. Playback also isolates the sound stimulus from other confounding variables, such as the presence or behaviour of the bird itself, and gives the experimenter a great deal of control over the experimental variables present in the sound stimulus. By holding all other variables constant, the feature under investigation can be varied and the responses measured in a variety of ways. With playback to territorial males, approach to the speaker is common and so measures such as latency of approach, nearest distance

and time spent within a prescribed radius of the speaker tend to be used. McGregor (1992b) reviewed the different measures that can be used in experiments on males. Females are elusive in the field and so playback to captive females is the main technique used. The number of sexual displays that females make is the main response measured, and Searcy (1992b) discussed this and other techniques used on females.

There are now many different types of playback experiment. Although early ones used only one speaker, two-speaker designs can also be used to present subjects with a choice between stimuli or, for example, with the separate male and female components of a duet (Rogers *et al.* 2004). More modern techniques attempt to treat birds as more than just passive receivers. Weary (1992) has reviewed experiments based upon operant conditioning. With this technique, birds are rewarded when they respond to playback of certain song types, and their powers of discrimination between different song types can be subsequently investigated. But perhaps the most exciting recent development has been with interactive playback (e.g. Dabelsteen & McGregor 1996) and this is discussed further in Chapter 6. In a standard playback experiment, the signal is presented in a rigid predetermined fashion, irrespective of how the receiver responds. This is a very different situation from natural communication, where the signaller may well modify his signal according to the reactions obtained from the receiver. In interactive playback, there is an attempt to modify the signal and so achieve a more natural interaction. This is clearly an impossible experiment to do with conventional playback equipment involving tape recorders. What is needed is equipment with a fast reaction, which can quickly respond to changes from the receiver.

Dabelsteen & Pedersen (1991) were the first to develop such a system, and equivalent ones, based on laptop computers, are now widely used. With them the experimenter can interact with a singing territorial male, for example by switching to match its particular song type (see Peake *et al.* 2000). Dabelsteen (1992) described interactive experiments with both European blackbirds and great tits, where responses are very different when compared to non-interactive playback (see also McGregor *et al.* 1992b). This technique has been applied increasingly over the past decade (e.g. Otter *et al.* 1999, Burt *et al.* 2002b) to give new and more realistic insights into the interactions between singing males.

Various playback designs and the results obtained from such experiments feature heavily in later chapters. But variations in the design of

playback experiments can lead to problems in their validity and interpretation. In a series of papers, Kroodsma (1986, 1989b, 1990b, Kroodsma *et al.* 2001) has criticised past playback designs, mainly for using a restricted range of song stimuli, and has also suggested improvements for future designs. Kroodsma pointed out that too few stimuli can lead to 'pseudoreplication' problems. Technically, pseudoreplication is the use of inferential statistics to test for treatment effects with data from experiments where either treatments are not replicated or replicates are not statistically independent. In song playback experiments, it can occur when only a restricted range of song stimuli are used to test a much more general hypothesis about the functions of song.

Kroodsma's criticisms led to some lively exchanges in the bird-song literature (e.g. Catchpole 1989, Searcy 1989, 1990, Weary & Mountjoy 1992) and also to a special workshop and book on the design of playback experiments (McGregor 1992a). All the participants, the above authors amongst them, were able to agree upon the general principles of good playback design, and their detailed advice was published as the first chapter of the book (McGregor *et al.* 1992a). But there is no overall, general recipe, as there are many different designs and hypotheses to be tested. However, the fine details of how to avoid pseudoreplication are clearly set out, and are essential reading before even contemplating a first experiment. Several other important points emerged from the workshop. Pseudoreplication is a very widespread and general problem in the scientific literature and is not peculiar to playback studies on birds. It is not ubiquitous in playback studies, nor does it impose a constraint upon their usefulness. The heart of the matter lies in clearly specifying the hypothesis to be tested and in using an appropriate number of different song stimuli. A restricted number of stimuli can be used with appropriate statistics, but in this case the hypothesis being tested can only be a restricted one. Finally, designing an elegant experiment is all very well, but there are even more problems in its execution, especially in the field. Yet controlling for all other potentially confounding variables, or holding them constant, must be attempted if our playback experiment is to be a valid one. A list of some of these more important variables is shown in Table 1.2, and the reader may well think of even more.

In spite of the problems of designing and carrying out playback experiments, especially in the field, there can be no doubt of their importance to the scientific study of bird sounds. But, important as playback experiments

Table 1.2. Some potentially confounding variables affecting the execution of playback experiments, particularly in the field

Experimental recordings
Background noise
Distortion
Degradation
Sound level
Sound per unit time
Total amount of sound
Playback equipment

Experimental subjects
Location in territory
Distance from speaker
Stage of breeding cycle
Motivation
Other behaviour
Neighbouring subjects
Predators

Environmental conditions
Time of year
Time of day
Weather
Vegetation
Background noise
Position of speaker
Position of observer

Modified from McGregor *et al.* 1992a.

are, we should not place all our faith in just one of them, no matter how well designed or executed. What about observation, context and correlation studies, and the comparative method? There are also new techniques such as radio tracking (Naguib *et al.* 2001) and the use of microphone arrays to pinpoint the locations of singing birds (e.g. Burt & Vehrencamp 2005). Whilst each technique has its particular advantages and disadvantages, if integrated and used together they permit a whole suite of related predictions to be tested (Catchpole 1992). An integrated approach is surely a much more powerful form of hypothesis testing and it is particularly encouraging when results from the laboratory support those obtained in the field. As we shall see throughout the book, those studying bird song have been ingenious in developing a whole variety of techniques, sometimes in the most unlikely situations.

CHAPTER TWO

PRODUCTION AND PERCEPTION

Marsh wren

A light broke in upon my brain
It was the carol of a bird
Byron *The Prisoner of Chillon*

2.1 Introduction

This book will show that birds produce an astonishing variety of vocal sounds. These range from short, monosyllabic calls, to some of the longest and most complicated sounds known to science, their songs. As we have seen in Chapter 1, any communication system needs a sender, a signal and a receiver. A bird sends its vocal signal by using a special sound producing organ the syrinx. In many cases the signal is an elaborate song, which originates and starts its journey in the brain. How this takes place is one of the most remarkable stories in biology, involving an understanding of some complex neural circuitry within the brain itself. The story also has some unusual twists, such as the growth of new neurons in the adult brain, as well as asymmetry and sexual dimorphism in brain structure. Hearing and perception is at the other end of the communication system, and we will see that the brain also contains highly selective neurons which only respond to specific songs. A more recent discovery is that the listening brain also responds selectively to songs by expression of immediate early genes (IEGs) in various parts of the song system. Neurobiologists are making exciting discoveries about how the brain works, and bird song is a favourite model for their theories and experiments. There is enough material in this particular field for a whole book, but in this chapter we can only hope to outline some of the most important developments whilst pointing the reader to more detailed reviews.

2.2 Sound production

Before we start it is advisable to remind ourselves of some of the special characteristics of sound and some of the common terms we use to describe it. Sound waves are alternating changes in the pressure of the medium, which in the case of bird songs is always air. The volume of the sound is related to the height or amplitude of the sound waves. The waves are measured in microbars, but a more familiar unit of sound volume is the decibel (dB), a logarithmic scale of pressure ratios. The pitch of the sound varies with wavelength, which measures the length in millimetres of one complete wave cycle and the number of cycles per second is known as the frequency. Frequency is measured in thousands of cycles per second or kiloHertz (kHz), and gives an indication of how high or low the sound is pitched.

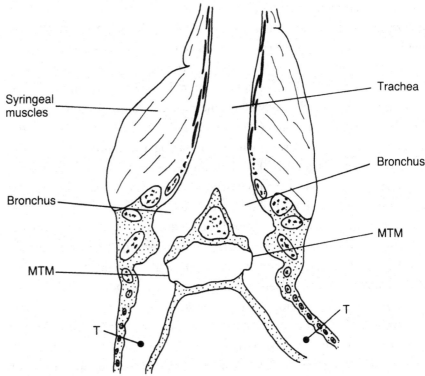

Fig. 2.1. Section through the syrinx of a brown thrasher, showing the positions of the two thermistors (T) used to record airflow separately through each bronchus. Also note the positions of the two separate medial tympaniform membranes (MTM) (from Suthers 1990).

Labels on figure: Trachea, Bronchus, MTM, Syringeal muscles, Bronchus, MTM, T, T

2.2.1 The syrinx

The syrinx (Fig. 2.1) is the very special sound producing organ in birds (Brackenbury 1980, 1982, 1989), and the equivalent of the human voice box or larynx. Like the larynx, the syrinx contains special membranes which vibrate and generate sound waves when air from the lungs is forced over them. The most important appear to be the medial tympaniform membranes (MTMs) situated on the medial walls of the bronchi (see Fig. 2.1). This was established in early work by Ruppell (1933) who dissected out the syrinx from herring gulls and suspended it in an airstream. He found that the MTM vibrated when extended into the bronchial lumen under increased air pressure. This pressure normally builds up in the interclavicular airsac and, as we will see in Chapter 6, if this is punctured a male bird cannot sing. Endoscope observations on a number of non-songbirds have now confirmed that the MTM does indeed vibrate in airflow (Larsen & Goller 1999). However, this is not the case in songbirds. Instead, it seems that connective tissue forming the labia at the end of each bronchus are adducted into the syringeal lumen

and these then vibrate. Indeed, after many years of speculation about their importance, surgical removal of both MTMs has now confirmed that they are not essential for song production (Larsen & Goller 1999, Suthers 2004).

Another complication is the sets of syringeal muscles, which have a number of functions. Songbirds have the most complex syrinx, with five pairs of muscles, which may well be one explanation for the complexity of their songs. Larsen & Goller (2002) have used an endoscope to observe the motion of syringeal muscles whilst stimulating the control areas in the brain of anaesthetised birds. The dorsal muscles play a key role in operating a pneumatic valve at the upper end of each bronchus and this may be important in the timing of phonation. The ventral muscles appear to be more concerned with the control of frequency. Although syringeal muscles are important in controlling such aspects of singing as timing and frequency, exactly how this is done, through movements of the labia and membranes in the lumen, is still poorly understood.

As every human being knows, speaking has to be somehow co-ordinated with breathing for respiratory purposes, and birds face the same problem. This may be particularly acute in birds such as canaries, which sing in long bursts. Calder (1970) suggested that one solution is for the bird to synchronise the two, and sing whilst taking a series of mini-breaths. Singing canaries produce syllables at the rate of up to 27 per second, making a mini-breath strategy unlikely. However, Hartley (1990) used a system of electromyogram (EMG) analysis of abdominal expiratory muscle activity, combined with airsac pressure recordings in singing canaries. In general, she found that each syllable was accompanied by a pulse of air pressure and a burst of EMG activity. Furthermore, pressure fell below zero during each inter-syllable silence, supporting the mini-breath hypothesis. Mini-breath patterns of inspiration and expiration have also been clearly shown in zebra finch song (Wild et al. 1998, Franz & Goller 2002).

The nature of the syringeal mechanisms by which songbirds produce sound has been a major growth area and been the subject of continual investigation and modification in recent years. Yet, in spite of this, the precise details are still far from understood and remain the subject of considerable debate. The following is only a selection of the many reviews published since the first edition of this book: Gaunt & Nowicki (1998), Doupe & Kuhl (1999), Suthers (1999), Suthers et al. (1999), Goller & Larsen (2002), Podos & Nowicki (2004), Suthers (2004).

2.2.2 Two voices

Whereas the human larynx is situated at the top of the trachea, the syrinx is much lower down, at the junction of the two bronchi. This means that there is another important difference, and one which may well help to explain the extraordinary complexity of bird songs. Being located at the bronchial junction means that the syrinx has two potential sound sources, one in each bronchus. The sounds are then mixed when fed into the common trachea and buccal cavity. Anatomical inspection confirms that there are indeed separate tympaniform membranes on the medial walls of each bronchus (see Fig. 2.1). Not only does each side have an identical set of membranes, they are also separately innervated by the descending branches of the left and right XIIth cranial (hypoglossal) nerves.

Greenewalt (1968) initially suggested what has become known as the 'two-voice' theory of song production. He proposed a model which embodied three major features: that the MTMs vibrated to produce sound, that they were functionally independent, and that they were the sole source of any modulations. As we have already seen, there is now evidence which disputes the role of the MTMs, but we will now go on to see whether he was right about lateral independence. The first line of evidence from Greenewalt (1968) relied upon observations from sonagrams. Many complex syllables contain harmonics, but these are multiples of the same fundamental frequency produced from a single source. Greenewalt pointed out that some syllables contain structures which are not related to each other harmonically, and must presumably originate from separate sound sources.

This observation was soon supported by experimental evidence from Nottebohm (1971) who sectioned the branches of the hypoglossal nerves supplying the left or right sides of the syrinx. In both canaries and chaffinches he found that sectioning the right side produced hardly any effects, only one or two syllables dropping out from the song. However, sectioning the left side produced dramatic effects, and most of the syllables disappeared from the postoperative song (see also Nottebohm & Nottebohm 1976). A similar asymmetry has been found in other species, except the zebra finch where the right side is dominant (Williams *et al.* 1992, Wild *et al.* 2000). As well as confirming the two voice theory, these experiments presented an intriguing new finding. Although two sound sources are used to generate the final song, one side of the system appears to dominate the other.

Fig. 2.2. Sonagram of a complete grey catbird song. Below are sonagrams and records of airflow obtained separately from the right and left side of the syrinx during the production of the same song (from Suthers 1990).

The two voice theory has since been confirmed in a variety of song-birds, and has been supported by a particularly elegant series of experiments carried out by Suthers (1990, 1999, 2004). Instead of eliminating the contribution of one half of the syrinx, Suthers was able to monitor the activity of each side directly, and relate this to the composite structure of the final sound. For the first time, we can literally see what are the relative contributions from each half of the syrinx as a living bird sings its song.

To do this Suthers (1990) implanted tiny bead thermistors (miniature sensors) in the centre of each bronchial lumen close to the MTM (Fig. 2.1). By synchronising recordings in the syrinx with those from the singing bird, he was able to pinpoint which half contributed which sound to the final song. For example, in the grey catbird the contribution of each half of the syrinx varied from syllable to syllable (see Fig. 2.2). Some were produced only from the right, some only from the left, and others by a

combination of right and left. Catbirds are noted for their elaborate sing-
ing patterns, and this was reflected in the complexity of their syringeal
compositions. Unlike chaffinches and canaries, catbirds and brown
thrashers did not show lateralization, both halves of the syrinx were
equally involved. When both sides of the syrinx contribute simultaneously
to a syllable, both may generate the same sound, or each side generates a
different part, but when the syllable is repeated the same mechanism is
always used. The timing of the contribution from each side was controlled
by opening or closing the syringeal lumen. Each syllable seems to be
represented as a unique combination of stereotyped motor patterns, con-
trolling the muscles of the two halves of the syrinx as well as those of the
respiratory system. Clearly this complex interaction between the two
halves requires great co-ordination from the central nervous system, so
that the respiratory and syringeal muscles act together in a complex dance,
controlled and choreographed by the brain (see review by Suthers 2004).

2.2.3 Modulations in the vocal tract

So far we have dealt with the possible sources of sounds, but there is still a
considerable distance between the syrinx and the point at which the song
escapes into the outside world. In between is the vocal tract, and there has
long been speculation about whether or not it plays an important role in
shaping the final structure of the song (see review by Podos & Nowicki
2004). One of the main features of bird song is its pleasant 'tonal' quality
(Marler 1969, Nowicki & Marler 1988, Nowicki et al. 1992). Tonal quality
is achieved by the production of 'pure' sounds within a restricted fre-
quency range, relatively free from harmonics or overtones (Fig. 2.3).
But is tonal quality the result of the sound source itself, or does the vocal
tract somehow help the bird to sing a more beautiful song?

If the latter is true, then there are two main possibilities. One is that the
vocal tract acts as a simple resonator, rather like the long tube of a wind
instrument. The rigid requirements of coupling the source and resonator
make this extremely unlikely. The more plausible alternative is that, as in
human speech, the vocal tract acts as a complex, variable filter, emphasis-
ing some frequencies and attenuating others, literally tuning the final song.

The fact that the two sides of a bird's syrinx can operate independently
appears to argue against resonance playing an important role. The vocal
tract could not be acting as a coupled resonator and still support two

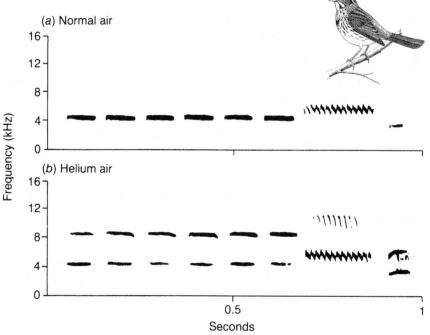

Fig. 2.3. Sonagrams of the song of a song sparrow produced in normal air and then in helium air. Note the pure tones produced in normal air compared to the harmonics produced in helium air (from Nowicki 1987).

harmonically unrelated sounds. Acoustic resonances are determined by both the properties of the resonant cavity, and by sound velocity. The density of the propagating medium also has a direct effect upon resonance frequencies. If the nitrogen in the air is replaced by less dense helium, the velocity of sound increases considerably. The net result of this is that when a human being speaks in a helium atmosphere, the voice sounds high and squeaky. This is not due to a change in the fundamental frequency of the voice, but to an increase in the higher frequency overtones which the vocal tract normally filters out. These are the equivalent to the harmonics often seen in bird songs (Fig. 2.3). It follows that if acoustic resonances play no role in the production of bird song, then the song should be unaffected when a bird sings in a helium atmosphere.

Nowicki (1987) recorded the songs of nine species in both normal and helium atmospheres, and showed quite clearly that they were affected. The most obvious difference, found in species like the song sparrow (Fig. 2.3), was the addition of harmonic overtones to each syllable. In no case was there a shifting of the fundamental frequency, ruling out the possibility that resonances control the vibrations of the sound source. In species that produce more broad band sounds, such as chickadees, the higher

frequencies were emphasized, as in human speech. It seems that, just like humans, birds use their vocal tract as a selective filter to modify the final sound.

The human vocal tract is noted for its ability to vary its resonant properties, mainly by changing the shape of the buccal cavity. Westneat *et al.* (1993) pointed out that birds also appear to adjust their vocal filter in a variety of ways. The vocal tract itself can be lengthened and shortened by stretching or retracting the neck. When birds sing the throat is often expanded and the bill frequently opened and closed. Westneat *et al.* (1993) studied videotapes of synchronised head and beak movements in singing birds. They found clear correlations between the degree of bill opening (gape) and the frequencies of simultaneously produced sounds, as did Hoese *et al.* (2000) when they experimentally manipulated beak movements.

In order to produce a high frequency sound, a bird must open its bill widely, but to produce a low frequency sound it must close the bill down. Therefore to produce a syllable with a wide frequency range, as in a trill, the bill has to open and close the bill over a wide angle. As there must be some limit to how fast this can be done, it follows that there must be a trade-off between bandwidth and repetition rate. The wider the bandwidth of the syllable, the lower is the maximum repetition rate at which it can be produced (Podos 1997). Evidence for the existence of such a constraint comes from experiments on song learning in swamp sparrows. Podos (1996) tutored young males with songs whose trill rates had been artificially increased. The young birds were unable to learn them in this form, and instead often left gaps in the trills. Podos (1997) also found evidence of the trade-off from a comparative study of 34 New World sparrows. When he examined the relationship between bandwidth and trill rate for these species he found that syllables with a narrow bandwidth could be produced at any rate. However, syllables with a wide bandwidth were only produced at a slow rate. When plotted as a graph (Fig. 2.4) this gives a triangular distribution with an upper-bound regression line which indicates the performance limit for any given trill rate. Songs close to the line are at the upper limit of performance whereas those further away deviate most from high performance.

But why should males struggle to produce the highest quality songs they can? As we will see in later chapters, male songs are used in signalling to rival territorial males and to attract females for mating. Ballentine *et al.*

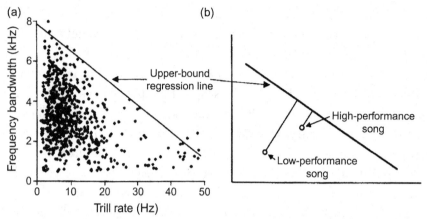

Fig. 2.4. Vocal performance in New World Sparrows studied by Podos (1997). a) The triangular distribution of frequency bandwidths plotted against trill rates from the songs of various species, and b) a schematic representation of how high and low performance songs can be identified (from Searcy & Nowicki 2005).

(2004) used this measure of performance in selecting swamp sparrow songs to play back to females. They found that females consistently displayed more to songs closer to the performance limit, suggesting that females may well use measures of song quality in female choice. As we will see later in Chapter 7, females are extremely selective when choosing a mate, and female choice for song quality may have played an important role in the evolution of male song.

2.3 Hearing

Sound production is only one half of the communication system, the other being hearing, the ability to detect, identify and discriminate incoming sounds. The ear is the main organ concerned and after that there is the auditory pathway leading to the main receptive fields in the brain. As we shall see, within the brain are auditory neurons which respond selectively to a variety of sounds including specific songs. One of the ways in which they respond is in the expression of immediate early genes (IEGs) which may result in eventual changes to neurons and their synapses. As the review by Dooling (2004) points out, it is fair to say that much more is known about sound production than about hearing in birds. Yet we do know that birds can hear as well as humans and that in some aspects, such as the fine discrimination of temporal patterns, their hearing abilities are superior to our own. In Chapters 6 and 7 we will see how they put their remarkable hearing abilities to good use in discriminating between species, sexes, and individuals in their communities, even eavesdropping on their neighbours' songs.

2.3.1 The auditory pathway

The outer ear of birds is not very obvious, due to the absence of an external pinna, and the opening of the external auditory meatus is covered by protective feathers. The main features of the peripheral auditory system in birds have been reviewed by Saunders & Henry (1989) and Saunders *et al.* (2000). The meatus leads to the tympanic membrane, which vibrates due to changes in pressure. In birds these vibrations are transmitted to the inner ear by a single bone, the columella. The columella is held against the fluid-filled cochlea of the inner ear by a complex system of ligaments. The cochlea is a tube with a much folded roof, the tegmentum vasculosum, which covers the basilar membrane. The basilar membrane contains the sensory hair cells, which, as they are deflected, act as transducers to change pressure into nerve impulses. It has long been suspected that different sound frequencies excite different regions of the basilar membrane. In support of this view, the tallest sensory hairs are located at the distal low frequency end of the cochlea, and the shortest at the high frequency end. However the exact mechanism of peripheral frequency analysis is not yet fully understood.

Sensory output passes from the cochlea along the eighth nerve to a collection of cochlear nuclei. From here they progress to a dorsal nucleus in the mesencephalon, and then to the nucleus ovoidalis in the thalamus. Finally, the nucleus ovoidalis projects to the major auditory area of the forebrain called field L (see Fig. 2.8), which itself projects towards the main song control centres. Adjacent to HVC is HVC shelf and adjacent to RA is RA cup and both these appear to play some role in song perception. Recently, two areas near to field L called NCM and CMM (Fig 2.8), have received considerable attention as during playback of song the neurons not only fire, but are also extremely active in producing immediate early genes such as ZENK.

2.3.2 Hearing ranges

There are several methods for determining the hearing ranges of birds. Konishi (1970) developed a neurophysiological method by playing sounds to anaesthetized birds and then recording directly from auditory neurons in the cochlear nuclei. He was able to determine thresholds of response by finding the sound level and frequencies at which the neurons were most sensitive.

(a) Audibility curves

(b) Power spectra

Fig. 2.5. Audibility curves of song sparrows (open circles) and swamp sparrows (filled circles) compared to the power spectra of their songs (from Okanoya & Dooling 1988).

The more common method is to use a behavioural measure, obtained by conditioning birds to peck a key when they hear a sound. Konishi established that the two methods gave very similar threshold curves for the starling. Dooling (2004) has reviewed a number of such studies, and found that most species of birds have rather similar curves. Hearing is generally best in the 1–5 kHz range and most sensitive between 2 and 3 kHz.

A more recent technique is to use the auditory brainstem response (ABR) recorded from electrodes positioned under the skin of the scalp. Results so far show that hearing thresholds in nestling canaries and budgerigars reach adult level at 20–25 days of age (Dooling 2004). These results are particularly interesting as early isolation experiments often use young taken from the nest before this stage, and there has been speculation about whether or not they are capable of hearing and learning adult song in the nest.

There is no evidence from any study that birds can hear in either the ultrasonic (above 20 kHz) or infrasonic (under 50 Hz) range. One of the reasons why birds are so obvious to us is that they tend to communicate primarily within the same frequency range that we use for language and music, although in general their sounds are pitched rather higher.

If birds are transmitting and receiving sounds to and from each other, then we would expect a good match between sound production and hearing capabilities. Okanoya & Dooling (1988) measured both of these capabilities in song and swamp sparrows (Fig. 2.5). In song sparrows, the audibility curve showed peak sensitivity at 2 kHz compared with 4 kHz

in the swamp sparrow. When the power spectra of their songs were analysed, the two distributions showed a similar separation in the same direction. In general, there was a very good match between song production and hearing range within the two species.

The ability to detect a sound is obviously important, but so too is the ability to localize sound. As we will see in later chapters, there are many instances when it is important to know where the sender is positioned. Localization is theoretically possible when there are two receivers to compare the relative position of the sound source. Although birds have two ears, in small animals such as most songbirds, the ears are very close together. This makes comparison of time differences in arrival at the two ears very difficult. The size of the head is also too small to form an effective sound shadow, when sound waves are reflected from the receiving side, again setting up a differential between the two ears. As we will see in Chapter 4, both these methods have been suggested as the basis of an efficient localization mechanism in birds.

Finally, birds do have a rather unusual physical structure, which might also explain their apparent ability to localize sounds – a hole in the head (Lewis & Coles 1980). The hole is really just the joining of the two middle ear passages via the Eustachian tubes and other air passages in the skull. These interaural pathways are capable of transmitting sound arriving at one ear through the skull to the other. This provides a delay mechanism between the two ears, but also sets up a differential pressure which could be equalised by orientating the head towards the sound source. This is the basis of the pressure gradient system for sound localisation in birds proposed by Coles *et al.* (1980) and Hill *et al.* (1980). However, no later work has followed this up and so it remains untested and highly speculative.

2.3.3 Hearing and discrimination

Hearing is only a first step, and so we now turn to consider how auditory information is received and processed, and how it is used in discriminating between the many sounds that birds receive, particularly those from other individuals. In his review, Dooling (2004) points out that in general birds are better than humans at discrimination of small differences in the temporal patterns of sounds, about the same at frequency discrimination and slightly worse at intensity discrimination. In Chapter 6 we will see many examples of species discrimination and how males in particular learn to

recognize territorial neighbours on the basis of their song structures. In Chapter 7 we will also see that females have extremely fine powers of discrimination when it comes to choosing a mate on the basis of his songs.

First, we will briefly consider where sound signals are processed centrally in the brain. Auditory field L (Fig. 2.8) is the main receptive area, although nearby areas such as NCM and CMM are also involved. The neurons contained within field L were investigated in a series of experiments by Leppelsack & Vogt (1976). They studied starlings, and played back calls and songs to anaesthetized males while recording from microelectrodes in area L. They found that some individual neurons acted as rather simple frequency filters, whereas others only responded selectively to certain combinations of parameters. The latter type act as feature detectors which may be able to recognise specific sounds contained within complex songs. Hausberger (1993) completed a more detailed series of experiments on European starlings using simple whistles. No neurons in area L responded to the whole whistle, but many responded selectively to a part of a whistle. Neighbouring neurons responded in similar and often complementary ways. It seems that in auditory field L, there are groups of neurons which work together to encode particular sounds. This could be a first step in a categorisation and recognition process, which then projects into the main song production and learning pathways (see Fig 2.8).

Margoliash (1983, 1986) implanted microelectrodes into the HVC of white-crowned sparrows and played back whole songs to them. He found 'song-selective neurons' which responded much more strongly to their own songs than to the songs of other individuals. Margoliash & Fortune (1992) later investigated this selectivity in the zebra finch. They found highly selective neurons in HVC, which only responded to particular syllables (Fig. 2.6). But song-selective neurons are not only found in the main motor pathway. Doupe & Konishi (1991) also studied the zebra finch, and found song-selective neurons in all three of the nuclei (Area X, LMAN, DLM) which form the song learning pathway. Clearly, the pathways which control song production and learning are closely associated with the perceptual processes involved in recognising and responding to individual songs.

We have already seen that there is some evidence for asymmetry of control in the peripheral part of the song control pathway. Williams *et al.* (1992) investigated central control by unilateral lesions of HVC in the zebra finch. Although the results were not as clear as with syringeal

Fig. 2.6. Activity from two selective neurons in the HVC of a zebra finch when particular syllables are played. Total output from the neurons is shown above and sonagrams of the syllables below (from Margoliash & Fortune 1992).

denervation, there was some evidence that the right side of the brain was more dominant. Cynx *et al.* (1992) used the operant technique on zebra finches to investigate hemispheric differences in discrimination. Auditory input to the right or left forebrain was first disrupted by operation. The birds then attempted to discriminate between their own song and that of a cage mate. In this task, right-side lesioned birds did better than left-side lesioned birds. Another experiment tested for discrimination between two versions of the same song, one with a slightly altered syllable. In this task the left-side lesioned birds did better. It seems that, in the zebra finch, although there is some evidence for central asymmetry of control and perception, the direction of cerebral dominance may vary from task to task.

There is now a great deal of behavioural evidence that birds are capable of very fine discrimination in the recognition of their songs. Chapter 6 will deal in considerable detail with how birds recognise the songs of their own species, as well as those of other individuals such as their neighbours. Chapter 7 will then go on to show that females also show fine discrimination between the songs of individual males. Meanwhile, in the laboratory, operant conditioning techniques have been used to investigate particular aspects of song perception and discrimination. Weary & Krebs (1992) trained male great tits to discriminate between different songs from two individuals. They then tested the males on quite unfamiliar songs from the same individuals, and they were still able to discriminate. This suggests that some birds are capable of categorical discrimination, based upon individual vocal signatures, not just the learned structure of particular songs. However, Beecher *et al.* (1994a) carried out similar experiments on song sparrows, but found no evidence in this species for a general voice trait or signature carried within the song types of each individual.

There are also other ways of testing whether birds do share the human ability to categorise and recognize distinctive patterns of sound. Humans routinely segregate sounds that occur together into separate auditory objects, as in the 'cocktail party effect', where one voice can be picked out from a background noise of others. To test whether birds can also do this Hulse *et al.* (1997) trained starlings, again using operant conditioning techniques. They found that birds were indeed capable of discriminating the song of one species from others presented at the same time. They then went on to show that starlings can even pick out the songs of one individual starling from others presented at the same time (Wisniewski & Hulse 1997). There seems no doubt that birds do indeed possess remarkable abilities to discriminate between different sounds in much the same way humans do.

2.3.4 *Gene expression in the brain*

More recently, investigations into the auditory pathway have started to study song induced changes in gene expression (reviewed by Bolhuis & Eda-Fujiwara 2003, Jarvis 2004, Bolhuis & Gahr 2006). Playback of song can have many effects, but any lasting ones must presumably involve molecular and cellular changes within the brain brought about by gene

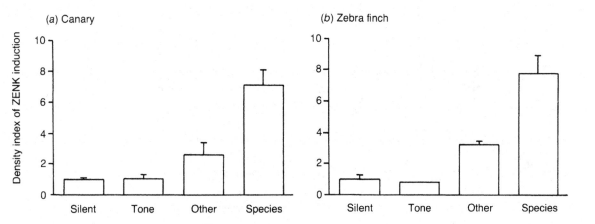

(a) Canary

(b) Zebra finch

Fig. 2.7. The induction of an immediate early gene (ZENK) in the brains of zebra finches and canaries, by playback of their own species song (species). Controls are the other species song (other), another noise (tone) and no playback (silent) (from Mello et al. *1992).*

expression. This initially involves the expression of an immediate early gene (IEG) encoding a transcriptional regulator. The IEG is detected by measuring the increase in specific messenger RNA, and this is done by image analysis of brain sections hybridised to RNA probes. The resulting index of density (Fig. 2.7) gives a measure of IEG induction.

Mello *et al.* (1992) investigated the effects of playback on induction of an IEG called ZENK in the auditory pathway of zebra finches and canaries. Although we now know that several areas are affected in this way, NCM seems to be particularly important. The experiments involved playing songs to birds for 45 minutes, and then sectioning their brains soon afterwards. The results showed that high ZENK mRNA levels occurred in birds exposed to their own species song, when compared to those exposed to the song of the other species or controls (Fig. 2.7). It has also been discovered that the species-specific IEG response is actually learned. When young zebra finches were raised in acoustic isolation, playback of their species song did not induce ZENK expression (Jin & Clayton 1997). Furthermore, when zebra finches were cross-fostered and raised with canaries, the most ZENK expression in the adults was induced by playback of canary song (Jarvis 2004).

ZENK synthesis is also influenced by familiarity with the playback songs, it is reduced after many repetitions but increases with the presentation of new songs (Mello *et al.* 1995). Because of this and other evidence there has been speculation that IEGs may be involved in the learning and memory of songs (Bolhuis & Eda-Fujiwara 2003, Jarvis 2004, Bolhuis & Gahr 2006). Cellular analysis suggests that the level of ZENK induction

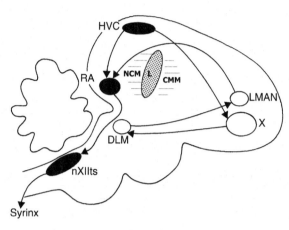

Fig. 2.8. A simplified, schematic diagram of the main song control nuclei and pathways in the brain. Some nuclei (dark) form part of the descending motor pathway, others (light) play a special role in song learning. The primary auditory receptive field L (stippled) is also shown as well as two adjacent areas NCM and CMM.

may reflect the proportion of neurons recruited to express the gene. However, IEGs are also thought to regulate the expression of other genes, and we still do not know precisely how ZENK affects neuronal structure and function. Nevertheless, there is now molecular evidence that even a brief exposure to song may set in motion both transient and permanent changes affecting the perception, learning and eventual production of bird song.

2.4 Singing in the brain

Having dealt with the peripheral areas involved in song production, we now move on to consider the role of the brain itself. Neurobiologists have discovered that there is a complex but discrete brain pathway which controls the production of songs, and another which is more involved in the mechanisms of song learning. However, the two pathways are interconnected and the system is also involved in the hearing and perception of song. Some of the areas which are involved in the auditory pathway have only recently been discovered. Anatomically discrete clusters of neurons in the brain are called nuclei, and axons from these project to other nuclei to form recognizable pathways. There are many separate nuclei that have now been identified as playing some role in the song system. Most of these are in the forebrain, but there are also others in the midbrain and hindbrain (Fig. 2.8).

For many years the avain brain was considered to be very different in form from the mammalian brain, but recent work on homologies and a new system of nomenclature now make it clear that the two are much more similar in both structure and complexity than previously thought (Jarvis *et al.* 2002, Jarvis 2004). There are also homologies between the song system in different groups of birds. It appears that the song system has evolved independently in parrots and hummingbirds as well as in songbirds, and these are also the three groups of birds that learn their songs during development (see Chapter 3). The neural pathways controlling song production and learning is a highly technical field which is rapidly advancing, and there are several more detailed recent reviews which can be used to supplement the brief outline which now follows (e.g. Nottebohm 2000, 2005, Brainard & Doupe 2002, Jarvis 2004, Bolhuis & Gahr 2006).

2.4.1 The motor pathway

The principle motor pathway for song production is also known as the posterior vocal pathway. The main nucleus here is now known as the high vocal centre (HVC) and this projects directly to the robust nucleus of the arcopallium (RA). Nearby is the main auditory field L, and this projects towards both of these motor centres. From the RA there is a projection to the tracheosyringeal portion of the hypoglossal nucleus (nXIIts), from which branches of the hypoglossal nerve control sound production in the syrinx itself.

Nottebohm *et al.* (1976) were the first to demonstrate the importance of this main motor pathway, using a variety of anatomical and behavioural techniques on the canary. Bilateral destruction of HVC stopped normal song production, although a singing posture and some faint sounds were produced. Lesions of HVC, RA or the hypoglossal nerve caused severe deterioration of adult song, compared to control lesions elsewhere. They also found by making lesions on one side only that, like humans, canaries were left-side dominant in their vocal production. From these early studies it seemed clear that the main motor pathway is essential for the production of song.

McCasland (1987) used a quite different technique to investigate the role of the song nuclei, recording from the brain as male northern mockingbirds sang. He was able to confirm that HVC and RA were both active when birds actually sang. Single unit recordings showed that various

motor neurons in HVC were specialised and only fired for particular song syllables. They also found that neurons in RA fired after those in HVC. The importance of HVC and RA was confirmed in experiments on zebra finches by Simpson & Vicario (1990). However, they found that, whereas HVC and RA were needed for song production, they were not essential for the production of simpler, unlearned calls. Instead, there was a separate pathway for these sounds, located lower down in the brainstem.

Zebra finches have become the favourite model in song neurobiology, and one reason is that they sing only one, short, simple song which they repeat over and over again. Thus the song system and the individual neurons within it can be related to a relatively simple song structure. Yu & Margoliash (1996) studied the firing patterns of neurons in singing zebra finches in both HVC and RA. They found complex patterns whereby HVC neurons fired one burst to produce particular syllables and those in RA fired in synchrony but much more. Hahnloser *et al.* (2002) recorded from HVC neurons which projected directly to RA and found the same firing pattern every time the bird sang. A common inter-pretation of these results is that HVC is the dominant, controlling nucleus and functions as the 'conductor' of the 'musicians' in RA. However, the neurons in RA do have some rather special properties. Dave & Margoliash (2000) recorded bursts from RA in zebra finches while they sang and made a remarkable discovery. When the birds fell asleep the same cells still fired and showed the same pattern of activity as when the fully awake bird sang his one simple song. Although the significance of this last finding remains obscure, it does seem that some birds may even sing in their sleep.

The activity of HVC and RA during singing has also been confirmed by molecular studies which detect the synthesis of immediate early genes (IEGs). IEGs produce mRNA followed by later protein synthesis and there is speculation that this causes changes in the neurons and synapses of the song system. For example, Jarvis (2004) suggests that IEGs such as ZENK may help to replace proteins used up during singing and maintain song motor memories in the pathway. Singing certainly produces a dra-matic increase in the production of ZENK, which appears within 5 minutes and peaks after 30 minutes (Jarvis & Nottebohm 1997). The amount of mRNA being produced is stable if the bird sings at a constant rate and the amount is also proportional to the number of songs produced. Future molecular studies using IEGs may eventually help us to understand the

neural changes that underlie both the formation and maintenance of singing in the brain.

2.4.2 *The learning pathway*

While HVC and RA are necessary for song production, the role of the more anterior forebrain nuclei has been more difficult to establish. Indeed, the enigmatic nature of one nucleus to earlier researchers was shown by the naming of it as 'area X.' However, what is now known as the anterior vocal pathway, seems to play an important role in song learning rather than song production. The pathway consists of area X and LMAN, which are connected indirectly through a dorsolateral nucleus in the thalamus (DLM) (see Fig. 2.8). Area X receives projection from HVC, and LMAN projects to RA. Bottjer *et al.* (1984) demonstrated that, whilst lesions in LMAN did not disrupt song production in adult zebra finches, they did impair song learning in juveniles. Similar effects of area X lesions have been demonstrated by Sohrabji *et al.* (1990). It seemed at first that regardless of age, the anterior vocal pathway is needed for new song learning, but not for song production. However, studies on IEG expression by Jarvis & Nottebohm (1997) have shown that ZENK synthesis occurs in LMAN and area X in both juvenile and adult zebra finches during singing. Also, Hessler & Doupe (1999) have reported that neurons do fire in LMAN and area X during singing. It now seems that the anterior song pathway also plays at least some role in song production and its importance has been confirmed in a number of experimental studies (Brainard & Doupe 2002, Jarvis 2004). However, there is another rather puzzling feature about its design.

As seen in Fig. 2.8, the elaborate circuitry means that there are two different pathways from HVC to RA before a song can be produced. One is direct (HVC, RA), but the other is extremely indirect (HVC, area X, DLM, LMAN, RA), and its significance remains something of a mystery. Williams (1989) has drawn attention to this curious design, and has called the indirect pathway the 'recursive loop'. Nottebohm *et al.* (1990) pointed out that it could provide a possible brain mechanism for recognising or comparing song structures stored in different locations. The importance of comparing songs with a stored neural template is an important aspect of models of song learning discussed further in Chapter 3. Such models require that the neural pathway contains centres which would be able to

recognise and respond to specific song structures. As we will see later, several studies have found 'song-selective neurons' in various locations (e.g. Margoliash 1983, 1986). Song-selective neurons have also been found in area X and LMAN in the zebra finch (Doupe & Konishi 1991). This study confirmed that there are indeed two song-selective pathways (direct and recursive) from HVC to RA.

The properties of such neurons in juvenile zebra finches are very different; they are not song-selective. It appears that selectivity, like songs, must be developed through a process of learning. The timing of song-selectivity in the direct and recursive pathways also differs in zebra finches (Brainard & Doupe 2002). The neurons in the direct pathway do not project and innervate until 25 days after hatching. However, the recursive loop seems to be connected as early as 12 days, which is more consistent with sensory learning. It is tempting to speculate that one pathway is tuned to an auditory template, and the other to song production, so that the young bird can really compare the two as it goes on to develop its final song structure.

It seems that the true songbirds, the oscine passerines, have developed a special forebrain system for the production and learning of their elaborate songs. Further evidence for this view comes from comparative studies on non-oscine passerines. These do not learn their songs and neither do they have the special forebrain circuit we have described here. As several reviewers have pointed out (e.g. Brenowitz 1991a, Jarvis 2004), there are only three groups which learn to produce their vocalizations, songbirds, parrots and hummingbirds. It seems that, although they all have a similar forebrain song system, it may well have evolved independently three times, although there are alternative hypotheses (reviewed by Jarvis 2004).

2.4.3 Neurogenesis in the brain

Neurogenesis literally means the birth of neurons, and biologists generally assume that this takes place in the developing embryo or very young animal. Everyone knows that human adults cannot replace or regenerate adult brain cells which have been damaged by injury. Therefore there was considerable surprise amongst neurobiologists when Goldman & Nottebohm (1983) claimed that they had detected new neurons in the brains of adult canaries. They injected canaries with radioactive thymidine, which is only taken up by new cells. After thirty days they found to their surprise that as many as one percent of the neurons in HVC were

labelled. When they examined brains after only one day, there were no labelled neurons in HVC, but many in the ventricle just above. It seemed that the new neurons originated in the ventricle and then somehow migrated to the HVC.

Alvarez-Buylla and Nottebohm (1988) were able to follow the paths of labelled migrating neurones, and observed that they used radial glia to reach their final destination a few millimetres away. The migratory phase lasts for only a few weeks, and many cells may lose their way or simply die. When they get there, about a third of the migrating cells develop into HVC neurons and make synaptic connections (Burd & Nottebohm 1985). The labelled cells are then recruited into fully functioning HVC circuits as confirmed by intracellular recordings (Paton & Nottebohm 1984). Neuron death and replacement in the adult HVC has since been found to occur on a regular basis and a large proportion of neurons are replaced each year (Kirn *et al.* 1991, Nottebohm 2002).

There may well be parallels with development of neural circuitry in young birds. Nordeen & Nordeen (1988) found that in young zebra finches many new HVC neurons grew long projections to RA. This occurred particularly between 25 and 60 days, the sensitive period when young zebra finches are known to learn their songs. A similar pattern has now been reported for adult canaries. Alvarez-Buylla *et al.* (1990) followed the projections of new neurons in the HVC by backfilling them with fluorogold, and traced their axons some 3 mm to RA. Unlike zebra finches, canaries continue to learn new song syllables from year to year, and this takes place in autumn. At this time of the year the rate of neurogenesis in HVC was found to be six times higher than at other times. In the field, wild canaries even change their songs throughout the year, but without significant changes to the volume of HVC or RA (Leitner *et al.* 2001).

In the canary, about half of the neurons in HVC project to RA, about a quarter to area X and the remainder are interneurons (Nottebohm *et al.* 1990). Most of those that project to area X are in place at hatching, and very few are added later (Alvarez-Buylla *et al.* 1988). The converse is true for those that project to RA, where the peak occurs after hatching. Song development occurs later, and by this time HVC is starting to gain some control over RA.

Even though it is tempting to conclude that neurogenesis is associated in a simplistic way with song learning, the situation is far from clear cut. Even female canaries, which do not sing, show some degree of

neurogenesis in HVC. Adult zebra finches only learn one song type yet continue to show some neurogenesis after this (Ward *et al.* 2001), and just the act of singing itself can increase neurogenesis in the adult HVC (Li *et al.* 2000). However, the pattern of neuronal replacement is not uniform throughout the song system. Although neurons in both HVC and area X are continually replaced in adulthood, those in LMAN and RA are not (Kirn *et al.* 1991, Jarvis 2004). Why new neurons are needed to replace the death of old ones in some areas but not others is not known. Clearly, the existence of neurogenesis in the adult brain is no longer in doubt, but its role in the song pathway, particularly in relation to song learning, continues to cause considerable speculation (Nottebohm 2002, Gahr *et al.* 2002, Jarvis 2004, Bolhuis & Gahr 2006).

2.4.4 Sexual dimorphism in the brain

Neurogenesis in the brains of adult vertebrates was one of many surprises researchers on bird brains were to spring upon their fellow biologists. Another was the apparent existence of sexual dimorphism in brain structure. Indeed, one of the earliest findings about the song control centres in the brain, was that they were sexually dimorphic. Nottebohm & Arnold (1976) reported that HVC and RA in male canaries and zebra finches were three to five times as large as in females. This correlates broadly with singing behaviour, as in most species of songbirds only the males sing. In spite of such differences in behaviour, up until that time it was widely believed that, in vertebrates, there were no sex differences in brain structure. But the differences were not just in size; later work (reviewed by Arnold 1990, DeVoogd 1991) went on to demonstrate that not only did males have more neurons in these areas, but that they also had larger cell bodies and longer dendrites.

Sexual dimorphism in the song system develops with age. Kirn & DeVoogd (1989) were able to trace the number of neurons in the developing nuclei of young zebra finches. At hatching there are relatively few neurons, but high levels of neurogenesis, migration and differentiation lead to rapid brain growth. Large sex differences in the size of both HVC and RA then occur, but it is important to realise that this is caused just as much by cells dying off in females as by new cells appearing in males (Fig 2.9).

There are several groups of birds in which females sing as much as the males and this is particularly true of those that duet. Brenowitz & Arnold

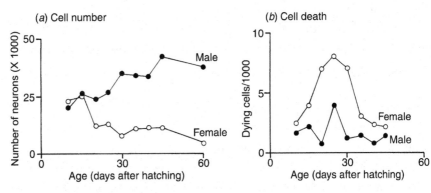

Fig. 2.9. Sexual dimorphism occurs in the song control centres of the zebra finch brain. The developmental changes in the HVC of male and female zebra finches involve not only the production of new neurons but also neuron death (from Kirn & DeVoogd 1989).

(1986) studied two different species of tropical duetting wren species, the bay wren and the rufous-and-white wren. Male and female bay wrens have similar sized repertoires, and there were no significant differences between their HVC and RA sizes. In rufous-and-white wrens, females have smaller repertoires than their males, and their HVC and RA sizes are rather less than in the males. Ball & MacDougall-Shackleton (2001) have reviewed the many studies since, and although there are some cases where the relationship is less clear, in general the hypothesis that sex differences in the song system have co-evolved with sex differences in singing behaviour seems to hold.

But what controls the sex of a developing brain, so that in general males have a larger song control system and tend to sing more? The main culprit appears to be the different pattern of sex hormone secretion and in particular the production of testosterone in males. Young female zebra finches, 'masculinized' by treatment with steroids, developed both male song and a larger song control system (Gurney & Konishi 1980). Such experiments suggested that the natural secretion of sex hormones in young birds plays a key role in the development of the song system. However, the relationship between testosterone and the size of song nuclei is more complex than previously thought, as discussed at length in Chapter 5.

Furthermore, the genetic sex of brain cells may also play an important role in the development of sexual dimorphism in the song system. This view has recently received support from the discovery of a remarkable gynandromorphic zebra finch (Agate *et al.* 2003). This bird arose by an unusual mutation which made it genetically male on one side and genetically female on the other. Male plumage was on the right side of the body which contained a testis, whereas female plumage and an ovary were on

the left side. When the song system was investigated, the right, male side was clearly larger than the left female side, supporting the view that genetic differences also play an important role in the development of sexual dimorphism in the brain.

2.4.5 Sexual selection and female choice

The very existence of sexual dimorphism in the song system suggests that sexual selection and female choice may well be behind the evolution of both song complexity and the underlying song control pathways. If so, then we can predict that, as song complexity increased, so too did the volume or complexity of the relevant nuclei in the song system. We will now review the various studies which have looked for correlations between song complexity (repertoire size) and the volume of key nuclei in the song control pathway, such as HVC and RA.

The most obvious approach is to look at individual species. Having established that males have larger brain nuclei than females, Nottebohm *et al.* (1981) went on to investigate the relationship between repertoire size and the size of HVC and RA in male canaries. In both cases they found a strong positive correlation with repertoire size. There were no correlations between repertoire size, overall brain mass or the size of brain areas not involved in the song control pathway. Leitner & Catchpole (2004) also studied male canaries and found a slightly different result. Although they found no overall relationship between repertoire size and song nuclei, there was a correlation between the repertoire size of 'sexy syllables' and the size of HVC in older males. As we will see in Chapter 7, these sexy syllables have a special, complex structure which females find particularly attractive (Vallet & Kreutzer 1995, Vallet *et al.* 1998) and may be important in female choice.

Airey *et al.* (2000a) studied a population of sedge warblers in the field. As we will see in Chapter 7, the sedge warbler has an elaborate song used in female attraction. The singing males were ringed and recorded, and then monitored as they started the breeding cycle. The males were then caught a second time, anaesthetised and their brains perfused in the field. Later analyses of song and brain structure revealed a strong correlation between repertoire size and the size of HVC. Males in the population that attracted females and paired successfully also had a larger repertoire size than those that remained single.

The zebra finch has a very simple song, and MacDougall-Shackleton *et al.* (1998) initially found no correlations between repertoire size and the size of song nuclei. However Airey & DeVoogd (2000) also studied the zebra finch and did find positive correlations between repertoire size and the volume of HVC and RA. Airey *et al.* (2000b) went on to complete an important breeding study on the heritability of song nuclei in the zebra finch. They found that heritability varied between the different nuclei in the song system. Heritability was highest in the motor pathway (HVC and RA) and much lower in the learning pathway (LMAN and area X). If sexual selection by female choice does drive the evolution of song complexity and the underlying brain nuclei, then we would expect to find higher heritability in the motor pathway rather the learning pathway.

Kirn *et al.* (1989) studied the red-winged blackbird, a species which adds to its song repertoire each year. It is tempting to speculate that any seasonal variation in song nuclei might be related to the tendency in canaries and red-wings to add new songs or syllables to their repertoires. However, although male red-wings had larger song nuclei than females, and these varied with season, there was no correlation between repertoire size and the sizes of song nuclei. Similarly, Leitner *et al.* (2001) investigated a wild population of canaries and found that, although there were seasonal changes in repertoire, they were not reflected in the size of the song nuclei. Brenowitz *et al.* (1991) studied the eastern towhee, a species in which song repertoires are stable after the first year. They also found seasonal fluctuation in song nuclei but again no correlations with repertoire size, as did Bernard *et al.* (1996) in the starling. At present there seems conflicting evidence for an association between repertoire size and HVC. However, Garamszegi & Eens (2004) have reviewed the various studies on different species and also conducted a meta-analysis. Overall they found a significant positive relationship between repertoire size and the size of both HVC and RA.

We now turn to a consideration of comparative studies on repertoire size and song nuclei. Canady *et al.* (1984) looked at variation in repertoire size and song nuclei in two distinct populations of marsh wrens. They found that eastern marsh wrens from New York had much smaller repertoire sizes (50) than western marsh wrens (150) from California. This was reflected in the size of both HVC and RA, which were significantly larger in the western population. Kroodsma & Canady (1985) decided to test how far this was really due to learning more songs, or conversely

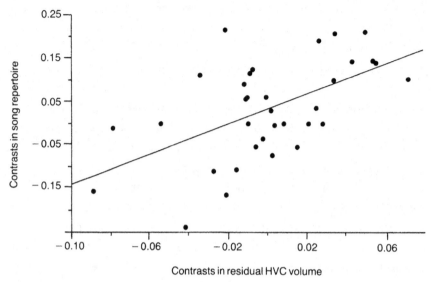

Fig. 2.10. The correlation between repertoire size and brain size in birds. The correlation shown here is between repertoire size and residual volume of HVC based upon 41 species of songbird and 34 independent contrasts (from DeVoogd et al. 1993).

whether it was caused by some basic genetic difference between the two populations. They took samples from both populations into the laboratory, and exposed young birds to 200 song types. The results were very clear, the California wrens learned on average over a hundred songs, but the New York wrens learned only about forty. It seems as though the New York wrens are indeed genetically constrained, and are only capable of learning up to about 50 songs however many they are exposed to.

We have already seen (Airey *et al.* 2000a) that in the sedge warbler a strong positive correlation was found between repertoire size and the size of HVC. An earlier comparative study had also looked for this relationship across closely related warbler species (Szekely *et al.* 1996). Using modern methods to control for variation in brain size and phylogeny, again there was a positive correlation across species between repertoire size and the volume of HVC. A broader comparative study had earlier attempted to examine this relationship across oscine songbirds in general. When DeVoogd *et al.* (1993) compared 41 species a significant correlation emerged between repertoire size and relative HVC volume (Fig. 2.10). These comparative studies suggest that, during the evolution of songbirds, selection for an enhanced song repertoire has consistently acted upon the size of HVC in the song control pathway.

Having established that, in general, the song system is smaller in females, it remains to consider what is its functional significance. Although

sexy | non-sexy

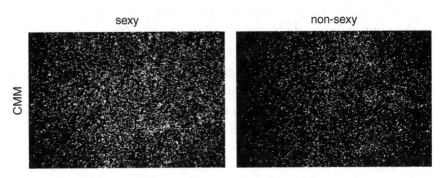

CMM

Fig. 2.11. Dark field photomicrographs of ZENK mRNA expression in the CMM of female canaries exposed to playback of canary songs. Significantly more gene expression occurs in response to songs containing sexy syllables (from Leitner et al. 2005).

females with a smaller system do not usually sing, or sing much less, as we will see in Chapter 7, they receive and respond to male song and may well be the driving force behind the evolution of song complexity by female choice. It follows that a likely role for the smaller female song system is in the hearing, perception and discrimination of male song. Until quite recently there were few studies that focused on the female song system, but this has changed in recent years, particularly with the advent of IEG studies.

Although females of several species have now been investigated, we will focus on canaries as they have now been exposed to a range of techniques. One of the first studies on the female system was by Brenowitz (1991b) who investigated the role of HVC. He found that females with lesions in HVC lost the ability to discriminate between playback of canary and heterospecific songs, a result later confirmed by Del Negro *et al.* (1998). Burt *et al.* (2000) have obtained a similar result from lesions in LMAN.

We have already seen that male canaries with larger repertoires have a larger HVC, and that males also produce special 'sexy syllables' to which females respond much more during playback. Leitner & Catchpole (2002) used this system to investigate the role of HVC and LMAN in song discrimination by females. First, they confirmed that females responded much more to playback of songs containing sexy syllables. When the brains were examined later, they found a positive correlation between female discrimination of sexy versus non-sexy songs and the size of HVC, but not with the size of LMAN. Furthermore, Del Negro *et al.* (2000) have shown that there are neurons in HVC which respond selectively to playback of sexy syllables. As outlined earlier, there are other areas outside the conventional song system which IEG studies have shown

to be activated after playback of song, and Leitner *et al.* (2005) also looked at these in female canaries. They found that ZENK expression in an area called CMM was higher in response to playback of sexy syllables than to playback of non-sexy syllables (Fig. 2.11). It does seem that parts of the female song system in canaries are used in the discrimination of male song quality. How female songbirds use this ability to select males of superior quality for breeding will be discussed at length in Chapter 7.

HOW SONG DEVELOPS

*Zebra finch male
and juvenile*

We can now manipulate experimentally all the components of the developmental equation. The complexity is awesome, but progress is being made.

Peter Marler *Nature's Music*

3.1 Introduction

As with any other aspect of behaviour, song does not suddenly spring up fully formed when a bird becomes mature. It has a developmental history. The development of bird song has probably been studied in more detail than any other aspect of animal behaviour. It has revealed some fascinating

insights into how nature and nurture interact with each other. Most people who study behaviour now accept that behaviour patterns cannot just be labelled as innate or learned but arise through an intricate interplay between the two. Studies of bird song have done more than any other studies to foster this realisation (Baptista 1996, Marler 2004).

Fig. 3.1. Sonagrams of a typical wild chaffinch song and of the song of a chaffinch reared in isolation (from Slater 1989).

3.2 The basic features of song learning

To describe how song is learnt, it is best to start with a case history and that of the chaffinch is an ideal example. In the wild, chaffinches hatch between May and July, and young males start to sing the following spring when they are setting up their territories for the first time. As Fig. 3.1 shows, their song consists of a trill followed by an end-phrase or terminal flourish. The trill is made up of a series of phrases, in each of which the same syllable is repeated a number of times. The flourish is a series of unrepeated elements which are usually longer and often cover a wider frequency range. The song is precise and detailed: the syllables are very constant in form and they always follow each other in exactly the same sequence. In the wild, the exact form of the song varies from place to place (Marler 1952), and this raised the possibility that its form might depend to some extent on copying from other individuals that the young bird has heard.

The chaffinch was in fact the first species in which song learning was studied in any detail, thanks to the work of Thorpe (1958), who hand-raised birds in captivity and examined how their experience affected the songs that they sang. Some of the young birds studied by Thorpe, and by Nottebohm (1968), were not exposed to any adult song. When they began to sing themselves their songs were very rudimentary (see Fig. 3.1). They were of about the right length (1.5–2.5 seconds) and in the same frequency range as normal song (1.5–7 kHz). They were also split up into discrete elements or syllables. But they lacked the detailed structure of the songs of wild birds. The syllables were very simple and tended to drift in form from one to the next throughout the song rather than being split into clear phrases. In many cases there was no obvious flourish at the end.

These results showed that hearing adult song was essential if young birds were themselves to develop normal songs. Other young ones were played recordings of adult songs at various stages in the first year of their lives, and subsequently they sang normal songs which were clearly based on those they had heard. Young birds reared in groups but without adults developed songs that were simpler than normal and also clearly similar to those of their group-mates. From results such as these, it was clear that the young birds were learning from what they heard. But it was not that they would copy just anything: Thorpe tried tutoring them with the songs of other birds, and even with a tune played on a tin whistle. The only success that he had was with tree pipit song. This is similar to the song of the chaffinch in many respects and the young chaffinches that heard it later produced songs that were obviously based on it.

Chaffinches do not start to sing until February or March the year after hatching when male sex hormones begin to circulate and bring them into breeding condition. A remarkable feature of Thorpe's results is that young birds would develop normal song even though the adult song that they had been exposed to was only heard several months earlier. They appeared to memorise the song and later match their output to it. This finding was confirmed by Slater & Ince (1982), who found birds copied songs heard both as fledglings and as young adults the following spring. These are the two stages when birds may learn in their first year, as adults do not sing between early August and late January (Bezzel 1988). Work by Thielcke & Krome (1989, 1991) suggests that young birds are also insensitive to learning during the autumn and early winter when adults are silent.

If birds were tutored after their first year, when they were already in full song themselves, Thorpe found that they would not modify the songs they

sang, even if these were very poorly developed. Their sensitivity, there-
fore, seemed to stop once they were in full song. Nottebohm (1969a)
provided an indication of why this may be from a study of one young
male which he castrated so that it did not start to sing at the normal time.
Then, when it was two years old, he gave it the male sex hormone
testosterone so that it started to sing. At the same time he exposed it to
adult songs and he found that it incorporated these songs into its reper-
toire, despite the fact that it was two years old. This suggests that the end
of sensitivity for song learning occurs either when the bird comes into full
song for the first time or when it achieves high levels of sex hormone. It is
certainly unusual, though not unprecedented (Nürnberger et al. 1989,
Goodfellow & Slater 1990), for chaffinches in the wild to change their
repertoires once they are developed. However, in the cases where this
has been described, there is no evidence that they incorporate new songs
that they have memorised when more than one year old. It is more likely
that each year they may use a slightly different selection from among
the songs they had memorised earlier, as has been found in nightingales
(e.g. Geberzahn et al. 2002). This general pattern of song learning in
chaffinches, with restriction to the first year of life and at least some of
it before the young birds disperse to take up their territories, means that
adult males often do not share songs with their immediate neighbours (see
Chapter 9).

Although it is difficult to generalise, because the exact pattern of song
learning differs between species, many aspects of these chaffinch results
have been confirmed by work on other birds. It was Peter Marler who
originally discovered local differences in chaffinch song in Scotland, rais-
ing the possibility that learning was an important influence (Marler 1952).
He subsequently went on to study a variety of North American species,
starting with the white-crowned sparrow (Marler & Tamura 1964, Marler
1970). His studies, together with others by Konishi (1964, 1965b, Konishi &
Nottebohm 1969), led to the development of what came to be called the
'auditory template model' of song development (Fig. 3.2). While more
recent studies have made it necessary to modify this in various ways, it
makes a useful starting point.

This model pictures young birds as hatching with a rough or crude
'template' defining the approximate characteristics of their own species
song. During a sensitive phase for memorisation, when they may hear
songs of many other species, only those that match this template are

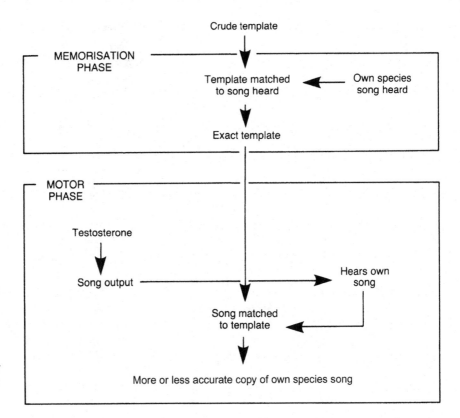

Fig. 3.2. The auditory template model of song development (after Slater 1983d).

memorised, at least in a form that will later be linked to their own singing. The crude template thus becomes an exact template: precise specifications for the song or songs they will sing. This may be achieved well before the bird starts to sing, or memorisation and production may overlap with each other, depending on the species.

When song production commences, as it does in most passerines at rather under one year old, once testosterone begins to circulate in their first spring, the young bird matches its output to the exact template it has formed. This does not happen instantaneously (Fig. 3.3). As song begins to develop, the young male goes through a period of subsong, which is quiet and highly variable, followed by plastic song, during which its output becomes louder and more normal in structure but is still not as stereotyped as full song. At this stage, for example, successive syllables within a chaffinch phrase may not be identical, and the bird may produce songs which are a mixture of two types. Slowly the song 'crystallises' from this into the full song typical of adults of the species. It is during this period, which may last some weeks in birds singing for the first time, that the

Sub-song
9 February

Sub-song
8 March

Plastic song
29 March

Plastic song
29 March

Crystallised song
28 April

Fig. 3.3. Stages of song development in a young chaffinch during his first spring, from sub-song, through plastic song to fully crystallised adult song. (Figure kindly provided by Katharina Riebel.)

young bird is thought to perfect its output in relation to the template it has developed. This change, from a highly varied output to an extremely stereotyped one, is very complex and has not been studied in chaffinches. The most detailed study has been that by Tchernichovski *et al.* (2001) on zebra finches, indicating how a combination of gradual and abrupt changes lead the young bird's song to home in on that of a tutor with which it is housed. In addition to these changes from day to day, remarkably the quality of the song in zebra finches has been found to deteriorate overnight and then to improve once more during the following morning, the copy the bird finally achieves being best where this deterioration was greatest (Derégnaucourt *et al.* 2005). This fits in with the suggestion by Dave & Margoliash (2000) that experience during sleep may be important for vocal development, and that some sort of 'off-line rehearsal' may be going on at night.

The evidence for the auditory template model of song learning came largely from hand-rearing experiments similar to those of Thorpe on chaffinches. These used three different approaches. First was the standard one pioneered by Thorpe of rearing young birds and exposing them to

recorded song at various stages. Second was that of depriving the young birds of adult song and seeing what effect this had on the song they developed. Third came the even more drastic experiment in which young birds were deafened. This could be done very early in life, before the birds had a chance to hear others singing, or later, when they would have memorised song but before they started to produce it. In either case, the effect was more extreme than simply depriving them of a song to copy, the syllable structure and patterning being badly distorted (Konishi 1964, 1965a,b). In some cases deafened birds have been found to produce songs that were little more than a screech (Marler & Sherman 1983). In the chaffinch, Nottebohm (1968) found the songs of birds deafened as juveniles to be virtually structureless, whereas birds deafened as adults in full song continued to sing normal song. The earlier the deafening during the period between these stages the more drastic the effect, showing that the ability to hear was important even before the birds started to sing themselves, as well as enabling them to match their own songs to ones they had heard. Yet feedback no longer seemed to be required once crystallisation had taken place. There may, however, be a long-term deterioration in the quality of the song which these early experiments failed to detect (see Nordeen & Nordeen 1992). This probably results from drift in the structure of song that accumulates in the absence of auditory feedback (Brainard & Doupe 2000). Using a different approach, Leonardo & Konishi (1999) used a computer to distort auditory feedback in intact adult zebra finches, playing back the sounds that they had produced after a brief delay, so that they heard both these and the genuine feedback. This led their songs to become distorted in various ways, again showing that feedback is important even in adulthood.

3.3 Variations

As the song development of more and more species has been studied, it has become clear that the auditory template model needs to be modified in various ways. For example, many birds have been shown to learn and respond to song that they do not produce themselves, so that the restriction is not strictly an 'auditory' one but is probably within the brain. It has also become clear that simply hearing song on a recording may not be enough to achieve learning and, even if they will learn from tapes, birds may respond very differently if given live tutors with which they can

interact. In this section we will examine some of these strong differences that exist between species.

3.3.1 Timing of memorisation

Like the chaffinch, many of the species that have been studied have been found to learn their songs as juveniles, long before they start to sing themselves. Perhaps the most striking example of such a sensitive phase at this stage is that described by Kroodsma (1978a) for the marsh wren. The histogram in Fig. 3.4 shows the number of songs learnt by the birds in this experiment, while the dotted line represents the total number to which they were exposed at different ages. It is clear that sensitivity rises at the start of the period and declines towards the end. Despite this decline, as in the chaffinch, marsh wrens can also be sensitive the following spring, when adults around them start to sing again (Kroodsma & Pickert 1980). Their learning programme is not, however, a rigid one. The more a young bird has learnt before the winter, the less it will learn next spring. Furthermore, even if the amount learnt is controlled, young that hatch late in the year are more prepared to learn next spring than those that hatched early because of an influence of daylength.

Several other species, including song and swamp sparrows (Marler 1987, Marler & Peters 1987), and nightingales (Hultsch & Kopp 1989), show this very early learning. In the song sparrow sensitivity extends to at least five months of age (Nordby *et al.* 2001); in nightingales recent evidence suggests that songs may also be memorised the following spring (Todt & Geberzahn 2003). It is typical to find a rise in sensitivity after hatching and there is, as yet, little evidence for learning in nestlings (see Gwinner & Dorka 1965 and Becker 1978 for exceptions). Perhaps this should not surprise us as the nestling period is only 2–3 weeks long in most song-birds, and the hearing of chicks is not fully developed at hatching (Khayutin 1985). However, some evidence points to accurate copying of sounds heard only a very few times. Song sparrows will learn a song type they have heard as few as 30 times (Peters *et al.* 1992). In nightingales, copying is not precise from 10 or fewer repetitions of a phrase but seems near perfect provided that a phrase is experienced 15 or more times (Hultsch & Todt 1989b, 1992) (Fig. 3.5). Despite this, there is no evidence of learning in birds trained before 13 days (Hultsch & Kopp 1989). White-crowned sparrows exposed to 120 repetitions of a song did not copy it, but two birds reproduced songs

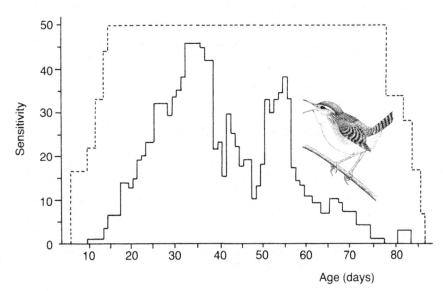

Fig. 3.4. The sensitive period for the marsh wren, a composite of individual graphs for each of nine males (from Kroodsma 1978a).

they had heard 252 times (Petrinovich 1985). While this is considerably more experience than required by a nightingale, it is still not great: an adult can produce that number of songs in less than an hour. Again no evidence of learning before 10 days of age has been produced (Marler 1970, Petrinovich 1985).

The zebra finch has proved a particularly successful species for studies on song learning because it can be easily kept and bred in the laboratory and produces a single, stereotyped and relatively short song. It is an opportunistic breeder from the Australian outback, and young birds mature very quickly, being able to breed when only three months old. Early experiments suggested that they would not learn from tape recordings (Immelmann 1969, Eales 1989), and their learning has largely been studied by moving young males through a series of tutors. They normally become independent from their parents at around 35 days of age, and their greatest sensitivity for song learning is between then and around 65 days old (Eales 1985). A tutor encountered between 35 and 65 days is much more likely to be copied than one the young bird meets earlier or later. However, the sensitivity of the young bird is pretty flexible. If no adult male is around in the 35–65 day period, the young bird will be more likely to copy sounds heard later, or to reproduce sounds heard from the father earlier (Slater *et al.* 1988, Slater & Mann 1991, Jones *et al.* 1996). Young birds isolated from adults but kept in groups will also copy each other

Fig. 3.5. How the number of presentations of a song type affects song learning in nightingales. (Figure kindly provided by Henrike Hultsch.)

(Volman & Khanna 1995), so end up sharing as many elements as if they had learnt from a tutor (Jones *et al.* 1996).

Some bird species can thus learn as juveniles, and others both as juveniles and young adults. There are two other possible patterns: learning as young adults only, and 'open-ended learning' so that the bird is prepared to learn much later in life. Both have been described in some species. Learning as a young adult is typical of indigo buntings, studied by (Payne 1981b). Payne (1985) has also described song in the village indigobird, a case where individuals can change their songs from year to year as adults. Males display in groups, which share a repertoire of different songs: the form of the songs changes from year to year in all these birds. In some cases, males may also change groups as adults and they also then change the songs that they sing to fit in with those of their new neighbours (Fig. 3.6). European starlings provide another example where songs can change from year to year (Eens *et al.* 1992, Mountjoy & Lemon 1995), as do canaries (Nottebohm & Nottebohm 1978, Nottebohm *et al.* 1986). In the last species, males add some new syllables when two years old: they also drop a few of those sung the previous year, although repertoire size does increase with age.

As remarked earlier for chaffinches, it is possible that changes in the songs of adult birds may be because of changes in production rather than because new ones are being memorised. Perhaps young birds learn a very large repertoire of song types and then only sing a small subsection of it each

Junction 1974 Song type

J14

J15

Cowpie 1975

B4

B11

Neighbour at Cowpie 1975

B4

B11

Frequency (kHz)

0 1
Seconds

Fig. 3.6. Song type repertoire change in a male village indigobird that dispersed from the junction neighbourhood to the cowpie neighbourhood. In 1974, male YGYG sang all junction song types (e.g. J14 and J15) but in 1975, after moving to cowpie, he sang all cowpie song types (compare B4 and B11 with the same types as sung by one of his new neighbours (from Payne 1985).

year. This is unlikely to be the case where birds change their songs to match those of neighbours very precisely but it may be true of some other species. Marler & Peters (1981, 1982b) have described how the songs of swamp sparrows, which consist of a very simple trill with only three or four elements in it, develop out of subsong. Until shortly before the full song crystallises, the young bird is singing around 12 different elements, but this number is dramatically reduced as the song homes in on its final form (Fig. 3.7). Just what leads some elements to be 'culled' and others to be retained is not yet known, but social factors are likely to be important (see below). In any case, the result suggests that birds memorise far more elements than they use. It is, therefore, possible that species which change songs in adulthood do so without the need for further memorisation at that stage. Good evidence of this comes from experiments on nightingales using interactive playback in which

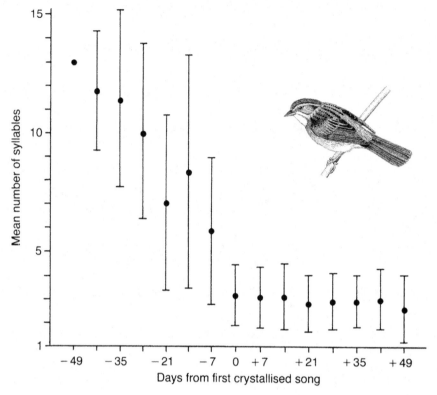

Fig. 3.7. Averaged syllable repertoires of 16 developing male swamp sparrows arranged with reference to the day of song crystallisation. Vertical bars indicate standard deviations (from Marler & Peters 1982b).

the birds were induced to sing songs they had heard earlier but not previously produced (Geberzahn & Hultsch 2003, Geberzahn *et al.* 2002).

3.3.2 Social influences

Many of the experiments described in the last section relied on training young birds with tape-recordings as a very convenient way of presenting them with songs in a highly controlled manner. Live tutors vary a great deal in their behaviour, including their song output, and there could be many reasons why young birds choose to copy from one rather than another. However, there may be cases where young birds will not learn from a tape-recording and, even where they will do so, a live tutor may be a more adequate stimulus. The importance of interaction in song learning is a matter of some controversy, but there is certainly more to song learning than just hearing the appropriate sounds. The issues occur at two different stages, that of memorisation, when the young bird commits the sounds that it hears to memory, and that of production, when some or all of the sounds it has learnt are selected to be produced.

MEMORISATION

Research on a wide variety of species has fuelled the controversy over whether or not interaction is important for memorisation. A good case history here is that of the white-crowned sparrow, originally studied by Marler (1970). He exposed young males to tapes of adult song at various ages and concluded that sensitivity was concentrated largely in the period 10–50 days of age. He also played them tapes of other related species, such as the song sparrow and Harris's sparrow, and found no copying. He concluded, therefore, that white-crowned sparrows are very selective in what they will learn and when they will do so. However, subsequent experiments, notably by Baptista and his co-workers (Baptista & Morton 1988, Baptista & Petrinovich 1984, 1986, Petrinovich & Baptista 1987), have suggested that these results were very much affected by the fact that song was presented on tapes. For example, it was found that white-crowned sparrows would imitate the songs of song sparrows if housed with them so that social interaction was possible. They would even learn the song of a strawberry finch, a quite unrelated Asian species, in these circumstances (Fig. 3.8). The timing of the learning was also broadened where social interaction was possible: while birds would not learn tape-recorded song when over 50 days of age, they would copy the songs of live tutors which they first encountered at that stage (Baptista & Petrinovich 1986).

These results, therefore, paint a very different picture from those in which tapes were used, and suggest that claims about timing and selectivity of tutor choice based on work only with recordings are questionable. However, Nelson (1998) has criticised these experiments. His main point is that the live tutors were presented for longer, so that young birds would have been able to countersing with them as their song production developed. They might have memorised the same amount from tapes but, lacking this countersinging experience, not produced so much of it. In another paper, Nelson (1997) has also criticised the whole body of work indicating that social interaction is important in song memorisation, on the grounds that there are too many confounding variables to allow such a conclusion. He argues that it is difficult to differentiate between late learning from live tutors, such as Baptista & Petrinovich claimed, and earlier learning (which might not require a live tutor) with subsequent selection of those songs that matched neighbours. We shall return to this latter idea below.

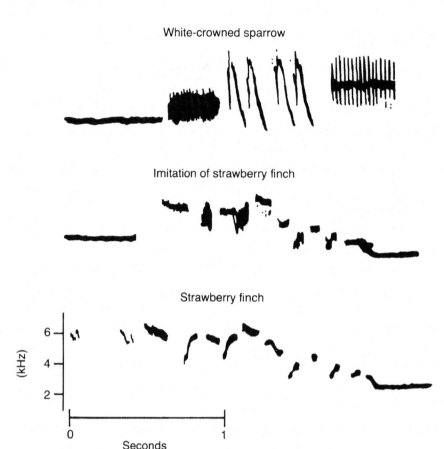

White-crowned sparrow

Imitation of strawberry finch

Strawberry finch

Fig. 3.8. Sonagrams of a white crowned sparrow song, a white-crown's imitation of strawberry finch song and the song of a male strawberry finch (from Baptista & Morton 1981).

Work by Marler (1987, Marler & Peters 1987, 1988b) also suggested that live tutoring may not be important in song and swamp sparrows. For example, swamp sparrows exposed to live tutors and to recordings showed very similar sensitive periods. However, caution is required here. The live tutors were distant from their pupils so that close interaction was not possible. It also remains possible that song learnt from tape recordings might be more easily over-written or superseded by others heard later than those copied from live tutors. Nordby *et al.* (2001) examined this possibility by exposing young male song sparrows to four different tutors when they were 30–90 days old and then, at 140–330 days, exposing them again to four tutors, two of which were not among those they had experienced before. These birds were chosen to have songs distinctly different from the others so that learning from them could easily be identified. All the birds developed several song types and in the majority of them some song was learnt from the two novel tutors. The evidence is therefore in favour of

song memorisation at a much greater age than identified from experiments with recordings. The learning is not, however, 'open ended', as adult males do not alter their song repertoires from year to year (Nordby *et al.* 2002). What they appear to do is copy songs from several tutors during their first year of life, tending particularly to adopt those they share with neighbours with whom they interact (Beecher *et al.* 1994b; Nordby *et al.* 1999); sharing with neighbours is important as it allows matched counter-singing, a topic to which we shall return in Chapter 8.

In zebra finches, early results suggested that young males would not learn from tape recordings (Immelmann 1969, Eales 1989), but that they might do so if they had operant control over the stimulus (Adret 1993). However, subsequent studies by Houx & ten Cate (1999) and Houx *et al.* (2000) found no difference between learning from tapes and from live tutors. Whether or not interaction with live tutors is important in this species may therefore depend on the exact conditions, but it does not appear to be crucial. Nevertheless, even if social interaction is not essential for song learning, there seems no doubt that it directs the attention of young zebra finches to one tutor rather than another, as they learn more from males that look or sound like their fathers (Clayton 1987, Mann *et al.* 1991), from paired males than single ones (Mann & Slater 1994) and from the more aggressive of two males with which they are housed (Clayton 1987, Jones & Slater 1996). In all cases it seems likely that the young bird learns from the individual that attracts its attention most, and with which it is therefore most likely to associate and interact.

PRODUCTION

Moving on to the production side of learning, perhaps the clearest cases are where young birds invent or improvise sounds rather than copying them from other individuals. King & West (1983) studied song development in male cowbirds. The brown-headed cowbird is a North American brood parasite and, as they are not reared by their parents, young males do not have an early opportunity to copy adult song. A particularly interesting finding is that females may influence male song development. There are two subspecies, one with an eastern distribution and the other more southerly, and these differ in song from one another. In their first study, King & West (1983) found that young males kept in the company of females of the other subspecies developed the song of that subspecies rather than their

own. They then went on to examine how females, which do not sing, were able to exert this influence (West & King 1988). It transpired that they have a brief display called 'wing-stroking' in which the wings are moved rapidly to and fro out from the body, and which they produce in response to certain male songs. Males are more likely to repeat songs to which their female companions respond in this way. They are thus trained to produce songs which match the preference of the females. Given the importance of song in many species for the attraction and stimulation of females, it is perhaps to be expected that other examples will come to light in which males modify their songs in response to female preference. The one other case so far is in zebra finches, although the evidence only favours a weak effect (Jones & Slater 1993).

The cowbird results suggest a rather different possibility for the way in which song develops from that put forward in the auditory template model. This has led Marler (1990, 1991) to propose a distinction between 'memory based learning' and 'action based learning'. He sees the former as the mechanism that has traditionally been put forward: the young bird memorises features of adult song and then, when it starts to sing itself, it matches its output to that memory. In action based learning, however, the selection may be taking place at the stage of production. As mentioned earlier, swamp sparrows sing far more elements in plastic song than in full song (Marler & Peters 1982), and song crystallisation involves rejection of elements. Indeed, the swamp sparrow has a comparatively limited number of element forms available to it, and Marler & Pickert (1984) argue that the process of song learning in this species is more a matter of selecting and recombining these into syllables rather than learning them as such. The most radical version of this hypothesis has been put forward by Marler & Nelson (1992), who argue that species-specific song structures in a variety of species may be 'pre-encoded' and that song learning may be primarily a matter of selection from among these rather than memorisation. The swamp sparrow is just one species where songs are built up from a limited number of syllable types suggesting that these may not have to be learnt; Baker & Boylan (1995) describe 129 species-universal song elements in indigo buntings and 122 in lazuli buntings. The idea that these may not need to be copied is an interesting one, but it is most likely to be relevant at the level of the syllable rather than that of the song type, as many species build up a huge variety of song types from a limited array of elements.

Recently, two examples have come to light of songbird species in which memorisation does not appear to be important as isolated birds develop songs as complex or more so than those exposed to tutoring. Kroodsma *et al.* (1997) found repertoire size in an untutored male catbird to be larger than that in other individuals that were played large or small song repertoires. In both field and laboratory, he also found that catbirds that could hear each other shared little. On the basis of this, admittedly small, sample these birds appear to build up their repertoire of sounds entirely by invention (in the terminology proposed by Janik & Slater 2000). The other example is the sedge warbler, studied by Leitner *et al.* (2002). Here isolated birds also developed song repertoires that were larger than in normal wild birds. In this species, adults stop singing once they are paired, so that first year birds returning to Britain from Africa have no experience of their own species song when they arrive and set up territories. What they appear to do is to develop a large number of elements but, in the few days after they arrive, cut down on these to retain mainly those that match their neighbours. This strategy of song development may be advantageous in a migrant that is only present on the breeding grounds for a short period and so must attract a mate as soon as possible. Achieving a match with neighbours in this way may also have an advantage in this species, as it is unusual among songbirds in often occupying different territories in successive years.

In several species where males are known to copy their songs from others, young birds have also been found to possess more songs early in development than they subsequently use. In white-crowned sparrows, males of the coastal subspecies *nuttalli* develop their songs rather earlier in the wild than do captive-reared ones and may set up territories in interaction with adults round about them as early as their first September. At this stage they can sometimes produce up to four song types, but the number later becomes reduced to one or two, and the ones that are retained are those matched to neighbours (DeWolfe *et al.* 1989). Hough *et al.* (2000) found that white-crowns raised in the laboratory could, in their second and later years, be stimulated by playback to re-express songs they had learnt as juveniles and then rejected. A similar picture occurs in the field sparrow: two or more song types are sung initially but, as the young bird settles on his territory, the number is reduced to one, and this is usually matched to that of his most actively singing neighbour (Nelson 1992). In nightingales, males tutored in the laboratory during their first year of life produce only a proportion of the songs they have heard in their

first singing season. However, if challenged during their second singing season with interactive playback using songs they had heard before but not produced, they will in many cases start to sing these. They had clearly been memorised but not incorporated into the repertoire for production (Geberzahn & Hultsch 2003). The action based learning idea does, therefore, have support in several species: rather than birds only memorising those songs that they later sing, many of them may copy or invent a greater variety and then select those for use during interactions with neighbours.

The role of social interaction in the song learning of many species is clearly well established. Taking such ideas a little further, Pepperberg (1985, 1990) has explored the application of 'social modelling theory', used by psychologists to understand how various social inputs affect learning in humans, to the mimicry of human speech by her parrot 'Alex'. She has illustrated how some interactive training regimes may lead sounds to be used in appropriate contexts and with reference to particular objects. She suggests that similar principles may operate where social interaction is important in the learning of sounds by wild birds, from their own as well as from other species. The way in which neighbours come to share song types through action based learning may well involve something akin to such 'social modelling'.

3.3.3 From whom do young birds learn?

The findings discussed in the last section suggest that social factors may be important in the choice young males in the wild make of which adult to copy and of which songs to use in adulthood, but they do not tell us exactly who they normally copy. This is a much more difficult question.

If young males copy in their first spring, this is likely to be from territorial neighbours with whom they actively interact. In line with this, Payne (1981b) found learning in captive indigo buntings to take place from individuals that they could see and engage in supplanting behaviour with at the time when territories would be being set up in the wild. Jenkins (1978) found that young saddlebacks, setting up their territories for the first time, adopted songs typical of that area rather than the one where they hatched. Although there are many other examples of birds singing songs typical of their breeding area (see Chapter 9), the extent to which they learn from neighbours rather than more distant individuals seems to vary

quite considerably. In some species, the similarity between the songs of birds on adjacent territories is no greater than would be expected by chance (Slater & Ince 1982). In other cases, neighbours have more different songs than birds further apart (Wolffgramm 1979, Rich 1981, Borror 1987), possibly because these individuals learn their songs before dispersing to their breeding territories.

If copying does occur earlier than territory establishment, the father would perhaps be among the most likely tutors. Indeed young birds hatching late in the year might have little opportunity to hear adults other than the father singing, as they would not disperse from their natal territory until the song season was ending. However, there is only very scant evidence that young birds do ever learn from their fathers. Field studies on indigo buntings (Payne et al. 1987, 1988a) and on great tits (McGregor & Krebs 1982a) have failed to find a tendency for males to have songs like those of their fathers. But there is evidence of sharing between fathers and their sons in some Darwin's finches (Grant & Grant 1996, Millington & Price 1985), and between parents and offspring in marsh tits (Rost 1987). The latter case is particularly interesting as the adult female sings, and the main time when she does so is just after the young fledge, at the stage when copying takes place. However, Rost's results are based on aviary studies and their relevance to wild birds remains to be established. We shall return to the Darwin's finch case in Section 3.3.5, as there is evidence that the cultural inheritance of song in these birds may be linked to female preferences.

To be sure that young birds learn from their fathers requires a rather controlled situation such as laboratory experiments allow, but the problem is that these conditions are also artificial. Two of the best known cases of learning from the father, in the bullfinch (Nicolai 1959) and the zebra finch (Immelmann 1969), were both based on studies of captive birds. If young male zebra finches are kept with their father until 60 or 70 days old they will produce a perfect copy of his song. But under normal circumstances, in the wild, they would have become independent from him at 35 days of age.

The zebra finch example is a particularly interesting one as young birds may well learn from their fathers in the wild (as Zann 1990 has argued on the basis of semi-captive studies) but, despite a great deal of work, the issue remains unresolved. The fact that independence occurs at 35 days and learning is primarily from 35 to 65 days would argue against it. But young

males prefer tutors who look like the father (Mann *et al.* 1991) and whose songs are similar to the father's (Clayton 1987). Thus, despite being independent, the young may seek out the father and learn from him. Independence is also not a sudden event in the wild, as it is when young are taken from their parents in captivity, but a gradual process during which the parents may drive the young away. Clayton (1987) found that young males given a pair of tutors learn more from the one that is most aggressive to them. Again, this might tie in with learning from the father.

Despite these results, there is a great deal of difficulty in determining exactly which individuals young males learn their song from. There are a few cases where experiments suggest the father as a likely tutor, and a few where they point to territorial neighbours, but the evidence is not particularly strong (Slater & Mann 1990).

3.3.4 How accurately do they learn?

The accuracy of song learning varies quite considerably. One factor in this may be the number of repetitions of a particular phrase that a young bird hears. As mentioned earlier, in nightingales, errors in copying are largely restricted to song phrases heard less than 10 times. Blackbirds may copy phrases heard 50 or fewer times (Thielcke-Poltz & Thielcke 1960). In the zebra finch, song output, beyond a very low minimum, is not a factor in tutor choice and does not seem to affect the accuracy of copying (Böhner 1983, Clayton (1987).

There is no doubt, both from laboratory experiments and from field observations, that the accuracy of song learning varies considerably, both within and between species. Individuals may have complex songs that are identical in detail to each other, indicating extremely accurate copying. But very often songs are found which are similar but not quite the same in various respects: a phrase may be missing, or elements may occur in a different order, or perhaps two different song types have been mixed up. In some species it is rare for two individuals to share closely similar songs, although they may still have many elements or syllables in common (e.g. house finch, Bitterbaum & Baptista 1979). In other species, large groups of perhaps up to 100 individuals (Baker & Cunningham 1985) may share the same song phrase or phrases, indicating very faithful copying.

Accuracy at the level of the song or song phrase may be a different matter from that at the level of the repertoire. Where individuals have

several song phrases, as in the chaffinch (Slater & Ince 1979) or great tit (McGregor & Krebs 1982b), it is normal for different birds in an area to have different combinations, indicating that whole repertoires are not copied from the same tutor. However, this is not always so. In the corn bunting all the individuals in an area may share the same two to three song types (McGregor 1980), and the males in a group of village indigo birds share more than 20 song types with each other (Payne 1985), suggesting that whole repertoires are accurately copied.

A higher level of copying is that of the temporal arrangement in which songs are sung. Nightingales will copy aspects of the sequence from a tape on which a series of songs are always produced in the same order, but will split the song up into groups or 'packages' (Hultsch & Todt 1989a,b) (Fig. 3.9). A package will typically consist of three to four songs which succeeded each other on the tape. They are then sung by the bird in close proximity, though not necessarily in the same order, one often being omitted or the order being reversed from that on the tape. Other songs which were close to this package on the tutor tape may also be learnt but belong to another package so that they do not appear close together in the output of the bird when it starts to sing.

Other aspects of temporal patterning may be modified by learning. Nightingales do not normally repeat song types but cycle through their repertoire, with many different song types between two utterances of the same one. If, however, a young bird is trained with a tape in which songs are repeated three or six times in a row, there is an increased tendency for it to repeat songs in its own output (Hultsch 1991). Chaffinches normally sing their song types in bouts, but there is no evidence that they will copy bout length from a tutor; some birds do, however, show an influence of the tutor on the sequence in which song types are sung (Riebel & Slater 1999a). Even if the tutor tape consists of exactly the same song repeated many times, birds that copy it still introduce the normal amount of variation in the number of syllables per phrase (Slater & Ince 1982), a feature also found in song sparrows (Nowicki et al. 1999).

The daily pattern of singing may also be modified by learning. Blue-winged warblers sing two different song types. One is produced largely at dawn and sung at a high rate, while the other is sung more slowly and later in the day. Training birds with the song types reversed in their speed and the time of day they were produced led young birds also to reverse them in their singing (Kroodsma 1988). A similar finding has been made by Spector

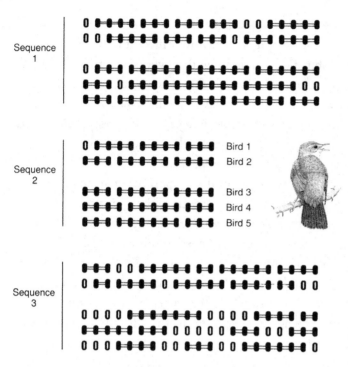

Fig. 3.9. Copying of song by nightingales. Each of five birds was played three different sequences of song types, represented by a row of symbols. Closed symbols indicate the song types that were copied; those that were associated in a package in the output of the young birds are linked by double lines (after Hultsch & Todt 1989b).

et al. (1989) in yellow warblers, with the added complexity that here the dawn song of each bird has several different types: trained with one of these at midday, the birds tended also to use it as their midday song.

3.3.5 What about females?

In many species of songbirds females sing not at all, or so infrequently that song has often been assumed to be a sign of hormonal imbalance rather than of any functional significance. However, female song is certainly more widespread than has generally been assumed (Langmore 1998), and has even been argued recently that it may have been the ancestral norm (Garamszegi *et al.* 2007). In species where females do sing, while their song is frequently described as simpler than that of males, it approaches it in variety and complexity, and is thus doubtless also learnt. Even where females do not normally sing, treatment with testosterone may make them do so and their songs often match those of males they have heard (e.g. Baptista & Morton 1982), indicating that they too have memorised songs even if they do not normally produce them. Their learning also tends to be subject to a sensitive phase (e.g. Riebel 2003a). As judged by calling in

response to song, that for song learning in female white-crowned sparrows ends earlier in females than in males (Nelson *et al.* 1997), and Yamaguchi 2001) also found this to be the case for the learning of songs that are later produced by female cardinals, a species in which both sexes normally sing.

While data on song production learning in female songbirds is sparse, there is rather more on the influence of learning on song preferences, though the results are somewhat mixed and there is a clear need for further work (see review by Riebel 2003b). The best evidence comes from zebra finches, where females have long been known to prefer the songs of their fathers (Miller 1979a, Clayton 1988) and, if cross-fostered, of the foster sub-species (Clayton 1990a). Even though females do not sing, early tutoring is equally influential on the adult song preferences of males and females (Riebel *et al.* 2002), and the preferences females develop are also repeatable (Riebel 2000); in addition, early tutoring is necessary if females are to prefer normal over abnormal song (Lauay *et al.* 2004).

By contrast with the zebra finch results, female Darwin's finches have been found to prefer males that sing differently from their fathers (Grant & Grant 1996). This contrast may not be as stark as it seems, however. A population of these finches includes several distinct song types, whereas wild male zebra finch songs have quite a lot in common and can differ in various ways from domesticated ones (Zann 1993). Thus choice of the father's song over a rather different one in captivity may be equivalent to preference for the local song over an alien one in the wild. In line with this, females of a number of other species have also been found to prefer the song they experienced as juveniles (e.g. Depraz *et al.* 2000, Hernandez & MacDougall-Shackleton 2004).

All these results point to the importance of early learning in the development of female song preferences, but they also point to a literature that is strikingly sparse in relation to that on male song development.

3.4 Mimicry

In the great majority of cases, birds learn only the song of their own species. The fact that they are reared by, and normally imprint upon and develop social relations with, members of that species is certainly one reason for this. But the 'crude template' may also involve constraints that channel young birds in the appropriate direction. For example, hand-reared white-crowned sparrows that are only 2–3 weeks old 'chirp' more

in response to songs of their own species than to those of various other species, indicating that they can discriminate at this early age (Nelson & Marler 1993, see also Whaling *et al.* 1997). There may also be learning rules that exclude the songs of alien species. For example, song sparrows and swamp sparrows, while they do not need social interaction to learn, do not learn each other's songs from tapes. The songs of the two species differ both in organisation and in the form of the syllables of which they are composed. By editing syllables from the two species into various patterns, Marler & Peters (1977) showed that each species has a filter that excludes the song of the other. Swamp sparrows will only learn syllables of the species-specific form, but they will do so even if they are edited into the song sparrow pattern. Song sparrows will learn their own syllables in the swamp sparrow pattern, or swamp sparrows ones on the song sparrow pattern, but changing both the syllables and the pattern makes them unacceptable (see also Marler & Peters 1988a). In white-crowned sparrows, the introductory whistle, which is always present at the start of a song, acts as a simpler filtering cue. While they normally only learn the song of their own species, young males will learn some sounds from other species if they are preceded by this whistle (Soha & Marler 2000).

Copying of songs between species is thus normally avoided and is a comparatively rare event, though with some striking exceptions which we will consider here. An unusual one, to which we shall return in Section 9.7 because of its relevance to evolution, is song learning in the brood parasitic viduine finches. Males learn to sing the song of their host, and females come to prefer it, thus ensuring that like mates with like. But this is one of the few cases where the functional significance of mimicry seems clear.

In a way, the subject of song mimicry is related to the question discussed earlier: from whom do birds learn? But the fact that a bird sings a song phrase from a species other than its own does not necessarily mean that it learnt from that species. Particularly if song learning is accurate, a phrase may have been passed down through many generations since it was originally copied from one species to the other.

Some of the most obvious examples of birds mimicking other species do not involve song at all. The striking capacity of grey parrots and hill mynahs to imitate human speech (as well as many other noises) is a case in point. Neither of these species sings. In hill mynahs, birds in an area have a small repertoire of 12 or so variable calls which they share with

local individuals of the same sex as themselves. However, the repertoires of males and females are different (Bertram 1970). As these calls are highly variable, perhaps the capacity to copy the wide range of patterns and frequencies they involve makes the birds incidentally capable of copying speech. As with the learning of song in many species, social contact may be an important factor here: studies of parrots (Pepperberg 1981) and of starlings (West *et al.* 1983, West & King 1990) copying human speech certainly point in this direction.

Mimicry in these captive examples may just be an emergent property of the birds' capacity to copy a wide range of different sounds and may be limited or avoided in the wild by the context in which the sounds are learnt. But it may occasionally be advantageous to one species to learn the call notes of another. Veerman (1994) reports that regent honeyeaters in Australia mimic the calls of several other honeyeater species, specifically those that are bigger and compete with them for food, suggesting that this may help them to compete. Goodale & Kotagama (2006a) have found greater racket-tailed drongos to imitate the alarm calls of a variety of other species and to use them in appropriate contexts. There may be advantages to doing so, because this species lives in mixed-species flocks. It also imitates the songs and other calls of species that flock with it and this attracts them, something that is likely to be advantageous because the association increases its foraging efficiency (Goodale & Kotagama 2006b).

Mimicry as a component of song learning may obviously also occasionally occur when song 'leaks' across from one species to another, as when a wild white-crowned sparrow learns a Lincoln's sparrow song (Baptista *et al.* 1981), or a song sparrow copies a white-crown (Baptista 1988). But such events are rare, testifying to the effectiveness of the filters that operate to ensure that young birds produce only the appropriate song. That these filters may not be at the level of memorisation is indicated by examples where young birds mimic sounds from other species in subsong which they do not later sing (Baptista & Morton 1988) and where they show territorial responses to other species which they do not mimic (Baptista & Catchpole 1989). However, the really striking examples of mimicry are those where all members of a species habitually incorporate many alien songs in their output. The phenomenon is shown by a variety of unrelated species: the European starling (Hindmarsh 1984), the Northern mockingbird in North America (Howard 1974) and the lyrebirds in

Australia (Robinson 1975, 1991) are perhaps the best known examples. The mimicry can be so precise that males of the mimicked species respond as strongly to it as to their own species song (Brenowitz 1982b, Catchpole & Baptista 1988); in Albert lyrebirds the mimicry of satin bowerbirds has also recently been found to be precise to the level of the local dialect (Putland *et al.* 2006).

One of the most remarkable cases of mimicry, which forms a good case history, is that shown by the European marsh warbler (Dowsett-Lemaire 1979). This species breeds in Europe and migrates down to East Africa. Young birds are thought to learn their song entirely in the first few months of life. They cannot base it on other members of their own species as adult males cease to sing before their chicks hatch. Instead, Dowsett-Lemaire estimates that each young male copies the sounds of many other species, on average 77, including those that they hear in Africa as well as in Europe. Many of the sounds incorporated are call notes (see Fig. 3.10), and the major limitation on what is copied seems to be whether the syrinx of a small bird like a marsh warbler can cope with the sound. The absence of deep sounds is not surprising. The song may well be built up entirely by mimicry but this is, as yet, uncertain: in such a widely travelled species, some sounds included in the song but of unknown origin may well be derived from other species which have not yet been identified.

Several suggestions have been made of the advantage that vocal mimicry may give (Dobkin 1979, Baylis 1982, Hindmarsh 1986). In some unusual cases it may simply result from mistakes in copying: Hindmarsh (1986) argues that this may be the fundamental basis of mimicry in cases where it is widespread too, although this is more an argument of last resort than an idea for which he could put forward good evidence. It has been proposed that birds may preferentially mimic competing species, so excluding them from their territories (e.g Harcus 1977), or predators, so that the area appears to be a dangerous one and other individuals are discouraged from entering it (Rechten 1978). Lyrebirds copy aggressive or predatory species preferentially (Robinson 1974), but there is little other evidence that such processes are in operation. Starlings appear largely to imitate those species commonest in their environment (Hindmarsh 1984), although certain species are represented very much more than their abundance would predict (Hausberger *et al.* 1991). However, no relationship to either competition or predation is obvious from the list of species involved.

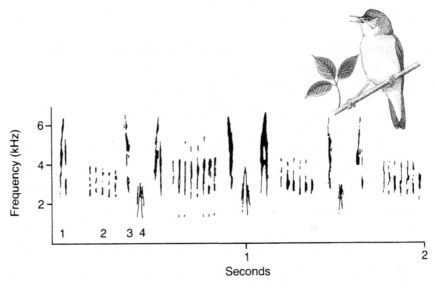

Fig. 3.10. The song of an adult male marsh warbler showing its complex construction. In this example, the elements that are repeated in the sequence shown are copied from four different African species: (1) tawny-flanked prinia;(2); African robinchat; (3) red bishop;(4) brown-headed tchagra (from Dowsett-Lemaire 1979).

A special case of mimicry is that of 'mixed singing', where one species commonly uses elements or songs from one other one in its full song (Helb *et al.* 1985). In most such cases, the two species involved are close relatives (e.g. common nightingale and thrush nightingale (Sorjonen 1986a); Eurasian treecreeper and short-toed treecreeper (Thielcke 1986), and it has been suggested, for example by Dobkin (1979), that the phenomenon is caused by vocal convergence in areas where the two species overlap and show interspecific territoriality. However, Helb *et al.* (1985) argued that most examples are either in isolated individuals or in small areas of overlap between species that are largely allopatric in their distribution. They suggest that misdirected copying is the most likely cause of the phenomenon, and that it is not, therefore, necessary to propose that it is of any particular benefit to the individuals showing it.

In many cases the functional significance of mimicry remains a mystery. It may pay honeyeaters or drongos to imitate the calls of other species with which they are associated, and brood parasites may attract an appropriate mate by imitating the song of their host. Miscopying seems the most likely cause in cases where it is only shown by a minority of birds within a species. However, where males habitually copy many other species, perhaps the most likely explanation is that it is produced by sexual selection as a means whereby they may build up large repertoires. In line with this, there is certainly no doubt that the most striking cases are in species with very elaborate songs.

3.5 Why all this variety?

Song learning varies in many different ways: in its timing, in its accuracy, in the relative importance of invention and imitation, in the variety of tutors that a young bird will accept. We are only just beginning to get some idea of why, but many of the differences undoubtedly relate to those between species in their life histories. Having a song repertoire may be advantageous, as may sharing with neighbours, but these are, at least to some extent, incompatible (Beecher & Brenowitz 2005). If intersexual selection is of prime importance it may favour the learning of very large repertoires, while intrasexual selection would be more likely to lead to the more modest repertoires males use in territorial interactions (see Chapter 8). The extent of sharing with neighbours will depend, amongst other factors, on the timing of song learning in relation to dispersal, on whether birds occupy the same territory from year to year, and on mortality. Many different factors interact.

Contrasts are most striking when they concern close relatives. The marsh warbler and sedge warbler both belong to the genus *Acrocephalus*, both nest in Europe and migrate to Africa, both cease to sing after mating so that yearlings returning from migration have had no opportunity to copy adult males of their own species. Yet, being migrants, they must set about breeding rapidly if their young are to fledge in time to move south and, to attract a mate, young males must thus generate a large repertoire of song elements very rapidly. Yet the two species do so in totally different ways, the marsh by imitation, the sedge by invention. Perhaps its migration route or wintering grounds give the sedge warbler less opportunity to copy other species, perhaps its less varied repertoire allows it to match neighbours more easily, perhaps differences in repertoire size are favoured as they label birds as belonging to one species rather than the other. At present we can only speculate on the reasons for the difference.

Whether or not a bird migrates has been identified as an important factor influencing song learning strategies, most impressively where there are differences between populations of the same species. The sedentary *nuttalli* subspecies of the white-crowned sparrow, kept in standard conditions, learnt over a broader range of ages but sang fewer song types in sub-song than the migratory *oriantha* (Nelson *et al.* 1995, 1996). This is in line with the migrant's shorter song season and greater uncertainty about

where it will settle: having several song types enables it to choose the appropriate one for the area where it does so.

A more extreme case is in the wrens studied by Kroodsma *et al.* (1999a). The marsh wren is sedentary in North America and birds develop song by imitation, so neighbours tend to share and songs to vary with distance. The same is true of resident populations of sedge wrens in Central America and in the Falkland islands (Kroodsma *et al.* 1999b, 2002). However, North American sedge wrens are nomadic, moving around during the breeding season and between breeding attempts. They do not share songs with neighbours, nor does song vary geographically; hand-rearing experiments have shown that each male develops a unique repertoire, by invention or by loose improvisation based on tutor songs. With their nomadic lifestyle, it appears the advantage is to have songs that allow them to communicate wherever they are, rather than being unique to a particular neighbourhood.

3.6 The distribution of song learning

Kroodsma & Baylis (1982) concluded that learning had a role in song development in every species of songbird studied up to that point. Broadly speaking this remains the case. The songbirds, or oscines, are a subdivision of the passerines, comprising some 4000 of the 9000 or so species of birds known to exist. The 1000 remaining passerines belong to the sub-oscines, a group which occurs primarily in Central and South America and includes ovenbirds, antbirds and tyrant flycatchers.

Sub-oscines have been less extensively studied, but the work that has been carried out on them suggests that their songs are usually not learnt. Kroodsma (1984) reared the young of two closely related species, the willow flycatcher and the alder flycatcher, under conditions where they were only able to hear tape-recorded song of the other species. Despite this, they produced normal species-specific song when they started to sing themselves (Fig. 3.11). The two species have quite simple, but very distinctive, songs, often represented as fee-bee-o for the alder flycatcher and fitz-bew for the willow. Kroodsma's result is striking, as it is very different from any obtained on songbirds. However, one point calls for caution: these species are also exceptional in starting to sing very young, a sound recognisable as song being recorded as early as 15 days of age (see Fig. 3.11). As the young Kroodsma studied were 7–10 days old when taken

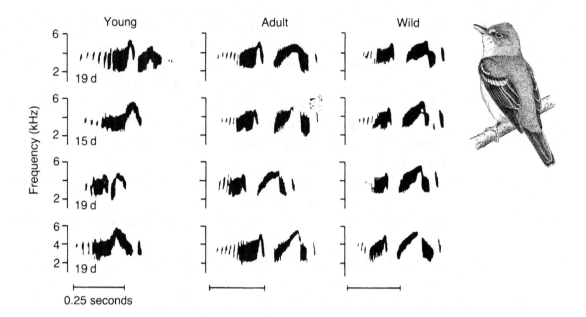

from the nest, it is possible that they do memorise sounds but also at a very early age compared with other birds, so that some learning would have occurred before they were taken from the wild.

Kroodsma (1989a) has also studied another sub-oscine, the eastern phoebe. He took young from the nest at 8–10 days old and exposed them to tapes of phoebe song, edited to give slightly different temporal structures, during the following two months. The young birds did not start to sing themselves until several months later and again, as in the flycatchers, there was no suggestion that their songs had been influenced by the tape. Birds housed together during the experiment developed songs that were no more similar than those of birds in different groups. Furthermore, Kroodsma (1985) found that young phoebes that were hand reared without the opportunity to hear adult song also sang normally. He suggested that this might be because they did not need to memorise song, but they still might require to match their output to their neural representation of song when they began to sing themselves. However, this can be discounted, as young males deafened at around 35 days of age were also found to develop normal song (Kroodsma & Konishi 1991).

These cases are striking because they are so dissimilar to any results from oscines. The clear distinction between these two groups is, however, less secure than it once seemed, once again due to work by Kroodsma (2004). This evidence comes largely from the three-wattled bellbird, a

Fig. 3.11. Development of the song of the alder flycatcher. The vocalisations of the four young laboratory reared birds between 15 and 19 days of age (left column) foreshadowed their adult songs (middle column), which in turn are very similar to songs of wild males (right column) (from Kroodsma 1984).

Three-Wattled bellbird

sub-oscine he has studied in Costa Rica. While this work is not experimental, it does add up to a convincing case. That the species shows dialects is not clinching as these might arise without learning if dispersal was very limited. However, during the course of 30 years the whistle frequency in the song of birds in the Monteverde district has declined by nearly 2 kHz, a change most likely to be attributed to cultural drift. Kroodsma (2005) also describes a captive bare-throated bellbird singing what appears to be a good imitation of a chopi blackbird. More research on this group (the Cotingidae) is needed, but the pointers certainly look strong.

Non-passerine birds do not normally produce the relatively long and complicated vocalisations we refer to as 'song'. In many cases, their rather simple calls appear to develop normally, even in birds that are deafened. This is true of turkey gobbling (Schleidt 1964), and the calls of chickens (Konishi 1963) and of doves (Nottebohm & Nottebohm 1971). However, there are clear cases among non-passerines where learning plays a substantial part in vocal development. Parrots and their relatives (order Psittaciformes) are an obvious example (e.g. Todt 1975b, Pepperberg

Fig. 3.12. Tree showing the phylogeny of avian groups that do learn their songs and related ones that do not (after Brenowitz 1991). In (a) the common ancestor is assumed not to have learnt (●) and this requires learning to have evolved three times (○). In (b) the common ancestor is assumed to have learnt (○), and this would mean that learning had been lost four times (●).

1994). There is now considerable work on budgerigars in captivity, showing that the call structure of individuals in groups converges (e.g. Bartlett & Slater 1999, Hile & Steidter 2000, Farabaugh *et al.* 1994); the same is true in pairs, with the call of the male being modified to match that of his mate (Hile *et al.*, 2000). The work of Wanker *et al.* (2005) on spectacled parrotlets even suggests that birds may use different forms of call to communicate with particular individuals. Field studies of parrots are less numerous, and largely restricted so far to the description of dialects (e.g. Kleeman & Gilardi 2005, Wright 1996). They are, however, consistent with the idea that vocal learning is widespread in this group.

A less expected example of song learning is among the hummingbirds (order Trochiliformes). The Anna's hummingbird was studied by Baptista & Schuchmann (1990), who hand reared several individuals and found them to develop more simple songs than those of birds in the wild. The normal song is quite complex and varies from place to place, though neighbours tend to share syllables and aspects of organisation; three birds hand reared in a group also developed similar songs. Therefore, despite its simple syrinx, the Anna's hummingbird has a complex song which develops in a very similar way to that of true songbirds. Other hummingbirds also have complex and varied songs (e.g. Gonzalez & Ornelas 2005, Ornelas *et al.* 2002, MacDougall-Shackleton & Harbison 1998), and song phrases have been found to be shared between species (Gaunt *et al.* 1994), suggesting that song learning may also be widespread in this group.

Vocal learning is thus well-established in three different groups of birds, and is thus usually considered to have evolved three times (see Fig. 3.12a). However, one should not be too hasty in accepting that conclusion. As Figure 3.12b shows, if the common ancestor of those three groups

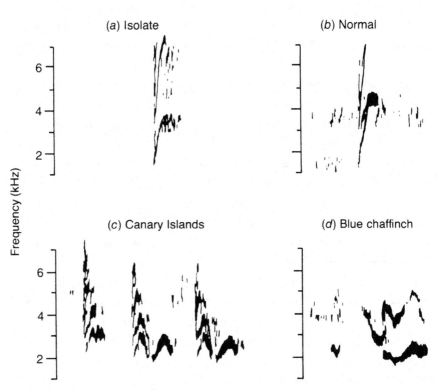

Fig. 3.13. Learnt variations in call notes illustrated by the 'chink' calls of chaffinches. In Britain and Europe this consists of a single call often produced in a series (b). Its form is similar in birds reared in isolation but lacks the pure tonal finish (a). In the Canary Islands' race of the same species it is distinctively different (c), and the equivalent call of the blue chaffinch, which also occurs on the Canaries, is again distinct (d) (from Slater 1989).

showed vocal learning, its current distribution could be accounted for by its loss on only four occasions. As loss of a trait is likely to be easier than its gain, this might be an equally plausible scenario. There is a need for studies on more bird groups, such as those by Kroodsma on bellbirds, as the more isolated the instances of song learning within a group the more likely it is to have arisen anew, making the search for its functional significance a more realistic possibility. It is to this functional significance that we shall now turn.

3.7 Why learn?

The final, and in some ways the most difficult, question concerns why so many birds learn their songs? A lot of ideas have been put forward, but there is as yet rather little evidence to support or refute any of them. The following are the main possibilities:

(1) Learning is a good mechanism for the transmission of complex information. Complexity in itself is probably not involved, as some learnt sounds (such as the 'rain call' and 'chink' call of chaffinches, see Figure

3.13) are very simple, and some unlearnt ones (such as suboscine songs) are quite complicated. However, learning does allow very accurate transmission, always provided that mistakes are not made.

(2) Learning allows social adaptation. Where birds learn from their neighbours, on territories or in groups, they will have songs that match each other. Matched countersinging is common between territorial neighbours, and sharing songs with neighbours may avoid conflict between them and could give rise to greater breeding success (as suggested by Payne 1982), presumably because less time is spent in territorial defence.

(3) Learning allows genetic adaptation. The idea that song learning allows birds to choose mates which are well adapted to themselves was originally put forward by Nottebohm (1972), who suggested that females might choose their mates on the basis of their songs to ensure that they were well matched. This idea has been particularly championed by Baker in his studies on white-crowned sparrows (see Baker & Cunningham 1985, 1987), and is discussed further in Chapter 9. The evidence in favour of it in this species is not strong, particularly as it appears that females often choose mates whose songs differ from those of the area where they hatched (Baptista & Morton 1982, Chilton *et al.* 1990), but female preferences for songs experienced early in life have been shown in a number of others (see Riebel 2003b). An alternative to the idea that song may ensure mating of like with like is that females might choose males singing rather differently and thus ensure a degree of outbreeding. This would be most likely where males learnt their songs from their fathers, thus giving females a guide to relatedness, and this is not often the case. Nevertheless, in the Darwin's finches where males do learn from their fathers, Grant & Grant (1996) did find females to choose males that sang differently from the way that their fathers had done. This was not true in first year birds, however, the authors argue because young females pair up later in the season and have less of a choice of mates.

(4) Learning allows adaptation to the habitat. This idea was put forward by Hansen (1979). He suggested that song learning may ensure that the song a young male develops is adapted to the habitat in which he sings. If song is learnt from other individuals singing some distance away within that habitat, the characteristics that carry best through it will be those most likely to be copied. Song should thus, generation by generation, become steadily honed to match the sound transmission characteristics of that habitat. As will be discussed in Chapter 4, there is evidence from several

Fig. 3.14. The cultural trap hypothesis. Males and females are symbolised as having filters, songs within the range of which may be developed by males and are acceptable to females. (a) In a population consisting largely of males with wide filters, their songs, symbolised here by the vertical line, will often be outside the acceptable range for a female whose filter is narrow. In the case shown here only the single male in the five territories with a narrow filter and two of the other males have a song she would find acceptable. As a result narrow filter females will have fewer mating opportunities and become rarer in the population. (b) In a population consisting largely of birds with narrow filters, broad-filter males will fare equally well because they are most likely to learn from a narrow-filter tutor and so have a song that conforms to the population norm (after Slater et al. *2000).*

species that song is matched in certain respects to habitat, although the effect is not often a striking one and it seems unlikely that this advantage of song learning would account for its widespread occurrence.

(5) Learning provides an honest indicator of male quality. Originally put forward by Nowicki *et al.* (1998a), and recently reviewed by Nowicki & Searcy (2005), this idea proposes that a range of learnt song characteristics including repertoire size and accuracy of learning may be deleteriously

affected in males that suffer from nutritional stress during development. Females might then use such cues in mate choice so that they select a male of high quality. Evidence in favour of this hypothesis is growing and, because of its relevance to mate choice, we will consider it in more detail in Chapter 7. That learnt song characteristics may provide a cue to male quality is clear, but what seems more doubtful is whether this would confer an advantage on song learning.

In addition to these main ideas, more speculatively, Nottebohm (1991) suggested that vocal learning might allow vocal flexibility so that birds could sing loudly without damaging their auditory system, Payne (1983) suggested learning might enable birds strange to a neighbourhood to blend in and avoid aggression, and Morton (1996) argued that it might assist birds in assessing the distance away of rivals (see Chapter 4). There are, therefore, several possible advantages of song learning. However, no one of them seems likely to account for all examples, although the ideas that learning is favoured because of social adaptation or adaptation to the habitat are perhaps the most likely.

Perhaps, however, the quest to find advantage in learning in every situation where it is found is a fruitless one. Given the widespread occurrence of learning among the oscines, it is probable that vocal learning was present early in the evolution of that group, maybe even before anything we would now recognise as song had evolved. Once learning had evolved, it may have been maintained through a variety of means: the need to match neighbours, the advantage of matching sound characteristics to the habitat and the attractiveness of variety to females may be just a few of these. A further alternative, put forward by Lachlan & Slater (1999) is that song learning may have no direct function as such but be maintained because the species showing it are caught in a 'cultural trap' (Figure 3.14). They carried out computer simulations that suggested learning was likely to persist under a wide variety of conditions simply because males that learnt were at no disadvantage and females were likely to be at an advantage if they were prepared to mate with males with learnt songs. The attraction of this idea is that it might account for the persistence of song learning even in the hugely varied range of circumstances in which it is found among the oscines. What is clear, however, is that the book is not yet closed on the reasons for vocal learning, which remains one of the most interesting and challenging questions in the study of bird song.

CHAPTER FOUR

GETTING THE MESSAGE ACROSS

Eastern towhee

And like an echo far away
A nightingale sang in Berkeley Square
Eric Maschwitz *Popular song*

4.1 Introduction

Any signal must obviously get through from the sender to the receiver if its message is to be understood and acted upon. In the case of song, many factors may conspire against this being achieved. The further from the source the hearer is, the fainter the signal is likely to be and the harder to pick up. The more noisy the environment, the more the signal will be masked or drowned out. In complex environments, such as dense forests, the exact form of the signal may be greatly distorted by echoes from tree trunks and from the canopy. These are amongst a whole array of factors that make it hard to get the message across.

The solution to problems like these is not just to shout as loudly as possible as close to the receiver as can be achieved. Research since the early 1970s has shown that more subtle choices may be involved and that natural selection has often taken animals down paths that yield signals which are maximally effective at getting through the particular environment that they inhabit. In this chapter we will explore some of the problems involved and the solutions to them that selection has generated.

4.2 The problems of transmission

There are two fundamental problems of transmission, that of attenuation and that of distortion or degradation. The theory underlying them was reviewed in relation to bird song especially by Wiley & Richards (1978, 1982, Richards & Wiley 1980). The two problems raise rather different issues.

4.2.1 Attenuation

Attenuation simply means that the further away from a source the receiver is the fainter a sound will be. This is partly attributable to a simple physical principle. If you imagine a bird calling on top of a tall tree, the energy that it puts into each call will spread outwards in all directions (see Fig. 4.1). At 10 m range, it will be spread over the surface of a sphere 10 m in radius (r), an area of 1257 m^2, while it is spread over four times that area at 20 m range. The intensity of the sound, or density of the energy in it, thus decreases in proportion to $1/r^2$. The unit of intensity used is the decibel (dB) and this is on a logarithmic scale. The basic expectation for 'spherical spreading' is that the intensity of the sound will diminish by 6 dB for each

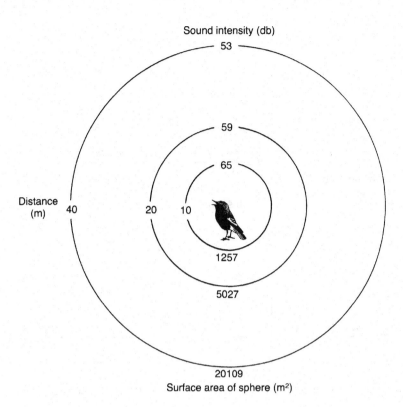

Fig. 4.1. Diagram to illustrate spherical spreading.

doubling of distance, as 6 dB corresponds to a factor of 1/4. Thus a sound measuring 65 dB at a range of 10 m from the bird should score 59 dB at 20 m and 53 dB at 40 m (see Fig. 4.1).

Spherical spreading assumes that we are dealing with a point source in the middle of an environment that is homogeneous in all directions (and that we are not too close to the source at that). Singing birds are not quite like this and there are several reasons why attenuation tends to be greater in the real world than spherical spreading would predict. This leads to a certain amount of what is referred to as 'excess attenuation', the attenuation observed in addition to that expected from spherical spreading. The amount of excess attenuation varies with habitat and weather conditions, but it also varies with frequency, some sounds attenuating more than others. Higher frequencies tend to show more excess attenuation, because they are more prone to being absorbed by the atmosphere, especially in hot or humid conditions. High frequencies also tend to be scattered more, bouncing off objects rather than bending round them, and their path is also more likely to be distorted by turbulence in the atmosphere. Sounds at lower frequencies, with their longer wavelengths, will

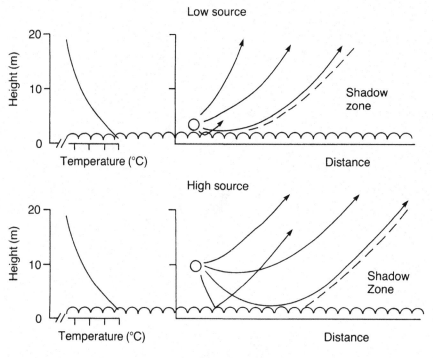

Fig. 4.2. How reflection off the ground and upward refraction caused by temperature gradients lead to sound shadows (after Wiley & Richards 1978).

only be affected by larger objects or greater differences in atmospheric conditions.

Among the objects that affect sound transmission is foliage, with greater attenuation the denser it is, particularly if the vegetation is broadleaved (Aylor 1972). Martens (1980) confirmed this but found that the effect depended on the frequency of the sound, with components above 4 kHz filtered out, while there was a tendency for those just below this level to be amplified.

Temperature and humidity are other factors that affect sound transmission, with their influence also depending on frequency, though in a rather complex way (see, for example, Larom *et al.* 1997). High humidity enhances transmission (Evans & Bazley 1956); the effect of temperature varies both with humidity and frequency (Harris 1966). For a frequency of 4 kHz, typical of many songs, the greatest attenuation is when humidity is low and temperature high. Temperature gradients also have an effect, and this is important as air tends to be warmer near the ground than higher up on sunny days (see Fig. 4.2). Sound travels faster in warm air and the gradient above the ground has the effect of bending an advancing sound wave upwards, leading to a 'sound shadow' near the ground some distance from the source (Michelsen 1983).

Fig. 4.3. Curves showing excess attenuation in decibels per 300 ft plotted against sound frequency for pure tone sound propagation at ground, 5 ft and 10–20 ft (1.5m and 3.0–6.0 m) elevations in forest habitat. The solid line at 0dB excess attenuation represents what the plot would look like if the frequencies were all attenuating at the rate of 6 dB per doubling of distance (from Morton 1975; © by the University of Chicago Press, 1975).

The arguments presented so far might suggest that animals should produce the lowest possible sounds if they are to achieve maximal transmission. But the real situation is not as simple as that. Low frequencies have problems of their own, especially for animals singing near the ground. Some sound travels direct from the signaller to the receiver, but some of it is reflected off the ground and these two different waves may interfere with each other. The results of this interference can be complicated. If the direct and reflected paths differ in length by half a wavelength, the two may cancel each other when they meet so that the sound does not reach the receiver. But if the two differ by one wavelength, they will sum with each other to amplify the sound. The exact nature of these effects depends on height of the singer and receiver above the ground, frequency of the sound and the characteristics of the substrate. Generally, however, the result of such interference is to disrupt the transmission of a band of low frequency sounds near the ground.

The fact that high frequencies are more prone to being scattered and low ones are more disrupted near the ground might suggest that signals at intermediate frequencies would travel best. In line with this, Morton (1975) found such a 'sound window' at around 1600–2500 Hz in woodland (see Fig. 4.3), though not in more open country. Subsequent studies suggest that such a sound window may occur in a variety of other habitats

89

as well, but that its lower limit may only be significant for animals vocalising near the ground (Marten & Marler 1977, Marten *et al.* 1977). Ellinger & Hödl (2003) found excess attenuation at a rainforest site to rise smoothly with frequency at heights of over 12.5 m, but also to show a low frequency peak at or below 2.5 m up, so that, on or close to the ground, there was a sound window at 0.5–1.5 kHz. These results are similar to those of Morton, though the sound window near the ground is at a rather lower frequency.

We can conclude from these results that, when a bird is singing more than a few metres above the ground, maximum transmission will be achieved by the lowest possible frequencies. Nearer the ground, interference disrupts low frequencies because of interaction between direct and reflected waves. However, to add a final complication, really low sounds are less affected by this interference, and sounds produced very close to the ground also propagate rather well, because of an interaction between the air above the ground and that in pores in the ground, setting up what is known as a 'ground wave'. Thus, the deep mechanical sounds of many grouse displaying on the ground may be heard at very long range (e.g. Hjorth 1976).

The reasons why songbirds do not usually produce very deep sounds lie in physical constraints on the bird itself rather than limitations set by the environment. First, deep sounds require a larger sound-producing mechanism, and this may not be possible in a small bird. Second, low-pitched sounds are more energetically expensive to produce, so that the benefits of increased range obtained by using them may be outweighed by the greater costs of production (Ryan & Brenowitz 1985).

Aside from such physical limitations, and the problems of sound transmission, one might perhaps imagine that all birds should sing as loudly as possible. However, this may not be true. In contrast to some earlier claims, recent evidence suggests that producing sounds is not costly in terms of energy expenditure (e.g. Oberweger & Goller 2001, Ward & Slater 2005a). However, it may have other costs, such as the possibility that a predator may be attracted, or that of heat loss incurred by sitting on an exposed perch (Ward & Slater 2005b). Such costs may lead to compromises so that birds produce sounds that only transmit as far as it is beneficial for them to spread. If a song is used in mate attraction it may indeed be best to transmit over as wide an area as possible, but if song is used primarily in interactions between territorial neighbours there would

be little advantage in it travelling more than two territory widths at most (Brenowitz 1982a, Shy & Morton 1986, Lemon *et al.* 1981b).

4.2.2 Degradation

Song does not just become quieter with distance; it is also subject to degradation or distortion. The scattering referred to above, which is particularly true of high frequency sounds, is one reason for this degradation. It is especially problematical in dense cover, such as forests, where reflection off the ground or canopy and echoes off the tree trunks may make it very difficult to distinguish successive elements or to detect the form of the individual elements as they become slurred and run into one another. These problems, collectively referred to as reverberation, have a profound effect, especially on the form of rapidly modulated signals. The exact form of a trill (an amplitude-modulated signal) or of sounds that sweep up and down over a wide frequency range (frequency-modulated) is easily distorted by reverberations. By comparison, pure whistles remain pure whistles of the same frequency when they bounce off objects: they thus travel rather well through forests, and reverberation may indeed enhance both their length and loudness (Slabbekoorn *et al.* 2002). Despite this, animal signals do not often consist of pure tones. This may be partly because pure tones can be so badly attenuated through interference, for example when they are reflected off the ground, so that some modulation may be beneficial. But, perhaps even more importantly, pure whistles can be easily masked by a second individual whistling at the same frequency. In contrast, if animals use a frequency modulated sound, the second animal has to be in perfect temporal synchrony with the first to achieve complete masking, and this is highly unlikely.

If reverberation distorts a trill so much by running the elements together, then perhaps the answer might be for the elements to be less rapidly repeated, so that one has died away before the next is produced. But this has a difficulty of its own, that of low-amplitude fluctuations caused particularly by wind. Two elements produced in quick succession will transmit to much the same extent, but the longer the interval between them the more different conditions will be and the more likely that the amplitude of one at the receiver will be different from that of the other. This is another reason why coding information in changes in amplitude is not advantageous, as the environment, particularly if it is a turbulent one,

will impose its own amplitude modulation on the signal so distorting it to the extent that it may be unreadable.

Atmospheric turbulence does not distort frequencies to any great extent. So, as with reverberation, the coding of information in patterns of frequency would be predicted as best for transmission. This does indeed appear to be the way in which most birds code their signals.

4.3 Does practice match theory?

Having outlined some of the theory of sound transmission, we can now consider how the song and singing behaviour of birds are adapted to overcome the problems of getting the message across.

4.3.1 Singing position

One expectation, almost regardless of habitat, is that small birds (which cannot produce very low frequencies) should sing well clear of the ground so that problems of interference are minimised. Of course, singing on exposed perches high in the vegetation also makes birds more vulnerable to predation: it may not always benefit them to occupy the best possible position for sound transmission if predation is an important consideration.

Krams (2001) found chaffinches to sing high in the trees but that those with song perches at the very top tended to move to sing rather lower after being exposed to a hawk model. Corn buntings have been observed invariably to choose the higher of two singing perches, even if the next highest is only a few centimetres lower; so important are song perches to this open-country bird that they appear to be an important factor in the distribution of territories within the habitat (Møller 1986). In a study of the songs and habitat characteristics of *Acrocephalus* warblers, Jilka & Leisler (1974) suggested that all the species they studied would benefit from high singing positions. In line with this, the great reed warbler sings from the tops of the reeds, and the sedge warbler uses frequent song flights. The former species uses much lower frequencies than the latter, and these transmit better through the reed beds. The behaviour of the reed warbler, by comparison, does not seem so well matched to sound transmission, as it sings lower in the vegetation; however, the intermediate frequencies typical of its song do transmit well at that level. A further member of this genus, the rare Seychelles warbler, has been studied by

Catchpole & Komdeur (1993). It has a more melodious song, with a narrow frequency range, which contrasts sharply with that of its European relatives and seems well adapted to the fact that it occupies dense forests rather than reed beds.

An interesting variant of behaviour matched to the need to broadcast sound widely is shown by the blue-black grassquit, a Neotropical grass-land dwelling species. The males leap clear of the grass when they display, and this is probably an adaptation for sound transmission as the call that accompanies the display travels very much better above the grass than through it (Wilczynski *et al.* 1989). Less remarkably, many other species sing in flight, thus increasing the broadcast area of their songs. Evidence from the bluethroat suggests that this song is primarily used in territorial defence (Sorjonen & Merila 2000).

These various means of singing from high up are just as one would expect to maximise the transmission of sound, although singing in flight or from an open perch may also give the bird a good all-round view so that evasive action can be taken should a predator appear. But skylarks, Shelley's blithe spirits, with their beautiful aerial songs, take this one step further. In winter, a time of year when they would not normally sing at all, they sometimes actually sing in the presence of a predator. Cresswell (1994) watched many instances of birds being chased by merlins. Those that sang well were chased for a shorter time than those that sang poorly or not at all, and those that produced full song were also less likely to be caught. This rather unconventional use of song appears to act as a pursuit deterrent signal, informing the merlin of the bird's condition and thus the improbability that it could be captured.

Returning to earth, in a study of how European blackbird song trans-mits through deciduous woodland, Dabelsteen *et al.* (1993) argue that birds should be able to discriminate between songs across at least two territories. This is based on the transmission characteristics of the purer initial part of the song rather than the terminal twitter, which transmits less well and is thus likely to be used in shorter-range communication (see Dabelsteen & Pedersen 1993). An interesting additional finding here was that transmission did not depend on loudspeaker height, although (at 3 m) the lowest height used was well clear of the ground, but was substantially affected by that of the microphone. This suggests that singing birds may use high perches because they can hear responses better rather than because they are better heard. In interpreting these experiments, we have

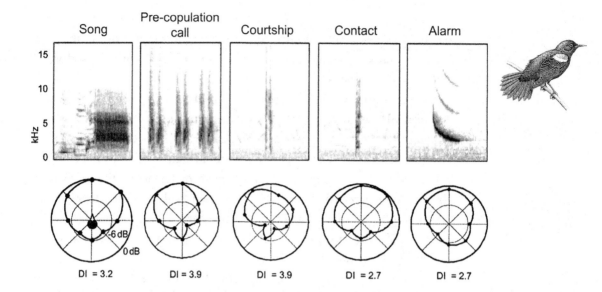

Fig. 4.4. Sonagrams of five different red-winged blackbird vocalisations produced in the contexts indicated above them. The polar plots below are single examples of the relative amplitude of the sound at eight microphones around the calling bird, recorded with its beak facing the upper one. The directionality index (DI) is a measure of how focussed in direction the sound is, the precopulatory and courtship calls being the most so and the alarm and contact calls the least (after Patricelli et al. 2007).

to be careful not to generalise too much: they were carried out when the leaves were off the trees and, as we shall see below, precise habitat characteristics may have a substantial effect. Nevertheless, it is important to look at things from the perspective of the receiver too. Winter wrens played degraded songs move to higher perches, thus enhancing both their ability to hear and be heard by their distant rival (Mathevon et al. 1996).

Sound does not necessarily spread equally in all directions. Thus, Larsen & Dabelsteen (1990) found that, while low-frequency sounds in European blackbird song radiated in all directions, those above 5 kHz were more strongly beamed in the direction the bird was facing. Using an array of microphones and video cameras around the song perches of male red-winged blackbirds, Patricelli et al. (2007) found some quite sharp divergence from simple spherical spreading. Interestingly, courtship and copulation calls were most directional, while contact and alarm calls were much less so as one might predict from the functions of these vocalisations (see Fig. 4.4). Song was intermediate, in keeping with its role in both advertisement and courtship. To compensate for directional biases, singing birds often move their heads from side to side or face in different ways. Breitwisch & Whitesides (1987) found unmated northern mockingbirds to show greater variability of singing direction within a bout of song than mated ones and suggested that song early in the season is primarily a mate-attractant signal, with changes of direction insuring that it broadcasts over as wide a range as possible. During normal singing behaviour nightingales

move their heads from side to side, so broadcasting the song in different directions. In response to playback, however, they face the speaker and show fewer head movements, thus addressing the particular source of the intrusion (Brumm & Todt 2003).

Another intriguing, but somewhat perplexing, aspect of singing behaviour was described by Hunter (1989), who discovered that Himalayan birds of various species face uphill when singing. Perhaps this aids in sound transmission, their songs being carried towards the ground rather than up into the air above the trees. But there may also be anti-predator advantages. Hunter suggests that they may be able to keep a better look out for predators approaching from beneath; another possibility is that their more camouflaged upper parts are less likely to be spotted by hawks flying above the canopy.

In looking for more subtle matches between song and habitat, two different approaches are possible. One is a comparative survey across species to see whether those occupying particular habitats have song features in common which are matched to the acoustic characteristics of their environment. The other is to look within a species and see whether song varies between habitats in predictable ways.

4.3.2 Comparisons between species

Morton (1975) surveyed a large number of Neotropical bird species for evidence of links between habitat and song features. He found that species in more open country used rapid frequency modulation more than those in forests, as one would expect given the disruptive effect that trees can have on signals of complex structure. By comparison, more pure tones were found in forests, where such sounds transmit well. The frequencies of the songs of forest birds appeared well matched to the 1600–2500 Hz sound window that he had found in his transmission experiments. An equivalent analysis was carried out on the songs of European passerines by Sorjonen (1986b), who also reported more whistles in forest habitats and more trills in open ones. Wiley (1991) came to similar conclusions on the basis of a comparative study of North American songbirds: the most striking association he found was that between habitat and the temporal properties of songs, which seem adapted to minimise reverberation in forest habitats. In a more recent study, which controlled for phylogenetic history, Badyaev & Leaf (1997) examined the songs of 30 species of leaf and tree

warblers, and also concluded that species in closed habitats avoided rapid frequency modulation, as well as producing shorter and more spaced out elements, just as one would expect for avoidance of reverberation. In addition they found that smaller species sang at higher frequencies and produced shorter notes than larger ones, in line with constraints that size puts on sound production.

As mentioned earlier, the importance of the sound window is doubtful except for birds singing close to the ground. A good test of the idea is to examine species singing in marsh and grassland areas, as the differences in substrate between them lead to a ground effect in grassland but not in marshes, which absorb the reflections that cause ground interference. In line with this, Cosens & Falls (1984a) predicted that marsh birds would use lower frequencies in their songs than those in grassland and, using a small sample of species from each habitat, they did find a significant difference.

In other respects the match between bird sounds and habitat is less straightforward. Both Morton (1975) in central America and Chappuis (1971) in equatorial Africa reported forest birds to average lower frequencies in their songs than ones in open habitats (Fig. 4.5). But Richards & Wiley (1980) found no such effect in a survey of passerines in North Carolina. A possible reason for this was put forward by Wiley & Richards (1982), who pointed out that tropical forest species often sing on the ground, at which level only low sounds transmit well, whereas temperate species, even those that forage on the ground, tend to sing high up.

Although sound windows may lead to transmission advantages of singing in a particular frequency range, most other considerations suggest that low frequencies will usually be best: they attenuate less, they are less disrupted by objects and they are the only frequencies that transmit well at ground level. Size is probably the major factor in restraining birds from using them to best advantage: the smaller the bird, the less capable it is of producing low-pitched sounds (Wallschläger 1980, Sorjonen 1986b, Tubaro & Mahler 1998). Indeed there may be an interrelationship between body size and habitat: larger species tend to live in forests and this may be partly responsible for the lower frequency sounds Morton and Chappuis found to be used there (Ryan & Brenowitz 1985). Larger-bodied birds also tend to have larger territories and are capable of producing louder songs.

Another idea, put forward by Heuwinkel (1990), is that birds may actually use the acoustic characteristics of their habitat to enhance

Fig. 4.5. Histograms showing the frequencies emphasised in the sound of bird species (to the nearest 500 Hz), arranged according to the habitat in which the species occurred. The arrow shows the mean frequency emphasised for each of the four habitats shown. This is the sum of the frequencies emphasised for all species in that habitat divided by the number of species in that habitat (from Morton 1975; © by the University of Chicago Press, 1975).

transmission. He found the songs of *Acrocephalus* warblers to transmit particularly well through the spring vegetation of the reed beds in which they live (as had Jilka & Leisler 1974), and argued that the predominant frequencies in their songs are matched to the resonant frequencies of the reeds. In this way sounds may cost less in energy input for a particular distance of travel. Likewise, in dense forests, narrow frequency band notes may be both extended in length and amplified by reverberation so that the habitat works to their advantage (Slabbekoorn *et al.* 2002).

4.3.3 Comparisons within a species

As mentioned in Chapter 3, Hansen (1979) suggested that learning the songs they sing within the habitat that they occupy may enable birds to match their songs to the acoustic characteristics of that habitat. If learning takes place at a distance, those features which transmit well through the habitat will be learnt whereas those that are lost in the transmission process

will not. Thus song should become progressively honed to the features of that particular environment. Where a species occupies several different habitats, the songs of individuals might thus be expected to match the characteristics of the one in which they are found.

In line with this proposal, Gish & Morton (1981) found the songs of Carolina wrens recorded in two different locations to transmit with less degradation in the place where they were recorded than in the other one. In a subsequent experiment, they also found young birds to learn more from songs that were undegraded, having been recorded close to the singer, than from ones that had been degraded by passage through the habitat (Morton *et al.* 1986).

The evidence on whether such processes lead songs within a species to vary between habitats is mixed. No differences could be found in the song of the chaffinch (Williams & Slater 1993) or that of the American redstart (Date & Lemon 1993), but there is good evidence for several other species. Hunter & Krebs (1979) recorded great tit songs in a wide variety of locations in Europe, as well as in Morocco and Iran. Those recorded in forests had a lower maximum frequency and frequency range, as well as fewer elements in a phrase, than those in more open woodland (Fig. 4.6). As discussed above, songs of lower frequency and with more spaced out elements are better adapted to transmission through dense vegetation, so their results were in line with predictions.

Another species in which differences between habitats have been examined in some detail is the rufous-collared sparrow from Argentina. Studies by Nottebohm (1969b) and King (1972) showed that songs vary between the different habitats in which this species occurs. Several features of song were found to be correlated with aspects of the environment, such as latitude, altitude and vegetation, by Nottebohm (1975). It is not easy to dissociate the influences of various related factors, but Handford (1981) suggested that the influence of height was largely through vegetation changes rather than altitude as such. Songs had higher frequencies in forests, at lower altitudes and in more tropical areas, which Nottebohm suggested may be because such frequencies transmitting best in very humid conditions. As with the great tit, he found trills to be slower in forested areas, but slow trills were also found in arid areas without any clear transmission advantage. Subsequent studies by Handford (1988) and Handford & Lougheed (1991) found that both trill rate and various other frequency and duration measures were most

Fig. 4.6. Great tits in forests sing songs with a narrower range of frequencies, lower maximum frequency and fewer notes than the songs of open-country birds (from Krebs & Davies 1993).

clearly related to the original vegetation that had been present in an area before agricultural incursion rather than to its current vegetation. This suggests a great deal of inertia in the system. Handford argues that this occurs because this is one of few species in agricultural parts of Argentina, and they nest at very high densities in these conditions, so that both acoustic competition from other species and the sound transmission characteristics of the habitat may have little influence on shaping songs. These, therefore, could have remained much as they were before the farmland was created. We will return to this example in Chapter 9, as the rufous-collared sparrow is one of the most interesting cases of a species showing dialects in its song.

Fine differences in song have also been found between habitats in several other species. In white-throated sparrows, open-country birds use higher frequencies in their songs than do birds in deciduous forests (Wasserman 1979, Waas 1988). As before, this correlates with the transmission problems that high frequencies encounter in dense foliage,

although it leaves unclear why open-country birds do not use low frequencies too. In summer tanagers, Shy (1983) also found a relationship between high tree density and lower frequencies in song, although this was not the case in scarlet tanagers, perhaps because this species sings in the canopy more often. In satin bowerbirds there is both local and geographical variation in the advertisement call up and down the east coast of Australia (Nicholls & Goldizen 2006). In line with the transmission characteristics of different habitats, in dense forests the calls are lower and show less frequency modulation, while those in more open habitats are higher and modulated more.

In addition to differences between them in their transmission characteristics, habitats may differ substantially in the frequency spectrum of their ambient noise (Slabbekoorn 2004), and this may be another selective force shaping vocalizations. There was little low frequency background noise in rainforest in Cameroon compared with that in nearby more open forest, and in line with this the songs of little greenbuls had lower minimum frequencies in the rainforest (Slabbekoorn & Smith 2002b).

There is some good evidence that song may even vary over very short distances where habitats are different. Anderson & Conner (1985) found differences in northern cardinal songs in Texas between forests of three different ages, ranging from saplings 18 years of age to a mature timber stand of 59 years of age with well-developed canopy. Songs in the two more mature stands showed little frequency modulation, which the authors suggest is because such patterning is disrupted by reverberations from canopy foliage.

A final example of differences within a species concerns the yellow-headed blackbird, a species in which males sing more than one sort of song (a phenomenon discussed in detail in Chapter 8). Cosens & Falls (1984b) found that the 'accenting' songs of this species travel further through their marshland habitat and appear to be used mainly in long distance advertising. On the other hand, 'buzzing' songs are distorted more during propagation and are elicited more by encounters with other individuals. Their use seems to be more in short-range interactions, for which distortion by the habitat is of little consequence.

Overall, there is good evidence in some species that song can vary with characteristics of the habitat. However, no such effect has been found in some other studies (e.g. Payne 1987, Williams & Slater 1993), and many of the influences found have been fairly minor, requiring statistical

demonstration rather than being obvious to the ear. But we must now turn to a topic we have so far only just touched on, but an increasingly important one: that of background noise.

4.4 Communication in a noisy environment

The world is a noisy place. Even if one discounts the effects of man-made noise, such as that of traffic or of aircraft, which is often infuriatingly obvious to those of us that record bird sounds, the natural world produces plenty of sound of its own. The sounds of waves on the shore, of wind in the trees or of rivers tumbling over rocks are all examples. Such sounds are, of course, relatively unstructured and they also tend to have most of their energy at low frequencies. This may be one reason why bird songs are normally quite high pitched, and also why they often incorporate regular rhythms such as trills and pure sounds such as whistles. These help to make the sound stand out against the background. But there is one source of sound against which such features provide no simple protection and this is that made by other animals, especially birds. In South American rainforests there are places where 400 species of bird occur within a square kilometre or two. Each must signal to its own kind against the extraordinary background noise made by other animals. In this context it is hardly surprising that the sounds that birds produce explore such a wide variety of tones, rhythms and more complex patterns, in addition to other differences such as those that lead them not to sing at the same time as each other (see Chapter 5).

But the problem remains: birds often have to signal against a noisy background. How can they maximise the chances of getting their message across in these conditions? The most obvious solution would be to increase the signal to noise ratio by singing more loudly. Brumm & Todt (2002) demonstrated such an effect in captive nightingales, and Brumm (2004) went on to show that it is also the case with males singing against the background of traffic noise in Berlin. Birds with territories in noisy parts of the city sang more loudly than those elsewhere. In a way, this result is surprising, as you might expect them to be singing as loudly as they could already, but then, if they are signalling to territorial neighbours or potential intruders, it would not be necessary to have a sound that travelled to more than one territory away. Under normal circumstances they might therefore not be singing at full volume. This raising of the amplitude of vocalisations in noisy conditions is known as the Lombard

effect, and has also been found in a number of other species, ranging from zebra finches in the laboratory (Cynx *et al.* 1998) to blue-throated hummingbirds in the field (Pytte *et al.* 2003). In a study by Potash (1972) the Lombard effect was shown in Japanese quail, indicating that amplitude regulation is possible even in species that do not learn their calls.

Another possibility is to raise the pitch of the sound so that it literally rises above the background noise. Slabbekoorn & Peet (2003) showed that great tits in noisy areas of the city of Leiden sang higher pitched songs than those in quieter districts, and Slabbekoorn & den Boer-Visser (2006) have gone on to show a similar contrast between the songs of this species in various European cities compared to woods nearby. Another, rather more curious, finding on great tits was made by Lehtonen (1983), following the earlier study of Bergman (1980). He analysed differences in their songs over the period 1947–81 in the Helsinki region. He found a shortening of the song phrase, with an increasing proportion of two- rather than three-syllable phrases, and the introduction of new one-syllable phrases. The effect was particularly noticeable in densely populated, sparsely wooded and noisy urban areas, and he argues that it relates in some as yet unde-termined way to the increasing noisiness of the environment. However, given the way in which the song types present in an area can change with time through cultural transmission (see Chapter 9), it pays to be cautious about this interpretation.

A third possible way of counteracting noise is to repeat the message more often so that the listener is more likely to hear it, either because one of the repetitions occurs at a quieter period or because the listener can extract more information from each successive song. Even in species with repertoires of different song types, it is common for birds to repeat the same type a number of times before switching to the next. This 'redun-dancy' is what one might expect if it was important to get each signal across. Chaffinches are among the species that sing with 'eventual variety' like this: as the redundancy idea predicts, in Scotland those close to noisy streams tumbling downhill sing longer bouts of a particular song type than those further away in the same area (Brumm & Slater 2006a).

One might also predict that, like great tits in cities, birds in such areas should sing at a higher pitch, as the sound of running water or of wind in the trees is concentrated at low frequencies, largely below 1 kHz, but with diminishing amounts of energy at higher frequencies as well (see Fig. 4.7). There was only slight evidence of this in the study by Brumm & Slater, but

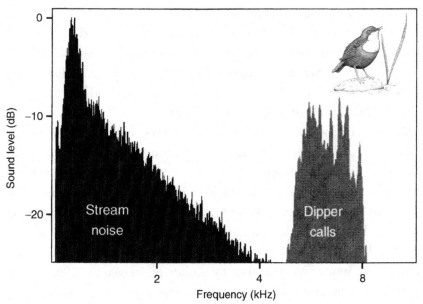

Fig. 4.7. Power spectra of white-throated dipper calls (shown in grey) and of the background noise of the fast running streams that form the bird's habitat. By calling at such high frequencies, dippers avoid the masking effect of this constant noise (from Brumm & Slabbekoorn 2005).

Martens & Geduldig (1990) found species of several families living close to Himalayan torrents to have loud and rather high-pitched songs, with the energy concentrated in a narrow frequency band, so that the elements used in their songs frequently consisted largely of whistles. These features seem ideal for sounds to stand out against such a noisy background, as Fig. 4.7 illustrates.

These then are some tricks that signallers can employ in noisy environments. The problems presented by noise are particularly great with advertising song, for noise can limit its active space severely. It may be because of this limitation that some social species, such as starlings (Eens *et al.* 1990) and brown-headed cowbirds (Cooper & Goller 2004) combine their songs with wing waving displays, so adding a visual component that must at least indicate which of a number of possible birds is singing. Where the target of song is closer and more focussed, as in the courtship song of zebra finches, noise is less likely to disrupt the message. Even here, however, as males have been found to sing more loudly to females that are further away (Brumm & Slater 2006b), it appears that birds modify their behaviour according to the situation.

So much then for the signaller: what of the receiver? How does it extract information from the cacophony of sound in its environment? If it is imperative that it responds, as for example to an alarm call, where the difference

may be between life and death, it may play safe and react to the slightest hint of a signal. In this case correct identifications may be combined with a lot of false alarms. But most song is not like that, as the odd missed signal is hardly a disaster and time may be wasted on false alarms. It seems more likely that the receiver listens out to hear if the message is repeated and approaches to maximise the chances of picking it up if it really is there.

Receivers do show a remarkable capacity to tease apart signal and noise, 'noise' in this broad sense often involving the songs of many other species. This ability is sometimes referred to as the 'cocktail party effect' in reference to the way in which we can pick out one conversation among many in a crowded room. Brémond (1978) first demonstrated it in winter wrens, played back song of their own species masked by that of several others, and there have been a number of other studies since (see Hulse 2002). One of the most impressive was that by Aubin & Jouventin (1998), who showed that king penguin chicks respond to their own parent's call, even when played back in combination with those of other adults played at higher amplitude. Selection has doubtless favoured this amazing ability because the chicks join together in enormous crèches and often move some distance while the parent is away at sea. The animals are able to extract one complex sound from a number of others even when the differences between them are very subtle.

4.5 Sound localisation

If song is to function well, either in advertising for a mate or in repelling rivals, it must include cues which enable the hearer to locate the singer. In the case of mate attraction, there is no doubt that giving accurate location cues is to the singer's advantage. As far as rival repulsion is concerned, this may also be the case, but we shall see that some theories suggest that it may pay singers to deceive rivals about how far away they really are.

Locating individuals involves two rather different problems, assessing distance and assessing direction, and we will discuss these in turn.

4.5.1 Distance

In theory, provided that the amplitude of a signal is known and the attenuation characteristics of the habitat can be taken into account, the distance away of a singer can be determined. However, several factors are

Fig. 4.8. Spectrograms (upper) and oscillograms (lower) of a normal eastern towhee full song and of the same song degraded (from Richards 1981a).

likely to make the estimation of distance in this way rather inaccurate. Even if song is always emitted at the same loudness, the source amplitude may be quite strongly dependent on the direction in which the singer is facing, as Fig. 4.4 shows (see also Hunter *et al*. 1986, Larsen & Dabelsteen 1990, Brumm & Todt 2003). Loudness also fluctuates with wind and temperature and will depend on reflections off the ground, vegetation and other objects in its path.

An alternative means of assessing distance is by analysis of degradation. Low frequencies transmit better than high ones, so the further from the source the listener is the more skewed downwards will be the frequency spectrum of the signal. Furthermore, the scattering of sound and its reflection off objects will lead the individual elements to be slurred and not so crisp and discrete as they are at short range (Fig. 4.8). That birds may use such changes to assess distance was first suggested by Richards (1981b). Working on Carolina wrens, he played back songs which had been recorded either close to or far away from the singer, so were undegraded or degraded, respectively. He equated playback volume so that any difference in response could not be attributed to those recorded close to the bird being louder. Despite this, there was a difference: the undegraded song elicited a strong response, the territory owner attacking the loudspeaker, whereas the degraded song received countersinging in response, as would be appropriate for a distant rival. This suggests that these birds do indeed detect the degradation of the signal and could use this to assess distance.

Morton (e.g. 1986, 1996) has taken this notion rather further to develop what he refers to as the 'ranging hypothesis'. He suggests that

selection may benefit individuals whose songs disrupt the behaviour of neighbours to the maximum possible extent. This will lead to songs which are well adapted to the habitat so that they show minimal degradation and are perceived as being close even when the singer is some distance away. He also suggests that a number of other features of singing behaviour may be accounted for by this hypothesis. For example, the listener will benefit from learning his neighbours' songs as this will enable him to accurately assess distance by comparing what he hears with the stored representation of the song. As a counter to this, it will benefit the singer to have a repertoire of song types so that he is likely to have songs which his neighbours are unable to range accurately. Thus, Morton sees his hypothesis as able to account both for learning from neighbours and for the existence of song repertoires, two features of song in many birds.

A greater response to undegraded than to degraded song has also been found in several other species (e.g. Fotheringham *et al.* 1997). In western meadowlarks (McGregor & Falls 1984) and in great tits (McGregor *et al.* 1983, McGregor & Krebs 1984b), familiarity with the song type in question has been found to be an important factor: if the birds were unfamiliar with the song concerned, then the response to its degraded form was not significantly different from that to the undegraded one. While this agrees with Morton's idea, some results, particularly those of McGregor & Krebs (1984b), suggest that familiarity need not depend on the bird having the song in its own singing repertoire. Great tits appear as able to discriminate between the degraded and undegraded forms of songs their neighbours sing as well as those that they sing themselves: they may after all be familiar with songs even if they do not sing them. Morton & Derrickson (1996) found that dusky antbirds also responded more to undegraded songs, though here the species is a suboscine and not thought to learn its song, so that all songs are rather similar and all likely to be familiar. By contrast, some other studies have failed to find that familiarity was an important factor. Carolina wrens can clearly assess the distance of songs rather well, as they fly beyond a loudspeaker that has just played a degraded song, and spend more time beyond it than if the song was undegraded (Naguib 1996), and they also do this with songs they are unlikely to have heard previously (Naguib 1997). Kentucky warblers can also assess the distance of playback songs recorded from strangers and which neither they nor any of their neighbours have in their repertoires (Wiley & Godard 1996). Overall, therefore, the evidence is mixed,

possibly because species vary in whether degradation cues vary from song type to song type or can equally be assessed from any song whether previously experienced or not. In one case, that of the black-capped chickadee, birds have been found to respond equally to degraded and undegraded songs (Fotheringham & Ratcliffe 1995): the authors suggest that degradation is a poor cue to distance in this species given the wide variety of habitats it occupies.

The idea that birds can correctly assess the distance away (range) of strangers' songs as well as of those of neighbours and those they sing themselves is damaging to Morton's hypothesis, based as it is on the suggestion that neighbours may disrupt each other's behaviour by singing unrangeable songs (McGregor 1991). The ranging hypothesis, therefore, remains controversial (see Morton 1998a,b, Naguib 1998, Wiley 1998). Although there is now no doubt that degradation is used as a cue in distance assessment, there is little evidence in favour of the suggestion that individuals can manipulate this cue to give false information which benefits themselves at the expense of their neighbours. Indeed, it is questionable whether individuals would gain from disrupting the activities of their neighbours. Such an action seems more akin to spite (the reduction of a rival's fitness, without any gain to that of the individual), and it is doubtful how behaviour of this sort might evolve, given that it appears not to be of any advantage to the animal that shows it.

4.5.2 Direction

There are three different ways in which an animal can estimate the direction of a sound making use of the differences between its ears (Marler 1955): it can detect differences in time of arrival, intensity and phase.

TIME OF ARRIVAL DIFFERENCES

If one ear is closer to the source of sound than the other, the sound will reach it first and this can be used to indicate that the source was on that side of the body. Given the speed of sound, the difference for a songbird will be very small (about 29 μs where the ears are 1 cm apart), even when the head is at right angles to the sound source. Of course, if the sound comes from immediately in front, behind or above the head, there will be no difference at all and it will not be possible to tell exactly where the sound is coming from without moving the head.

INTENSITY DIFFERENCES

Sound will also be more intense on the side nearer the source. This is not primarily because it is closer (with a song coming from several hundred metres away not much attenuation will occur over 1–2 cm!) but is the result of the 'sound shadow' created by the head. Just as sounds are quieter if one stands behind a tree, so the amplitude is lower in the ear beyond the head compared with that closest to the source. Again, the difference is greatest if the sound comes from the side, and there will be none at all if it is equidistant from the two ears, regardless of whether it is in front or behind. This method also depends on frequency. High-frequency sounds bounce off objects and leave a marked sound shadow, whereas low frequencies, with their long wavelengths, bend round them and so are more likely to stimulate the two ears equally. The larger the object the lower the frequencies that will give a sound shadow, but generally this means of location is best suited to high-frequency sounds. Because these are also disrupted by objects other than the head, it will not be very useful in dense habitats such as forests.

PHASE DIFFERENCES

Sound consists of waves of pressure changes and, with low-frequency sounds, the pressure in one ear will be different from that in the other. For example, if the wavelength of the sound is twice the distance between the ears and the sound comes at right angles to the head, one ear will experience peak pressure when the other experiences a trough. This difference can be used to determine the direction of a sound because the phase difference for a given frequency will vary with the angle to the head. The difference will be maximal with the sound source to the side and non-existent when it is immediately in front or behind the head. Here, however, localisation of the sound relies on it being of sufficiently low frequency. If the wavelength is less than twice the difference between the ears, ambiguities arise: the same pattern of pressure between the two ears can arise from sounds in more than one location.

LOCALISATION OF BIRD CALLS

These various possibilities for auditory localisation were first discussed in relation to bird sounds by Marler (1955). He suggested that the 'seeep', an

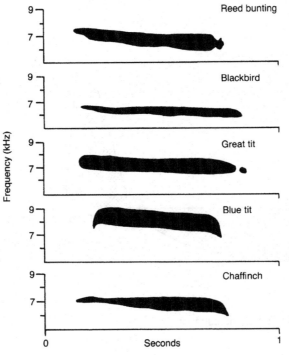

Fig. 4.9. The 'seeep' alarm call of several passerine species (from Thorpe 1961).

alarm call produced in almost identical form by a variety of European passerines (see Fig. 4.9), had converged on that form in different species because it was ideal to avoid localisation. Producing a call in the presence of a predator such as a hawk is a risky business, and selection might be expected to minimise that risk. The 'seeep' is an almost pure tone at around 8 kHz. This frequency is too high for the unambiguous use of phase differences and too low for intensity differences between the ears to be marked. Furthermore, the sound fades in and out so that it is hard to make use of time of arrival differences. It is certainly not at all easy for a human to locate the caller (to the extent that the sound has sometimes been referred to as 'ventriloquial'), and this is probably also the case for predators such as hawks (Brown 1982), assuming that they can hear it at all (Klump and Shalter 1984, Klump *et al.* 1986). As Fig. 4.10 shows, the hearing of the great tit is very much better than that of the Eurasian sparrowhawk at such high frequencies.

If a sound is to be easily located, it should have characteristics which are the opposite of those of the 'seeep'. One would, therefore, expect an abrupt onset and offset so that time of arrival differences were easy to detect. A broad frequency range would also ensure that cues were present

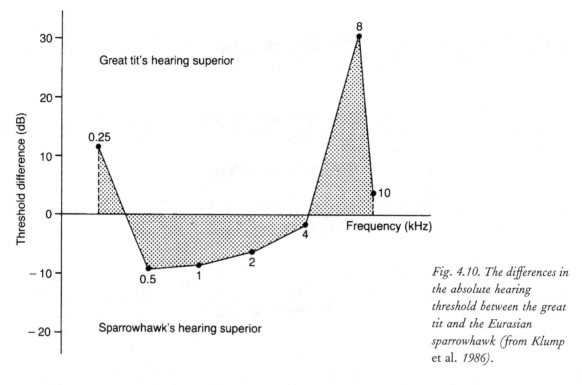

Fig. 4.10. The differences in the absolute hearing threshold between the great tit and the Eurasian sparrowhawk (from Klump et al. 1986).

both for phase-difference detection (low frequencies) and for that of intensity differences (high frequencies). This is true of many bird-call notes, such as mobbing calls or contact calls, which function to attract other birds to the caller and so must be easy to locate. It is also true of many of the elements in bird songs, which are, in addition, often repeated in a series. As noted above, each of the three mechanisms of localisation is of no use for sounds that are equidistant between the ears, as the hearer has no cues to tell whether it is immediately in front, behind or above. A single brief note may therefore be impossible to locate. But with such notes in a series, and sufficient time between them for the bird to move its head, such ambiguities can be removed and the sound located with precision. Anyone who has done playback experiments using bird songs will vouch for the remarkable ability of territorial males to detect and approach the precise location of a loudspeaker playing his species song. Part of this capacity may be because birds use a means of locating sounds different from that we possess. As mentioned in Chapter 2, there is evidence that air passages within the skull allow sound to reach each ear through the head as well as from the outside (Coles *et al.* 1980, Hill *et al.* 1980). Rather than relying on detecting differences between the ears,

each ear may, therefore, compare these two sources; this 'pressure-gradient' system may be a particularly effective way of localising sound for an animal with a small head (Larsen 2004). On the other hand, pressure-gradient detection may be rather poor when it comes to locating the 'seeep' alarm call (Brown 1982).

4.6 Conclusion

Song as a signal will only work if a recipient responds in some way that is to the advantage of the singer. Getting the message to such a recipient without it being hopelessly degraded by distance, distorted by features of the habitat or masked by environmental noises is quite a formidable task. The evidence suggests that a number of features of song and of singing behaviour are quite well matched to those that will give best transmission in the habitat of the species, and there is mounting evidence for finer and more subtle differences within a species in different habitats. The form of song seems well adapted to provide the listener with cues to the distance and direction of the singer, as might be expected particularly for a signal involved in mate attraction. The more controversial idea, that rival males may disrupt each other's behaviour by providing cues about distance which are confusing, has so far not received unequivocal support.

WHEN DO BIRDS SING?

*Northern
mockingbird*

It was the nightingale, and not the lark
That pierced the fearful hollow of thine ear
William Shakespeare *Romeo and Juliet*

5.1 Introduction

In this chapter we will consider in what context song occurs and, in particular, who does the singing and when it takes place. In most species songs are only produced by the male sex and song production is under the control of the male sex hormone testosterone. However, in several species females also sing and, as we will see in Chapter 8, some pairs even sing elaborate duets. Timing will be considered in some detail, as most species only sing at certain times of the year and in a particular daily rhythm. The context in which birds sing is important, as it can give valuable clues as to the possible functions of song. As we shall see in later Chapters, hypotheses can be generated by contextual studies, and then tested by observation or experiment. Everyone has probably heard of the dawn chorus, or at least been woken up by it! But why do birds sing more at dawn than at any other time? We shall see in this chapter that the answers to such apparently simple questions are far from clear, and that some ingenious and even unlikely answers have been proposed.

5.2 Song and the breeding cycle

In temperate zones, bird song is one of the most characteristic sounds which herald the return of spring. There is always some song throughout the year, but in spring the annual rhythm in most species reaches its peak. Resident birds are partly responsible, but their numbers and their noise are swelled considerably by the arrival of migrants on the breeding grounds. Males of most species can be observed singing as they occupy and defend their breeding territories. If they are successful in attracting females, mating, nesting and eventually successful breeding takes place.

In this annual cycle are our first clues as to the possible functions of song. There is a broad correlation between the seasonal production of male song, occupation of a territory and attraction of a mate for breeding. This has been largely responsible for the development of the 'dual function' theory, which suggests that male song functions both to attract females and repel rival males. We will be reviewing the evidence for this view in later chapters, but for the moment we will consider what evidence there is for a relationship between seasonal song production and the breeding cycle itself.

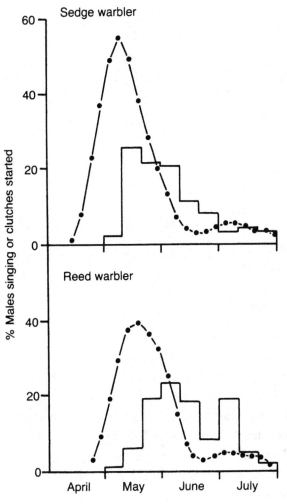

Fig. 5.1. The relationship between seasonal song (●) and breeding activity (histograms) in populations of two Acrocephalus species (from Catchpole 1973).

Although the early naturalists such as White (1789) were well aware of the seasonality of bird song, their accounts were descriptive rather than quantitative. Slagsvold (1977) has undertaken a detailed, statistical study of over twenty species of woodland songbird breeding in Norway. He quantified the amount of song in each species and related it to their breeding activities. Slagsvold was able to show that in each species peak song activity was correlated with egg-laying a few days later.

This general pattern has been confirmed by several studies on individual species, such as that by Lampe & Espmark (1987) on European redwings. Catchpole (1973) studied two marshland *Acrocephalus* species, the sedge and reed warbler, and again in both cases peak song production in the population was closely followed by egg-laying (Fig. 5.1). This annual cyclicity is most

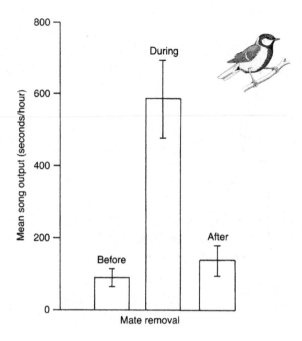

Fig. 5.2. Song output from male great tits, before, during and after mate removal (from data in Krebs et al. 1981b).

obvious in temperate regions where the seasons are clearly separated. The tropics lack such clearly defined seasons and any annual song cycles, if they exist at all, are much harder to detect. Breeding activity may be spread throughout the year, or be synchronised with other environmental variables such as rainfall or food abundance. Catchpole & Komdeur (1993) studied another *Acrocephalus* warbler which breeds in the tropics – the Seychelles warbler. They found that there was a more subtle annual breeding cycle, with two peaks of nesting activity in any one year. They also measured singing activity, and showed that over a period of two years there was a significant correlation between singing activity and nesting two months later.

5.2.1 Song before and after pairing

The European *Acrocephalus* species are migrants and the study by Catchpole (1973) also showed that male sedge warblers sang for between 6 and 12 days after arrival before attracting a female and pairing. As soon as they were paired, the males became silent and never sang again, unless they were deserted, when the diurnal pattern of song resumed once more (see also Fig. 5.8). Such contextual evidence suggests that in this species, song is particularly important as a sexual attractant. A more direct experiment to test this idea is to monitor song production from individual males,

and then remove their mate. This removal technique was applied by Krebs *et al.* (1981b) to mated pairs of great tits. Their results (Fig. 5.2) showed that male great tits increased their song output six-fold when their female was removed, and also that the rate returned to normal when the female was returned. Cuthill & Hindmarsh (1985) showed a similar result with their removal experiment on starlings. Again, the males dramatically increased song production after their females were removed but reduced it when the female returned, again suggesting that female attraction is an important function of song.

Kelsey (1989) studied another of the European *Acrocephalus* species, the marsh warbler. He found a similar picture to the sedge warbler, in that song production peaked in the unmated male in spring. After pairing, the diurnal rhythm was greatly curtailed, with much lower singing activity and small peaks particularly at dusk and dawn. This suggests that again the main function of song is sexual attraction, and that after pairing a much lower level of singing is needed to maintain territorial defence for the rest of the breeding season. In most studies it is impossible to rule out the possibility that song also functions in retaining the female, or even in attracting a second female in polygynous species. However, Kelsey continued his study on marsh warblers in their wintering areas in Africa. Here the males do not breed, but they do defend a winter feeding territory. He found that the amount and diurnal rhythm of song was similar to that of the paired male in Europe. Kelsey concluded that the intense burst of spring song was concerned with attracting and guarding a fertile female for breeding, and that the remaining song was used in defending a territory, whether in Europe or Africa.

5.2.2 Song cycles in mockingbirds

Monitoring song production from individual males throughout the breeding cycle was taken a stage further by Logan (1983) in a study of northern mockingbirds. She estimated the amount of song each male produced at different stages in the breeding cycle. Mockingbirds normally raise several broods a year, and Logan followed their progress right through the season. She found that during each breeding attempt, song activity peaked at the nest-building stage and then declined during the incubation and fledgling stages (Fig. 5.3). Logan suspected that song might have a stimulating effect upon the reproductive cycle of the females, and tested this in a series of

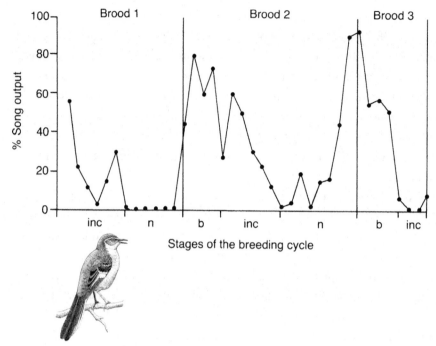

Fig. 5.3. Song output from a male northern mockingbird in relation to different stages in the breeding cycle. b, Nest building; inc, incubation; n, nestlings (from Logan 1983).

playback experiments (Logan *et al.* 1990). In the experimental design, half of the mockingbird breeding pairs had 140 minutes per day of recorded mockingbird song played in their territories through speakers. The other half served as controls and each day were exposed to playback of the same amount of brown thrasher song. In the experimental territories, the mockingbirds began re-nesting much earlier than the controls, suggesting that song does indeed have an effect upon their reproductive physiology.

5.2.3 Song and sperm competition

Although in most birds the peak of singing activity coincides with the onset of breeding in spring, there are some striking exceptions. Hiett & Catchpole (1982) studied yellowhammers in southern England, and found that song production peaked much later, in the summer months of June, July and August. Playback experiments showed that this was also when territorial males were most aggressive, although territorial behaviour and pairing occur much earlier during March, April and May. Møller (1988) studied a yellowhammer population in Denmark, and also described a late season song peak. He found that, like mockingbirds, yellowhammers lay several clutches in succession. He also established that peak male song rates

correlated with the fertile periods of the females. This is the time when males mate guard their females, to stop rival males from obtaining extra-pair copulations. Møller (1988) also found a correlation between mate guarding and song activity within the fertile period in yellowhammers. He suggested that part of the function of male song is to deter rival males from entering territories when females are fertile and vulnerable to extra-pair copulations.

Møller (1991) has extended his ideas on sperm competition and song from the yellowhammer to songbirds in general with his 'fertility announcement hypothesis.' The hypothesis suggests that male birds sing more during the fertile period of their female as a mate guarding strategy, to prevent rival males from invading and obtaining extra-pair copulations. It also suggests that males may sing as a strategy to maximise their own success in obtaining extra-pair copulations from other females. From a review of 49 species he found that 71% show a song peak within the fertile period of the female (nest building through laying). Of the examples we have considered in detail, the yellowhammer and the mockingbird are certainly species which come into this category. The sedge warbler clearly does not, and cases like the great tit and starling, where male song output increases dramatically when the female is removed, seem difficult to fit into his hypothesis.

Greig-Smith (1982a) studied seasonal song in the stonechat. He found that the main peak occurred in March, well before the fertile period, and was associated with territorial behaviour. Later, smaller peaks also occurred, and these were during the female fertile period. Greig-Smith's interpretation was that song functioned in territorial defence as well as attracting and retaining females during their fertile period, whereas Møller (1991) used this example in support of his fertility announcement hypothesis. Pinxten & Eens (1998) studied the starling in great detail and, although they found a significant peak of song in the fertile period, the males directed it towards their own females to obtain more copulations. In a detailed study on the dusky warbler, Forstmeier & Balsby (2002) also found a large song peak in the fertile period. However the males in this species do not mate guard their females, and nor was success in extra-pair paternity related to the amount of song. Other recent studies have simply confirmed what we discussed earlier in this Chapter (5.2.1), that in most species song peaks before pairing and not in the fertile period as Turner & Barber (2004) found in the song sparrow.

Gil et al. (1999) have produced perhaps the most compelling case against the fertility announcement hypothesis. First, they studied the willow warbler and found males sang very little during the fertile period

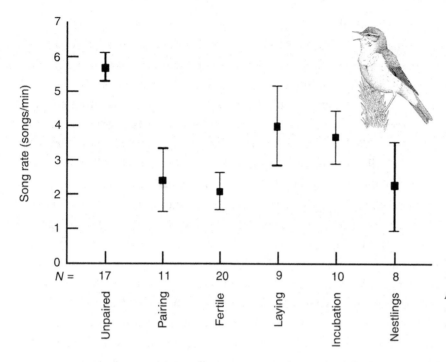

Fig. 5.4. Changes in song rate through different periods of the breeding cycle in male willow warblers (from Gil et al. 1999).

(Fig. 5.4), and male intrusions were not less common when resident males sang at higher rates. Gil *et al.* (1999) also reviewed a number of other studies and found that in most species where male song had been studied on an individual basis, males do not sing during the fertile period. This is quite a different conclusion to the review by Møller (1991), probably due to his use of population estimates of singing and fertile periods rather than using data taken from individual pairs. In most detailed cases now studied, it seems that song output and song rate does not peak within the fertile period. In any case, our more general dual function hypothesis merely predicts that male song both attracts females and deters males and is perfectly compatible with the idea that this remains important during the female fertile period.

5.3 Seasonal song and hormones

The seasonal song of male passerine birds has long been known to be under the control of gonadal steroid hormones (Arnold 1975, Nottebohm *et al.* 1987, Marler *et al.* 1988, DeVoogd 1991). Song production waxes and wanes with the seasonal cycle of testicular growth and secretion of testosterone which circulates in the bloodstream. In temperate zones, this is normally triggered by the increased photoperiod in spring, resulting in

testicular growth, secretion of testosterone and a consequent increase in song production. In autumn the reverse occurs and testicular regression is correlated with a decrease in singing activity. In adult males, song production decreases after castration and resumes with testosterone replacement therapy. In most species normally only males sing, but females implanted with testosterone can be induced to produce song (e.g. Cunningham & Baker 1983). The developmental and seasonal changes in song are accompanied by anatomical changes in the forebrain nuclei which are involved in the control of song, and are also thought to be under hormonal control (Nottebohm et al. 1986, DeVoogd 1991, Ball 1999).

However, the relationship between testosterone concentrations, enlarged song nuclei and singing behaviour may not be all one way, as it has recently been found that singing itself may promote cellular changes in the song nuclei (Ball et al. 2002, 2004). In canaries it has been shown that singing itself enhances the production of brain-derived neurotrophic factor (BDNF), a protein that increases the recruitment and survival of new neurons in HVC (Li et al. 2000, Alvarez-Borda & Nottebohm 2002, Alvarez-Borda et al. 2004). Furthermore, when male canaries are kept together, their singing and consequent development of HVC depends upon the amount and nature of their social interactions, particularly dominance relationships (Boseret et al. 2005). It now seems that both singing and social interactions may affect the development of HVC independently of circulating testosterone. These new findings may well help to explain apparent dissociations between levels of blood testosterone and singing behaviour in some species such as the European robin discussed below.

5.3.1 Hormone levels in wild birds

Most information on hormone levels has come from experimental studies on captive birds, but some have now shown how closely the seasonal production of testosterone is correlated with male singing activity in the wild. Rost (1990) studied the seasonal song pattern in male great tits in Germany. He found that song started to increase as early as January, and reached a peak by March, with a later, smaller peak in September. Great tits were caught and their blood sampled at monthly intervals to measure the levels of plasma testosterone. Rost found that the seasonal course of plasma testosterone was highly correlated with seasonal song production. The levels peaked in March, and there was also a later but smaller peak in September.

Fig. 5.5. The relationship between seasonal song and testosterone levels in male (solid lines) and female (dotted line) willow tits (from Rost 1992).

Rost (1992) has also related plasma testosterone levels to seasonal song production in willow tits (Fig. 5.5). In this study he also measured testosterone levels in females, although females do not normally sing. Again he found a strong correlation within data from the males, the main song and testosterone peaks were in April with smaller peaks in August. The level of testosterone in females showed no peak in spring, and females were rarely heard to sing at that stage. However there was a slight female peak of testosterone in August, similar to the male one. Again this correlated nicely with observed singing behaviour, as Rost found that the few individuals that sing at this time of year are juveniles of either sex.

5.3.2 Hormone levels in European robins

European robins are unusual in several respects, and provide a contrast with most species which are dependent upon testosterone for song production. Studies by Lack (1946) and Hoelzel (1986) have shown that robins are a resident species in which males sing throughout the year. In spring the male defends the breeding territory, but in autumn and winter both sexes occupy separate winter feeding territories and defend them with song.

Schwabl & Kriner (1991) found that, just like other passerines, male robins have elevated levels of testosterone in spring, but not in winter. In this case there is not a simple correlation between song production and testosterone, as male robins sing throughout the winter. Whether spring song has a more complex structure as suggested by Lack (1946) is not yet clear, but Hoelzel found that male song was more complex than female song. Schwabl & Kriner (1991) implanted male robins with flutamide, an androgen antagonist which negates the effects of any circulating plasma testosterone. They found this had no effects upon song production in either spring or winter. This suggests that in the robin, song production itself may have become emancipated from a direct dependence upon circulating testosterone (see discussion by Ball *et al* 2002, 2004), and would certainly help to explain the continued production of song throughout the winter months.

Kriner & Schwabl (1991) also investigated the relationship between testosterone and song in female robins, and found that females singing in winter had elevated levels of plasma testosterone. Flutamide implanted females sang less, and those that received testosterone implants sang more. At least in female robins, seasonal song production appears to be largely under the control of testosterone.

5.4 Females that sing

The robin is one of a number of species in which it is usual for females to sing. Even in species where females do not normally sing it is quite common to find occasional records of female song. In a recent survey, Garamszegi *et al.* (2007) found that, even in Europe, female song has been described in over 100 species of songbird. As we have just seen, song is thought to be largely under the control of the male sex hormone testosterone, and when plasma levels become elevated, in most cases the probability of song production is greatly increased. Females normally have low levels of circulating testosterone, but if these are increased then females will often produce song. Cunningham & Baker (1983) capitalised upon this in their study of song dialects in white-crowned sparrows. They implanted captive females with testosterone and were able to check whether females had learned the same song dialect as their mates.

What is probably happening in many cases of females singing in the wild is that their hormone levels occasionally rise above the threshold level for song production. In these isolated cases there is no need to seek a

general, functional explanation for female song. Arcese *et al.* (1988) investigated the incidence of female song in a population of song sparrows, and found that only 12 out of 140 females produced song. However Ritchison (1983a) reviewed the literature and listed some 40 species where female song was relatively common. Both Langmore (1998) and Riebel (2003b) in their reviews also emphasise that female song is often overlooked. One reason for this may be that female song in the tropics is much more common than in the much-studied temperate zone (Slater & Mann 2004). We shall now look more closely at female song and see whether females use their songs in similar or different ways from males.

5.4.1 Female song in red-winged blackbirds

The polygynous red-winged blackbird of North America has probably been studied as much as any bird, and female song is no exception. Females sing throughout the breeding season, and Beletsky (1983a) found that they produce two distinct types of song, quite different in structure from the normal male song (Fig. 5.6). Type 1 song is produced in the presence of males and seems to function in communication between the mated pair. Beletsky (1985) showed that when females are exposed to playback of male song, they invariably answer with type 1 song. Beletsky & Orians (1985) also found that when females left the nest they gave the type 1 song.

Type 2 song is produced when other females appear, and seems to be an aggressive signal (Beletsky 1983a). Male red-wings are polygynous and attract several females to nest on their territories. Each female defends her own small section of territory, and Beletsky (1983b) used playback experiments to show that females use the type 2 song in response to simulated invasion by other females. Beletsky (1983a,b) found that whichever type of female song he played back to females, they replied with the aggressive type 2 song. It seems that the type 2 song of female red-wings is the counterpart of male territorial song used to repel rival males (Chapter 6). When Beletsky (1983c) played back both kinds of song to males, they were able to discriminate between different females. However, female red-wings were not able to discriminate between each other in a parallel series of playback experiments (Beletsky 1983d). It may be that is important for males to recognise the different females in their harem, but less important for females to recognise each other as individuals.

Fig. 5.6. The different
male and female songs of
red-winged blackbirds
(from Beletsky 1983a).

Yasukawa *et al.* (1987) monitored the production of the two types of
song through the breeding season. They found that type 1 songs were
given fairly constantly, but that females who received no help from their
males gave more. The rate of type 2 songs gradually declined through the
season, and they concluded that these songs were produced to deter later
arriving females from settling and breeding.

It seems that female song in the red-winged blackbird is unusual in several respects. Not only is it quite commonly produced, but there are two structurally distinct forms, one for communicating with each sex. As we will see in later chapters, there are some striking parallels with the male song of some other species, although in most cases the dual function of repelling rivals of the same sex and attracting members of the opposite sex are contained within the same song.

5.4.2 Other examples of female song

We have already considered one other well studied example, the European robin, where song is a normal and important part of female behaviour (Lack 1946). Female song in this species is clearly related to winter territorial behaviour, and is shorter and simpler than male song (Hoelzel 1986). When males were played back songs from males or females, they produced more complex songs when responding to females. This suggests the shorter and simpler songs are used by both sexes in territorial defence, and that males use a longer and more complex version of song for female attraction.

Baptista *et al.* (1993) found that female white-crowned sparrows produce a song which is structurally similar to that of the male, but usually shorter (Fig. 5.7). The song was produced regularly during autumn, winter and spring prior to breeding, but then ceased. As well as producing spontaneous song, the females also responded with song to playback of male song. Another species in which female song appears to be shorter and simpler than in the male is the black-headed grosbeak. Ritchison (1985) found that female songs were very variable, and that they were used when females and fledglings were together in family groups. He confirmed with playback experiments that young grosbeaks were indeed able to recognise and respond to the songs of their parents (Ritchison 1983a). Grosbeak families often leave their territories in search of feeding areas, and females sing to maintain contact with their wandering fledglings. Ritchison (1983b) also found that females occasionally produce a longer song which is indistinguishable from male song. This is usually produced when the male has been absent during incubation for some time and is overdue to relieve his female on the nest. Ritchison suggests that the female is deceiving the male and tricks him into returning by giving the impression that a rival male is trespassing on his territory.

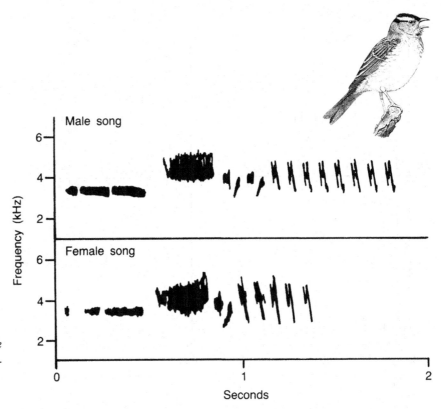

Fig. 5.7. The similar male and female songs of white-crowned sparrows (from Baptista et al. *1993).*

Finally, we have examples from species with more unusual mating systems. Cooney & Cockburn (1995) studied the cooperatively breeding superb fairy-wren in Australia. Playback experiments revealed that females were more likely to sing in response to songs of female strangers than to songs of female neighbours, a result reminiscent of classic male neighbour-stranger experiments (see Chapter 6). In two accentors with complex mating systems, female song seems to have a more sexual function. In the alpine accentor, Langmore *et al.* (1996) used playback to provide the first experimental evidence that female song attracts males. In polygynandrous groups, females are in competition with each other for males and use song to attract males for copulation. Females in the closely related dunnock also live in complex mating systems and show a similar pattern in their use of song to attract males (Langmore & Davies 1997).

In conclusion, females of many species do produce song and appear to use it in much the same way as males, suggesting that both intra and intersexual selection have acted to shape female song for the same basic functions of territorial defence and mate attraction. Females also join

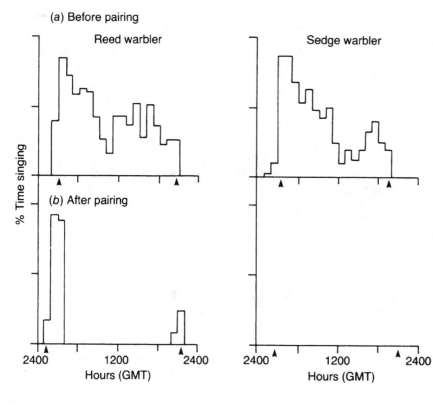

Fig. 5.8. Diurnal rhythms of song in a male sedge and reed warbler before (top) and after (bottom) pairing. Arrows indicate approximate times of dawn (sunrise) and dusk (sunset) (from Catchpole 1973).

in with males to produce song duets, but this is discussed later in Chapter 8.

5.5 The dawn chorus

Anyone who has studied bird song soon finds out that there is one time of day which is particularly good for observing, recording and carrying out field experiments – the early morning. Male birds show a marked peak of singing activity around dawn, and because so many individuals of many species join in, this has become known as the dawn chorus. Although many observers have noted this effect, surprisingly few have quantified the pattern in any detail. Catchpole (1973) studied the diurnal rhythms of song production in two European *Acrocephalus* species, the reed and sedge warbler (Fig. 5.8). In the unpaired male, both species show a well defined pattern of intense song production around dawn, which gradually declines towards midday. Observations showed that males were gradually spending more time feeding and less singing as the day progressed. A second peak occurred towards dusk and then silence for the relatively few hours of

darkness. After pairing, sedge warblers cease singing completely, but reed warblers like most other species continue to produce a strong peak at dawn and a lesser one at dusk.

Although most species contribute to the dawn chorus, it has often been noticed that they tend to start at different times, some earlier than others. For example in an English woodland, robins, blackbirds and song thrushes are 'early birds' whereas chaffinches and blue tits join in later. Armstrong (1973) noticed that species which were earlier tended to have larger eyes, and speculated that as they could see better at low light intensities they tended to start daily activities earlier. This has now been tested by Thomas et al. (2002), who actually measured eye size and pupil diameter as well as the time at which different species started to sing. They found that species with larger eyes did start to sing earlier, supporting the hypothesis that visual capacity at low light levels determines when a particular species will start to become active and sing in the morning light. Berg et al. (2005) completed a similar comparative study on a wide range of species in the tropics. They also reached a similar conclusion but, although species with larger eyes sang earlier, foraging height was the best predictor with canopy species starting earlier than those lower down in the darker conditions of the forest floor.

This is obviously a causal explanation and there has been much more speculation as to the possible functions of the dawn chorus and a long list of possible explanations (reviewed by Staicer et al. 1996). The dawn chorus of the great tit has been studied in particular detail, and has given rise to many of the ideas about why it may be best to sing at dawn. Kacelnik & Krebs (1983) combined these various ideas into one functional hypothesis which considered three main advantages of singing at dawn. First, sound transmission is particularly good at dawn; second, birds are not able to see sufficiently to forage, and finally, some territories may have become vacant overnight and new prospecting males may be moving in. We will now consider these as well as other ideas, for example that dawn song may also defend a fertile female during the egg-laying period.

5.5.1. Sound transmission at dawn

Henwood & Fabrick (1979) produced a model of sound transmission which suggested that conditions at dawn are particularly favourable. Certain microclimatic conditions occur around dawn, and the most significant for sound transmission are reduced wind and air turbulence.

Henwood & Fabrick calculated that songs broadcast at dawn could be 20 times more effective than when broadcast at midday. This may be somewhat optimistic, judging by the detailed studies on sound transmission at dawn in American sparrows carried out by Brown & Handford (2003). They found no difference in the quality of song transmission between dawn and midday, but did find that transmission quality was more consistent at dawn. The importance of sound transmission has already been discussed in detail in Chapter 4. If bird songs really do propagate most effectively at dawn, then it clearly makes sense to utilize this particular time slot, although this may not be the only explanation for the widespread occurrence of the dawn chorus.

5.5.2 Feeding conditions at dawn

We have already seen that one reason why sedge warbler males sing less after dawn is that they start to feed more. Kacelnik & Krebs (1983) suggested two reasons why feeding conditions might be bad at dawn so that singing becomes the more favoured option. First, light intensity is low making it difficult to hunt by sight, particularly for cryptic prey. Secondly, relatively low temperatures at dawn may reduce the activity of invertebrate prey, again making it more difficult to detect. Kacelnik & Krebs (1983) also reviewed experimental evidence which does suggest that decreases in both light intensity and temperature reduce the foraging success of captive great tits.

Cuthill & Macdonald (1990) tried supplementing the food supply of paired male European blackbirds. They found that the extra food resulted in the males singing earlier, longer and at a higher rate in the following dawn chorus. Using an experimental laboratory approach, Mace (1987b, 1989) manipulated foraging profitability in captive great tits by varying their food reward from computer controlled feeders. The computer also recorded their daily feeding and singing activities. As seen in Fig. 5.9, when profitability was low, this particular male cut down on foraging at dawn and dusk, but still produced a dawn peak of song. Overall, there was no significant effect upon their singing patterns and they all continued to produce a dawn chorus.

The diurnal rhythm of song is clearly influenced by the demands of other behaviours such as feeding, but the nature of the relationship is not a simple one and still far from understood. One approach is to use stochastic

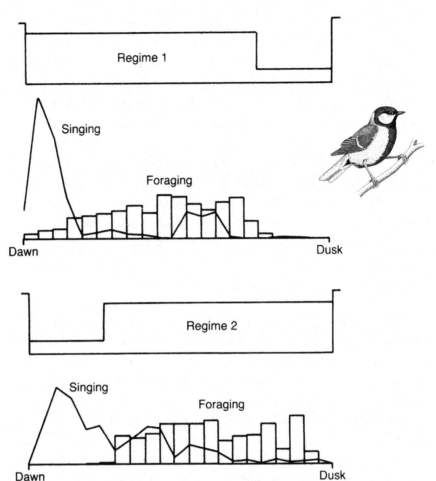

Fig. 5.9. The daily
routines of singing
(graphs) and foraging
(histograms) for a captive
male great tit on two
different foraging regimes.
In 1, foraging
profitability is high for
most of the day but low at
dusk, whereas in 2,
foraging profitability is
low at dawn but high for
the rest of the day (from
Mace 1987b).

dynamic programming (SDP) models (Houston & McNamara 1987, McNamara *et al.* 1987, Hutchinson *et al.* 1993). Their SDP models offer a general explanation of daily singing routines by predicting what the optimum behaviour of a bird should be at any one time in relation to just three possible activities: singing, foraging and resting. These behaviours are mutually exclusive and so there are trade-offs between the three to be made. For example, a bird cannot spend all its time singing or resting as food must be obtained in order to build up enough fat reserves to last through the night. Hence the model relates the relative costs and benefits of feeding and singing at different times of day to the size of the bird's fat reserves.

SDP models predict that if foraging success increases then so does song output, and Thomas (1999a) tested this in a population of wild European robins. He artificially increased foraging success with food

supplementation in territories and found this increased subsequent song output. The models also predict that stochasticity (unpredictability) will have an important effect upon the birds' daily singing routine. If unpredictability increases, birds should forage more at dawn as a buffer against bad luck, leaving less time for the dawn chorus. When Thomas (1999b) again manipulated the food supply, and made it less predictable, his robins responded by reducing their contribution to the dawn chorus. In a recent study, Thomas & Cuthill (2002) have gone on to confirm more predictions from the SDP model, for example that robins sing more at dawn when their body mass remains high. It is clear that a variety of experimental manipulations have now shown how food and foraging behaviour can influence the timing and duration of the dawn chorus.

5.5.3 Territory and vacancies at dawn

Kacelnik & Krebs (1983) argued that, due to overnight mortality, dawn is the time that any vacant territories first become apparent. This is when invading males would be most likely to secure a vacant territory, and are most likely to attempt invasion. Kacelnik & Krebs (1983) also presented some data which suggest that male great tits do indeed attempt to invade vacant territories in the early morning. However, the territorial prospecting of newly arriving males has been rarely studied in detail. Amrhein et al. (2004) have recently plugged this gap in our knowledge by using radio transmitters to follow individual male nightingales as they prospected for territories (Fig 5.10). They simulated newly arriving males by capturing them in one area then releasing them in another. The males made extensive excursions and visited the territories of several resident singing males at dawn, whereas after dawn they remained stationary. The resident males sang mostly at dawn, presumably to defend their territories against such intruders, and indeed Kunc et al. (2005) have concluded that male nightingales use dawn song mainly in territorial defence.

Amrhein & Erne (2006) also studied dawn singing in the winter wren, where once again territorial defence seems to be an important function. They used song playback to simulate intrusions at dawn then looked at song output the next day. The males who had suffered intrusions sang more than unchallenged controls at dawn the next day. It seemed that they remembered their previous experiences and responded the next day, perhaps anticipating more territorial challenges from rival males.

Fig. 5.10. The prospecting path of a single radio-tracked male nightingale released into a new area. Starting just before dawn he covered 1400 metres and visited 5 out of 7 of the territories occupied by resident males singing during the dawn chorus (black dots). (Figure courtesy of Valentin Amrhein, based on data in Amrhein et al. 2004.)

Defending territories from rivals at dawn is clearly important, even is species which may use their song primarily in female attraction. Such a case is the chipping sparrow studied by Liu (2004). Liu had previously observed that the normal diurnal pattern of song appeared to be used for mate attraction as males ceased singing after pairing. However, even paired males resumed song at dawn when interacting with rival males at the territorial boundaries. Liu removed neighbouring males and found that this stopped the dawn chorus of residents, but when the neighbours were put back dawn singing resumed. Removing females increased the daytime singing of males, but the dawn chorus was not affected. This series of experiments strongly supports the hypothesis that, in chipping sparrows, the function of the dawn chorus is to defend territory from rival males.

In conclusion, it certainly seems that in some species dawn is an important time for residents to defend their territories with song. Newly arriving males are a constant threat, exploring the area looking for vacancies and perhaps also assessing the resident males as they sing. Yet in some species, studies have found that dawn song can also be directed at females and separate functions may be more difficult to ascertain. For example in the great tit, Slagsvold *et al.* (1994) found that dawn song was used in both male and female contexts. We will now go on to discuss several examples

where the dawn chorus seems more important in attracting or guarding females.

5.5.4 Females and fertility at dawn

As we will see in Chapter 8, male song is an important signal to females searching for a mate and it may be that the dawn chorus is an important time and opportunity for females to assess then choose a mate. Otter *et al.* (1997) studied dawn singing in male black-capped chickadees during the breeding season and related it to male dominance in winter flocks. Higher ranking males sang earlier, longer and faster, so females could in theory assess the quality of prospective mates by listening to the dawn chorus. Indeed Poesel *et al.* (2001) studied a blue tit population, and found that females mated to males with higher song output at dawn bred and laid eggs earlier.

Mace (1986, 1987a) pointed out that in the great tit, the main dawn chorus does not take place until the female fertile period, and coincides with the period of egg-laying. Male great tits take up territory and pair well before this time. Mace observed that when the paired male awakes, he sings until the female emerges from the nest hole and then copulates with her. His song output then drops to a much lower level. Like many birds female great tits lay at dawn, and are then at their most fertile as the next egg starts its journey down the oviduct and can be fertilised by male sperm (Birkhead & Møller 1992). This would be the time that male paternity is at most risk from invasions by rival males and attempts at extra-pair co-pulations. Mace therefore suggested that the dawn chorus in the great tit has another important function, to protect male paternity by helping in mate guarding. Cuthill & Macdonald (1990) found that in the European blackbird the dawn chorus started earlier and continued longer as females reached peak fertility. They also emphasised the role of song in mate guarding during the period of female fertility. Møller has investigated the role of the dawn chorus in mate guarding in the European swallow, a species where extra-pair copulations are well known, and where male song activity peaks within the female fertile period. He extended the female fertile period by removing eggs, and found that males also extended their song peak (Birkhead & Møller 1992).

Not all studies support the idea that mate guarding is an important function of the dawn chorus. Part (1991) studied the collared flycatcher,

another species where extra-pair copulations are well known and mate guarding is clearly at a premium. Part found that although the diurnal pattern of the dawn chorus was similar to that in the great tit, there was no seasonal correlation with the period of peak fertility in females. Diurnal studies showed that dawn singing in males simply continued until the female appeared, and then stopped. Part concluded that male collared flycatchers are directing their songs at their own females, not rival males or other females.

Kempanaers *et al.* (1992) has shown that female blue tits often leave their territory at dawn seeking extra-pair copulations with higher quality males. It has also been shown that extra-pair paternity in blue tits is related to song output at dawn as more successful males sang longer songs during the dawn chorus (Kempenaers *et al.* 1997). Poesel *et al.* (2006) have now looked more closely at the relationship between dawn song and extra-pair paternity in the blue tit. They found that males who began to sing earlier at dawn had more mating partners and were more likely to gain in extra-pair paternity. These early singing males also tended to be older, so perhaps females are able to use dawn song as an indicator of male quality, selecting males who may carry good genes for viability.

It seems that there are indeed good reasons why birds should sing more at dawn than other times of day, and that in most cases these are associated with the basic dual functions of keeping out rival males and attracting females to mate. From the observations and experiments conducted so far it is difficult to interpret whether a male singing during the fertile period at dawn is warding off rival males, attracting and stimulating his female prior to copulation, attracting other females to obtain extra-pair copulations, or perhaps combining all of these important functions in his song.

5.6 Avoiding competition

So far we have considered how birds sing as if they were doing so in isolation, and were unaffected by the songs of other birds. But having considered the dawn chorus, the very term suggests that other males are also joining in. If this is so, then just as there is competition for space and suitable habitat, there may also be competition to get the message across in what may turn out to be a very busy and crowded acoustic channel. The advantages of avoiding competition could be considerable, any masking or interference would be minimised, and the efficiency of the communication

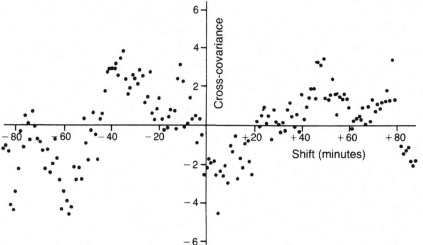

Fig. 5.11. Temporal song asynchrony in two species that thereby avoid competition for acoustic space. The dots show the cross-covariances as the wrentit is shifted forward (+) or backward (−) every minute with respect to the Bewick's wren (from Cody & Brown 1969).

system maximised. This would be particularly important where two species show similarities in song structure, such as sharing the same frequency range. There is now some evidence to suggest that the singing behaviour of males is not only effected by the songs of conspecific neighbours, but also the songs of other species with whom they share their habitat.

5.6.1 Competition from other species

One of the first studies to suggest that birds may adjust their song output in relation to other species was by Cody & Brown (1969). They studied the two most abundant species which occur in the Californian chaparral, the wrentit and Bewick's wren. In a detailed statistical study, they recorded which species were in song during every minute for two hours each morning during the early spring. They then used time series analysis by computer to test for any cyclical activity in the singing pattern of each species. Both species tended to cycle their song activity with a half-period of about 50 minutes, but the two cycles were markedly asynchronous. In other words, when wrentit song was at a peak, Bewick's wren song was at its lowest point, producing the asynchronous cycling effect seen clearly in Fig. 5.11. It is as though each species attempts to avoid acoustic competition with the other by subtly varying the diurnal rhythm of song production. How this is done is not clear, but Cody & Brown found some evidence to suggest that far from sharing equally, one species may control the other. The Bewick's wren is always the first to start in the morning, and

so initiates the cycle which the wrentit is then forced to follow if it is to avoid direct competition. Recently Brumm (2006) has used playback experiments to investigate whether singing nightingales avoid temporal overlap with six other species. He found that they did indeed avoid overlapping their songs with playback and sang mainly during the silent periods.

Several other studies have since revealed that this phenomenon may be widespread in other bird communities. Ficken et al. (1974) found that two forest species, red-eyed vireos and least flycatchers, tended to avoid singing in direct competition with each other. The singing patterns of individual males were recorded and analysed in great detail. Statistical analysis showed that flycatchers avoided beginning a song while a vireo was singing. The result was not so strong when they analysed vireo singing, again suggesting that one species tends to dominate the other. However, the study demands some caution: Planck et al. (1975) pointed out that there are difficulties in the analysis used as it assumes that the successive songs of an individual are independent of each other and this is very unlikely to be the case. Popp et al. (1985) went further by analysing a more complex system of four forest species. They found that each species attempted to avoid overlapping songs of the others, often by singing as soon as there was a gap. Each species sang more often and more regularly when the other species were silent and there was no need to modify their normal rhythm. Popp & Ficken (1987) were able to confirm this by playback experiments on one of the species, the ovenbird. When they played back the songs of other species, the ovenbirds adjusted their singing patterns to avoid overlap and interference.

5.6.2 Competition from other individuals

As well as varying the diurnal rhythm of their song, birds also sing in another rhythmic way referred to as cadence. Cadence involves not only length of the the song itself, but also to the regular pattern of silence between songs (intersong intervals). Hartshorne (1973) pointed out that whereas songs may often be quite short, the silent intersong intervals are usually longer. Beletsky (1989) has shown that intersong intervals can also be remarkably constant, and suggested that these rhythmic periods of silence may be an important and integral part of the song signal. Slater (1981, 1983a) has pointed out that the periods of silence may also be important

for the singing bird to listen and evaluate any replies to his own song. There are some studies which have attempted to see whether or not males adjust the rhythm of their own song in relation to that of other individuals, either to avoid competition, or even to mask or match their rivals message.

Wasserman (1979) applied the technique used earlier by Ficken *et al.* (1974) to individual white-throated sparrows. He showed that males avoided singing when other males were in song, and instead sang when their neighbours were quiet. Ficken *et al.* (1985) investigated this phenomenon in ovenbirds, using playback experiments. They found that males replied by singing during the first part of the quiet inter-song intervals, and so invariably avoided overlapping with the playback songs. In a sample of 250 songs, males only overlapped the stimulus song 10 times. Smith & Norman (1979) studied the natural way in which red-winged blackbirds responded to songs from their territorial neighbours. They found that males settled into a 'leader-follower' rhythm, one bird answering another in a regular pattern. Detailed analysis confirmed that the way one male sang always affected the other, and the two rhythms were not independent of one another.

However some workers have noticed that, far from avoiding overlap, individuals of some species seem to consistently attempt to overlap their rival's song. Todt (1981) found that, in the European blackbird, males avoided perches where their songs had been overlapped in playback experiments. In the nightingale, Hultsch & Todt (1982) found that individual males varied a great deal on how they replied to rivals in territorial interactions and used several different temporal strategies. Some, which they termed 'autonomous', showed no evidence of adjusting their timing to other males, whereas others always attempted to overlap with their rivals. Clearly, not all species attempt to avoid singing when their neighbours do, and some appear to deliberately overlap them as some sort of reply. There may be certain functional similarities between overlapping and matching a rival's song, and these phenomena are two of several complexities in singing behaviour discussed further in Chapter 8.

CHAPTER SIX

RECOGNITION AND
TERRITORIAL DEFENCE

*Great tit male
and rival*

The most important use of song to the robin in its territory
is to advertise possession to rivals and warn them off
David Lack *The Life of the Robin*

6.1 Introduction

In this chapter we move on to consider not only who does the singing, but to whom is the song directed? We have already discovered that males do most of the singing, and in this chapter we will find that males are also very much on the receiving end of bird song. The males of many species occupy and defend a breeding territory, and it seems that song can act as an effective territorial proclamation and first line of defence. However, obtaining direct evidence for a territorial function of bird song is extremely difficult, and has led to the design of some remarkable field experiments. One technique which continues to be developed is the play-back experiment, when the territory of a male is invaded by a loudspeaker broadcasting recordings of songs. Because males respond so readily to playback in the field, this technique has also been used with great success to investigate the features of songs which males use in species recognition. But males also recognise and sometimes respond to the songs of other species, such as closely related competitors. They also learn to recognise the songs of their neighbours, and can discriminate between neighbours and strangers of the same species.

At this stage we should point out that, although most of the experiments in this chapter demonstrate discrimination, it is more difficult to be sure that they also involve recognition. The problem is that recognition is an internal, cognitive process that we cannot observe directly, so many studies simply infer it from experiments designed to demonstrate discrimination. From our review it is clear that most workers still prefer to use the term 'recognition' and so, with the above caveat in mind, we have retained it in the book.

In recent years, research on territorial males has been extended to include more complex interactions within whole communities. Groups of territorial males live together in a communication network and may also 'eavesdrop' on their neighbours to gain additional information. This suggests that song, given mainly in a territorial context, has an extremely wide range of important functions, facilitating recognition of species, sex, kin, mate and many other individuals within the local community.

6.2 Territorial defence

In temperate regions, the marked seasonal rise in male song production coincides with the onset of male aggression and the occupation of

territory. In most species, a territory is an essential prerequisite for the attraction of a female and successful breeding. This relationship between territory and song in birds has been observed and emphasised by many early naturalists, such as Gilbert White of Selborne. White (1789) suggested that male rivalry for space was the most important function of bird song. With the publication of Eliot Howard's (1920) influential book on territory, the concept became widely known, and ever since then the view that bird song developed primarily as a territorial proclamation to rival males has held sway. In his major review Armstrong (1973) clearly regarded 'territorial song' as the most important type of song produced by male birds.

The term 'territorial song' thus came into common use, and with it the assumption that song has a territorial function. After all, male birds can be observed singing vigorously at rival males, and such vocal duals frequently escalate into overt aggression. Yet, until quite recently, there was no direct evidence that song actually functions in territorial defence. What was needed was some kind of experimental technique to test the hypothesis that song itself really does exclude or repel rival males. What evidence there now is comes from two different kinds of experiment; removing the ability of males to sing, and replacing males with speakers playing recorded songs.

6.2.1 Muting experiments

The first attempt to remove the ability of male birds to sing was performed by Peek (1972) on red-winged blackbirds. He trapped territorial males, anaesthetised them, and removed a small portion of the hypoglossal nerves which innervate the syrinx. This effectively muted the birds who were unable to produce normal songs. The operated birds quickly recovered, and when released back onto their territories appeared to be normal in all other respects. Control males were anaesthetised and operated on, but the nerves were left intact. Peek found that the muted males had their territories invaded much more frequently than the controls (Fig. 6.1), were involved in more fights and some even lost their territories.

Smith (1976) repeated Peek's red-wing experiment, but obtained less convincing results. He then went on to develop an improved muting technique (Smith 1979). Instead of severing the hypoglossal nerves, which can lead to respiratory problems and so confound the experiment,

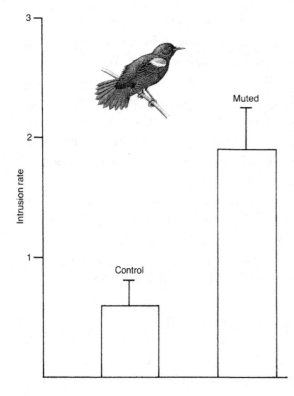

Fig. 6.1. The intrusion rates (intrusions per hour) of rival males into red-winged blackbird territories containing either operated (muted) or sham-operated (control) males (from data in Peek 1972).

Smith punctured the thin membrane of the interclavicular air-sac. This relatively simple operation produced more effective muting, as well as fewer side effects which might affect a male's ability to defend his territory. The males also eventually recovered their ability to sing when the hole healed, and so also served as their own controls. Smith confirmed that muted males suffered more frequent invasions from rivals. They also resorted to more visual signalling, using their red epaulets in an attempt to compensate for their lack of song. When the males finally recovered their ability to sing, they rapidly regained any territory which had been lost.

In another muting study, McDonald (1989) used the air-sac technique on the seaside sparrow. In this example, the experiment is more refined, as seaside sparrows only lose their ability to sing. Other vocalizations such as calls, which might also be important, are still produced. The experiment demonstrated that males incapable of song were slower to gain a territory, and like red-wings suffered more invasions from their neighbours. This is clearly a more precise test of the hypothesis that it is song alone which functions in territorial defence by keeping out rival males.

Westcott (1992) performed an air-sac muting experiment on a lek breeding species in Costa Rica, the ochre-bellied flycatcher. Song rates were correlated with the rates at which both males and females visited the lek, but when six males were muted their places on the lek were invaded by rival males. Five of the muted males lost their territories to floaters or satellite males, but none of the six intact controls nor the six sham-operated controls suffered this fate. It seems that song is also important for defending territories on leks, as well as those of the more conventional kind.

6.2.2 Speaker replacement experiments

An alternative technique to removing songs from males, is to remove the males themselves, and then replace them with speakers which play back recordings of songs. In muting experiments there is always the worry that, in spite of the use of controls, the operation may somehow impair the male's ability to defend his territory. Even with controls, there are still other variables in the system stemming from the males themselves, such as differing plumage, size or behaviour. A speaker replacement experiment is an attempt to isolate the effect of song itself, by completely removing the male and his confounding variables from the experiment. Any possible effects can only be attributed to whatever the experimenter plays through the speakers, giving much more control over experimental design.

A speaker replacement experiment was first attempted by Goransson et al. (1974) working on the thrush nightingale in Sweden. They replaced four males in a wood with speakers and recordings, and claimed that three of these territories were not invaded even after ten days of continuous playback. Control territories where the males had also been removed, were occupied by new males in under five days.

This basic technique was refined by Krebs (1977a) who used more subtle controls in his experiments on male great tits (Fig. 6.2). He removed all eight territorial males from a small wood in Oxford, and replaced them with a sophisticated system of speakers placed in trees. Several of the original territories were occupied by a system of four speakers and a tape recorder containing a continuous recording of normal great tit songs. Multiway switches ensured that the normal singing behaviour of a great tit singing in different parts of his territory was faithfully reproduced. The wood also had two control areas, one being left completely empty and

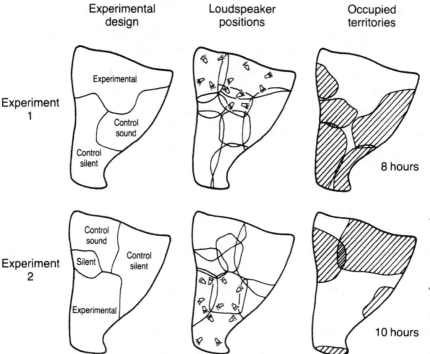

Experimental design Loudspeaker positions Occupied territories

Experiment 1

Experimental

Control sound

Control silent

8 hours

Experiment 2

Control sound

Silent

Control silent

Experimental

10 hours

Fig. 6.2. Excluding male great tits from parts of a wood by two speaker replacement experiments. The experimental design, positions of the speakers, and the resulting pattern of occupation after a few hours are shown (from Krebs 1977b).

silent, while the other area contained similar equipment and a control noise from a tin whistle. The first experiment took place in February, and after a few hours both control areas were occupied by prospecting great tit males, whereas the experimental area was avoided. As a final precaution, the experimental and control areas were changed and the experiment repeated in March, with much the same result. In both cases the experimental area was initially avoided, whereas the control areas were quickly occupied (Fig. 6.2). Eventually, after about two days, the whole area of the wood was reoccupied by singing great tits. Although song itself seems to deter intruders for several hours, there is no physical backup by real males, and eventually the bluff is called. What these experiments suggest is that song itself is indeed an effective first line of defence, but that a real male is still needed to reinforce the message and maintain territorial integrity.

Due to the obvious practical difficulties of such elaborate field experiments, there have been rather few attempts to repeat these pioneering studies. Falls (1987) used a similar design to Krebs on the white-throated sparrow in Canada, but with fewer controls. He replaced eight territorial males with speakers playing recordings of their songs, and removed males

from eight other territories as controls. The experimental territories were invaded later, and had fewer intrusions into their central areas. This pattern of results was also confirmed by Nowicki *et al.* (1998b) working on song sparrows.

The hypothesis that songs repel rival males has been extended to account for the tendency that many species show of increasing complexity in song structure and performance. Instead of singing one version of their species song, most individuals have a repertoire of different song types which they use. The central problem of song complexity and repertoire size, and why it has evolved in so many bird species, recurs at intervals throughout this book (see Chapters 7 and 8). But at this stage, we are concerned with the possibility that repertoire size might be important in male-male interactions, perhaps by enhancing territorial defence. But how could this possibly work?

The answer was suggested by Krebs (1977b) who provided an ingenious solution, inspired by the famous P. C. Wren novel *Beau Geste*. In the story, Beau Geste was in the French foreign legion, and faced with defending a desert fort alone. To foil the enemy he propped up dead colleagues, and ran behind them firing rifles to create the impression that the fort was well manned. His bluff worked, and Krebs suggested that by singing many different song types a male great tit might give the impression that a wood was fully occupied, so causing prospecting males to hunt for a territory elsewhere. Krebs *et al.* (1978) then designed a speaker replacement experiment similar to the previous one, only this time testing for the effects of repertoire size. As before, he divided the wood into three areas, but now one was defended by playback of a repertoire of song types, one by a single song type, and there was also a control silent area. The results showed that the control area was occupied almost immediately, the single song type area next, and the repertoire area was only occupied last of all (Fig. 6.3).

Although the Beau Geste hypothesis has some experimental support, Slater (1978) has pointed out that the way in which chaffinches sing (for example repeating song types rather than switching) is inconsistent with some predictions from the hypothesis. For example, to give a good impression that there are really several birds present, males should change perch when they switch song types. Dawson & Jenkins (1983) found that chaffinches did not synchronise changes of perch and song type in this way, and Bjorkland *et al.* (1989) found the same in great tits. There is only one

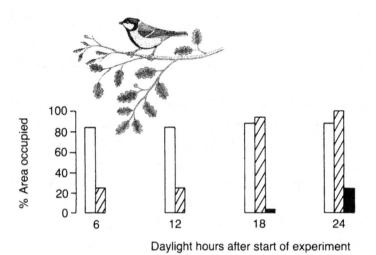

Fig. 6.3. Rates of occupation by male great tits of three areas of woodland during a speaker replacement experiment. The white area was a silent control, the shaded area contained speakers playing a single song type, and the black area contained speakers playing a repertoire of song types (from Krebs et al. 1978).

other example where the Beau Geste hypothesis has been tested directly by a speaker replacement experiment. Yasukawa (1981) studied red-winged blackbirds, and also found that a repertoire of song types was a more effective keep out signal than a single song type. He also found that the males in this study did tend to change perches when switching song types. However in a later study (Yasukawa & Searcy 1985) there was no evidence that prospecting male red-wings actually avoided areas which were more densely populated.

In some ways the speaker replacement experiment is a more satisfactory experimental design, as it isolates the effect of song itself from other confounding variables. Muting experiments cannot control for the additional effects of the individual birds themselves, including other behaviour such as visual displays. Nevertheless, by using each male as his own control, some of these effects can be minimized. Taken together the results from muting and speaker replacement now provide an important body of direct evidence that song itself acts as a first line of territorial defence. When song is present, prospecting males seem deterred from invading a territory. When it is removed, territories are quickly invaded and resident males have to fight considerably harder to retain their territory. It seems that, by singing, territorial males may have a number of advantages, such as reducing their chances of injury in fights. Instead of constantly chasing and fighting off rivals, they are able to conserve valuable energy and resources which can then be used to enhance future reproductive success.

6.2.3 Interactive playback experiments

The simplest kind of playback experiment places a speaker in the territory of a bird and plays back a tape containing a recording of the species song. The usual result is that the resident male sings back and approaches the speaker as though a real rival male had intruded. This has been taken as a first line of experimental evidence that song is used by males in territorial defence, and is the basis of the speaker replacement experiments described above. However, it is clearly a very artificial experiment, as the invading recording is passive in nature and quite unable to respond to what the resident might be doing or singing. What would happen if one male waits and replies to the other, or one tries to match the other by using a similar song? Observations have shown that there are many such subtleties in normal singing interactions between two birds. For this reason, modern experimenters have moved to interactive playback, where the experimenter with a laptop can respond by controlling what, when, where and how songs are presented to a rival male in territory.

In Chapter 1 we have already briefly outlined the early development of equipment for interactive playback, and some of the initial experiments were to compare how males responded when subjected to both passive (one-way) and interactive (two-way) playback. For example Dabelsteen & Pederson (1990) used both methods on European blackbirds and found that interactive playback produced stronger responses from their territorial males. Blackbirds have different song structures, some more aggressive in function, and by interacting with the males they were able to manipulate the nature of the territorial contests. McGregor et al. (1992b) worked with great tits and found that interactive playback affected the way resident males sang in response. They were able to show that resident males tended to match the song type played on the tape and sometimes attempted to overlap it when particularly aggressive. Matching and overlapping are ways in which territorial contests may be escalated and interactive playback has now enabled us to investigate these more subtle aspects of male-male signaling in considerable detail. Song type matching is discussed in detail in Chapter 8, so we will now focus here upon the use of interactive playback to study overlapping.

One bird that has been studied a great deal with respect to overlapping is the nightingale. Previous observational studies (reviewed by Todt & Naguib 2000) suggested that in male nightingales overlapping signals

increased aggression and threat, rivals tended to avoid males that overlap them and also tried to avoid being overlapped. When Naguib (1999) used interactive playback he found that males responded more when overlapped by singing back at higher rates. Recently, Naguib & Kipper (2006) gradually increased the amount of overlapping from 1 to 50% and found that the resident male nightingales increasingly interrupted their singing in response. All this suggests that males which overlap are more aggressive, so perhaps they might be more successful in acquiring territories and mates. Indeed a large proportion of territorial male nightingales remain unpaired throughout the breeding season. Kunc *et al.* (2006) used playback to screen all the males before pairing started and found that, after female arrival, those males which overlapped most were also those that paired successfully later.

Poesel & Dabelsteen (2005) studied blue tits and designed a complex interactive playback with multiple speakers (Fig. 6.4) in order to simulate song delivery from an intruder changing song posts. At any one time they had three speakers and could play back and interact between them as needed. They hypothesized that an intruder overlapping the residents'

Fig. 6.4. Design for a complex interactive playback experiment on territorial male blue tits. Speakers 1 and 2 are inside the territory and used to simulate an interactive dyadic playback where one 'male' overlaps the song of another. One 'male' can then move within the territory to speaker 3 inside, or move to speaker 4 outside (from Poesel & Dabelsteen 2005).

songs and moving inside the territory would be perceived as a greater threat than one also overlapping but then leaving the territory. Sure enough they found that residents approached more slowly when the two speakers inside territory overlapped them than when one of the speakers also overlapped but was located outside. A recent experiment on banded wrens (Hall *et al.* 2006) has confirmed that males tend to avoid the speaker more during playback from rivals who overlap them. A novel finding here was that males tended to remember this and avoided the same individual days later, even when the speaker played alternating and not overlapping songs. However, Hall *et al.* (2006) point out that it is not yet clear whether overlapping indicates motivation to escalate, or whether it is mainly a signal of male quality. While more complex interactive playback designs do reveal all manner of subtle interactions with song, interpreting the results and excluding alternative hypotheses can sometimes be difficult. Yet, on balance, most studies so far seem to show that overlapping does indicate an increase in aggressive motivation and a willingness to escalate a territorial contest.

6.3 Species recognition

It is common knowledge that experienced birdwatchers are able to identify species by their songs, and that this is particularly useful in closely related species with similar plumage. Indeed, Gilbert White of Selborne (1789) rightly suspected that distinct songs revealed the presence of several (not one) species of *Phylloscopus* warbler. Payne (1986) has reviewed the use of song in avian systematics, where it is of particular use at the species level, and we will discuss the phylogeny and evolution of song at the end of the book in Chapter 9. If songs can indicate species identity to us, it seems likely that one of the important functions of song is to enable birds to recognise their own species. As we will see in the next Chapter, it is particularly important for females to be able to recognise singing males of their own species. Clayton (1990b) has shown how females of the two subspecies of zebra finch use song as a reliable cue for assortative mating in captivity. However such studies on females are comparatively rare, and most questions regarding species recognition have been addressed to territorial males in the field.

We have already seen that a prospecting male seems to be repelled by playback of its species song. This suggests that responding males

recognise the song of their own species, although curiously no speaker replacement experiments have used the songs of other species as controls. We might predict that, if a prospecting male sings at the edge of a territory, then the resident would reply and also defend his territory by approaching and attacking the rival. Observations of territorial interactions between rival males confirm this, as does the very simple experiment of 'invading' a territory with a speaker playing back the species song. Ringers catching birds regularly use playback of tape recorded song to successfully lure males into their nets. The effect of a simple playback experiment is usually species-specific, males will only approach to playback of their own species song, not to controls of heterospecific song. Because this effect is relatively easy to obtain, the technique has been used extensively by those interested in what features of the song birds use to encode their species identity. Although such information is clearly of paramount importance for successful mating, females in the field are much less obliging in revealing their responses to playback (see Chapter 7). It is rather ironic that for purely practical reasons much of our knowledge about species recognition is based upon the responses of territorial males to playback of their species song.

6.3.1 Brémond's experiments

One of the first and most thorough experimental investigations into species recognition by song, was carried out by Brémond (1968) who studied the European robin. This was no easy task, as the robin has one of the most complicated of all bird songs. Each song is different from the last and consists of several different, complex phrases. Although a song may have only about four phrases, each robin has a repertoire of several hundred to choose from, and so the number of possible permutations is astronomical. Variability is the keynote and, as there really is no obvious, stereotyped robin song, how is species identity encoded? Brémond discovered a simple set of rules which appear to underlie the organisation of all robin songs. The first rule is that all phrases within a song are different, the second is that in a particular bout all songs are different, and the third is that successive phrases alternate in pitch between high and low frequencies (Fig. 6.5).

Brémond's main achievement was to successfully develop the technique of testing hypotheses using playback experiments with artificially modified

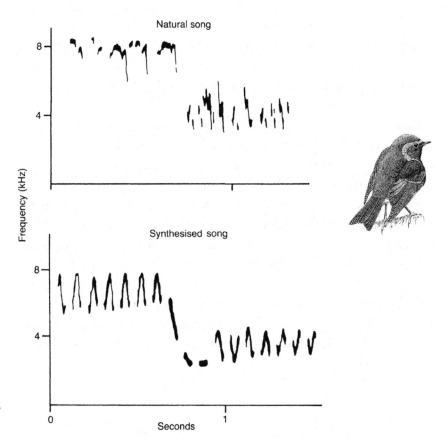

Fig. 6.5. Natural and synthesised European robin songs used by Brémond, showing the alternation of high and low phrases important in species recognition (after Brémond 1968).

song structures. By manipulating one variable at a time, while holding all others constant, it became possible to test the importance of individual features for species recognition. For example, to test the prediction that robins use the alternation of high and low song phrases as a cue, he spliced tapes of real robin song, creating experimental tapes with just high phrases, just low phrases, and the normal high and low phrase songs. He found that, whereas 90% of robins responded to the high and low controls (both natural and artificial), the response dropped to around 50% when only high or low phrases were used on the tapes. The alternation of phrases is clearly an important syntax rule which contributes to species recognition in male robins.

Brémond also pioneered the use of sound generators to synthesise completely artificial experimental songs (Fig. 6.5). He started with a continuous sine wave within the normal frequency range for robins, and modulated it to mimic the alternation of high and low phrases. This

extremely artificial stimulus obtained no response at all, perhaps because there are other important syntax rules within normal robin phrases, such as the repetition of the small separate elements. When this added complexity was introduced, the response obtained approached 70% when compared to natural control robin song.

That a relatively crude synthetic song containing no normal robin elements can still be relatively effective led Brémond to emphasise the importance of syntactical rules (alternation, repetition, etc.) within the normal frequency range of the natural song. He then turned his attention to species which had remarkably simple songs, consisting of mere repetitions of a single element. These songs are found among European *Phylloscopus* warblers, such as Bonelli's warbler, which has a song consisting of a single element type repeated about ten times to form a trill. Brémond (1976) found that interfering in any way with the structure of the element itself, for example reversing or inverting it, reduced the male response to as low as 10% compared to 100% with natural controls (Fig. 6.6). In Bonelli's warbler element structure itself, rather than syntax, seems more important for species recognition.

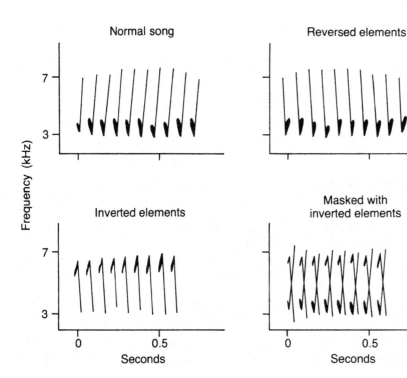

Fig. 6.6. Normal and modified Bonelli's warbler songs used in playback experiments by Brémond. The three modified songs obtained greatly reduced responses from territorial males (after Brémond 1976).

6.3.2 The important features

Brémond's experiments led to a tremendous growth of field playback experiments concerned with species recognition, particularly in the 1970s. In his review of these studies, Becker (1982) listed several features such as syntax, element structure, timing and frequency, which have been found to be important in various species. The review also reveals that the relative importance of these seems to vary considerably from species to species. In some, such as *Phylloscopus* warblers, element structure is vital, but in others this is either unimportant or less important than syntax or timing. In most species it appears that high redundancy from several cues are used to make species recognition an efficient and relatively fail-safe mechanism. For example, Ratcliffe & Weisman (1986) played back modified songs and found that structure, syntax and number of elements were all important to black-capped chickadees.

It may be that males have come to rely upon certain cues which are more reliable than others. For example Nelson (1988) investigated the several features used by male field sparrows, such as frequency, element duration, time interval between elements, and so on. He found that males were more sensitive to alterations in frequency, which normally varied much less than other possible cues. Weisman *et al.* (1990) returned to the 'fee bee' song of the black-capped chickadee and found that individuals varied considerably in the absolute pitch (main frequency) of their songs. However, the frequency interval between the 'fee' and 'bee' was relatively constant, and altering this resulted in much less response during playback (Weisman & Ratcliffe 1989). They suggested that pitch interval was important for species recognition, and that the variation in absolute pitch might be used in individual recognition.

Marler (1960) had earlier suggested that whatever features of the song are used in species recognition, those which are invariant (those which vary least and are relatively constant between individuals) are likely to be most reliable. This has now become generally known as the invariant features hypothesis. Emlen (1972) studied the indigo bunting, and found that timing was the most important feature. In his experiments, songs with the elements reversed were just as effective as control song, but if the time interval between the elements was altered, the response declined significantly. When he measured the time intervals from normal songs, he obtained remarkably low coefficients of variation. Emlen concluded that

invariant features were indeed the most important recognition cues, a view later supported by Becker (1982) in his wide-ranging review.

Later work on species recognition (Nelson 1989; Nelson & Marler 1990) has suggested that possible confusion from other species competing for the sound environment is also important. This has become generally known as the sound environment hypothesis, but distinguishing between it and the invariant features hypothesis is extremely difficult. For example, the frequency range of a species song may well show little variation, but this could just as well be due to pressure from competing species, resulting in each being constrained to a smaller sound niche with less scope for variation.

Brémond's (1976) earlier experiments on Bonelli's warbler are relevant here. In one experiment he altered the frequency range by transforming the song up or down by one kilohertz. Transforming it down had no effect upon territorial males, but transforming it up reduced the response to only 32%. Bonelli's warbler can be sympatric with the congeneric wood warbler, which has a similar song, but pitched one kilohertz higher. The evolution of reliable cues for recognition in one species is clearly constrained by the presence or absence of another. Where there are relatively few species, we might predict that competitive release will result in more variation, and this has indeed been claimed for the songs of island birds (see Chapter 9). Where there are many closely related sympatric species, we might expect much less variation and even examples of character displacement reflected in the song structures. One way to test this would be to compare song structures in allopatry and sympatry, and look for evidence of character displacement. This was attempted by Miller (1982); however the difficulties in attributing apparent changes in song structures to such evolutionary pressures are many, and the overall picture regarding character or variance shifts in song structure is still far from clear.

6.3.3 Recognising other species

Conspecific male birds are not the only class of individuals to whom species recognition is important. Although in general standard playback experiments have demonstrated that males respond only to conspecific song, there are some clear exceptions. For example, many species also defend their territories against closely related sympatric species, often congenerics who overlap in their ecological requirements. In such cases

we would expect interspecific territorialism to develop, with males also responding to heterospecific song. We have already mentioned the difficulties in detecting and interpreting differences in song structure between allopatry and sympatry, but there has been some progress in detecting differences in response to playback experiments. For example, Emlen *et al.* (1975) found that in allopatry North American indigo and lazuli buntings responded strongly to playback of conspecific song, but only weakly if at all to heterospecific song. However, in sympatry on the Great Plains, the two species showed interspecific territorialism and responded equally to the same playbacks of conspecific and heterospecific song. Prescott (1987) reported a similar pattern from his experiments on allopatric and sympatric populations of alder and willow flycatchers.

Catchpole (1978) obtained a similar result in allopatric and sympatric populations of European reed warblers, where instances of interspecific territorialism were also related to competition in sympatry. However close examination revealed that not all reed warblers in sympatry responded, only those which maintained interspecific territories. This suggests that far from being an inherited mechanism, individuals may learn to respond to heterospecific song during regular aggressive interactions with their territorial neighbours. This local learning hypothesis gains further support from another reed warbler experiment carried out by Catchpole & Leisler (1986). They played the same experimental tapes to two reed warbler populations only five kilometres apart, but only one of which had a few pairs of great reed warblers also breeding. They found that reed warblers only responded to heterospecific great reed warbler song at the sympatric location, and all of the responding males had great reed warblers as neighbours.

Learning to recognise ecological competitors, using multiple cues from both plumage and song together, was recently studied by Matyjasiak (2004) in the blackcap and garden warbler. These two are congeneric *Sylvia* species, both with brown plumage, but the male blackcap is clearly distinguished from the garden warbler by the black cap on its head. In the study population, male blackcaps defend their territories against the later arriving garden warblers. Matyjasiak used dual choice playback experiments accompanied by stuffed models. He found that while male blackcaps behaved aggressively to playback of garden warbler song, they also responded more to models of either species when the correct species song was played back.

In closely related species there are always some similarities in morphology or song structure, which might act as confounding variables, although most experimental designs have attempted to control for these. Reed (1982) studied an interesting case between chaffinches and great tits, where both song and plumage are completely different. In Scotland the two species maintained overlapping territories in woodland, and he found no evidence of interspecific responses to playback. However on adjacent islands, where space and habitat were severely imited, he found non-overlapping territories and strong interspecific responses to playback.

6.3.4 Kin recognition

In theory, song might also allow individuals to recognise and respond appropriately to kin. However a review by McGregor (1989) concluded that this is unlikely to be the case. As McGregor points out, the commonest pattern of song acquisition is by males learning from other non-related territorial neighbours, rather than their fathers. But the real problem is that only males normally sing, and so recognition of female kin by song could not occur. One possibility is that females might avoid incest by learning the songs of their fathers. McGregor & Krebs (1982b) studied a population of great tits and were able to compare the songs of a female's father with those of her eventual mate. They found that in general, females selected a mate whose song was not the same as their father's, but contained some similarities. This tendency to avoid the same, yet select a similar song, fits well with current hypotheses concerning incest avoidance and optimal outbreeding in birds.

In several species of Darwin's finches (Grant 1984, Grant & Grant 1996), it has been shown that sons learn and sing the song types of their fathers (Fig. 6.7). Furthermore, females avoid mating with males who sing like their fathers. This suggests a system based upon incest avoidance and mediated by song recognition.

Cooperatively breeding birds usually exist in groups based upon kin relationships, but it is not clear whether recognition is genetically determined in some way, or whether it develops through association with relatives. Payne et al. (1988b) have completed the only experimental study on recognition by song in a cooperative breeder, the splendid fairy-wren from Australia. They used playback to show that wrens responded equally to the songs of unfamiliar wrens, regardless of whether they were kin or

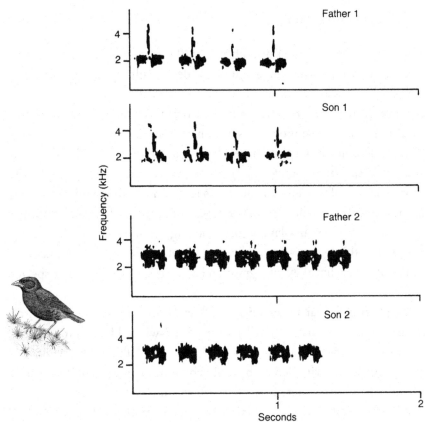

Fig. 6.7. Fathers and sons usually share the same song type in one of Darwin's finches, the large cactus finch from Isla Genovesa (from Grant 1984).

not. In general, they responded more to the songs of helpers from other groups than to the songs of helpers in their own groups. In this case, song appears to be used to recognize other wrens in a particular social group, rather than as a marker for kin recognition. Conversely, Price (1999) played back calls in the cooperatively breeding striped-back wren and found that males did recognize kin on the basis of patrilines, not on the basis of group membership. Finally, Hatchwell *et al.* (2001) working on calls in cooperatively breeding long-tailed tits, found significant differences in playback to kin and non-kin. They then used cross-fostering to separate the effects of relatedness from early experience. Adults who had been fostered as nestlings did not discriminate between genetic and fostered siblings in later years. This experiment suggests that kin recognition by calls in such species is learned during early development rather than genetically determined.

6.3.5 *Recognition by females*

This chapter is concerned primarily with the responses of male birds to song, and the responses of females are dealt with in the next chapter. However it is relevant to pause at this stage in the male story, as there are strong theoretical reasons to suggest that there may be sexual differences in species recognition. This was emphasised by Searcy & Brenowitz (1988) who pointed out that females have much more to lose than males if an error in species recognition is made. For a female, the error means the enormous cost of wasted reproductive success in breeding with the wrong species. The costs to the male of such an error are much less, perhaps just a waste of energy in displaying or attacking an intruder of the wrong species. For this reason we would expect females to be more discriminating than males in what they recognise and respond to as their own species song.

The first clue that males might be less discriminating came from an experiment by Brenowitz (1982b) on male red-winged blackbirds. Mockingbirds are superb mimics of many species, and Brenowitz wondered whether male red-wings could be fooled by mimic song. He used a standard field playback experiment on territorial males to demonstrate very clearly that the mimic song obtained the same strength of response as a control model song. Searcy & Brenowitz (1988) then repeated the same experiment on captive female red-wings, and showed that females displayed four times as much to the model as the mimic, demonstrating clear discrimination and a strong preference for the real species song. Sonagraphic analysis revealed that there are subtle structural differences between the model and the mimic songs, and we can conclude that females in this species have indeed been selected for superior powers of discrimination. How and why females use this ability to select males on the basis of their songs will be the subject of the next chapter.

6.4 Individual recognition

Species recognition is clearly one function of song, but there may well be other important information contained within the signal. As we shall see throughout this book, songs are extremely complex structures, and are unlikely to have evolved for one simple function. Marler (1960) has pointed out that there may be several conflicting selection pressures

acting upon song structure. Whereas invariant features are favoured to promote effective species recognition, some variation is also essential if other functions such as individual recognition are to occur. To demonstrate individual recognition by sound, two criteria must be fulfilled. First, there must be a consistent physical basis for recognition, and secondly an experimental demonstration that recognition occurs. There are also various categories of individual which it might be advantageous to recognise, and some of these are obvious, such as a parent or a mate. But there are also other individuals who it might be advantageous to keep track of, such as immediate neighbours who may have designs on your territory or mate. There are also new arrivals, strangers intent upon acquiring a territory of their own, who may pose even more of a threat than established neighbours. As we will see in this section, learning the songs of other individuals may have both obvious and more subtle advantages for the listening bird.

6.4.1 Parent–offspring recognition

Even relatively short, simple structures, such as the calls of seabirds, have been shown to contain enough information and variation to permit both species and individual recognition. Tschanz (1968) was the first to show this, working on a colonial seabird, the guillemot. One of the problems faced by colonial seabird parents is how to find their own young among thousands of others all packed together. From observations Tschanz suspected that parents and young recognised each other by calling out across the crowded colony. To test this out, he carried out playback experiments on young in the colony, using a two speaker design. From one speaker he played the call of a parent, and from the other a control call from another adult in the colony. He presented the calls both alternately, and together, and measured a variety of responses such as orientation, approach, begging for food and calling in return. The results were very clear: the young birds only responded positively to their parents' calls. Control calls from other adults were either ignored or even avoided. Tschanz also investigated the development of parent–offspring recognition by exposing guillemot eggs in incubators to playback of selected calls. In choice tests, the newly hatched chicks only approached the speaker transmitting the calls to which they had been exposed in the egg. Control chicks from an incubator with no playback approached both

speakers alternately. Later experiments on a variety of colonial seabirds, such as those by Beer (1969) on laughing gulls, provided further evidence for parent-offspring recognition.

Many penguin species breed in large colonies where there are few distinguishing features or landmarks in the flat, barren terrain. The parents have to travel long distances to catch food at out sea, and then have the problem of finding their own young in the crowded colony. King penguins compound the problem by having no nests; to be fed, the chick has to pick out its returning parent's call from all the others in the large colony. Jouventin et al. (1999) used playback to show that the parents' calls use a complex pattern of frequency modulations to form a distinct vocal signature. Other penguin species, such as Adelie and gentoo penguins, do have a nest-site to return to and Jouventin & Aubin (2002) have suggested that this is why they have developed rather less complex calls (varying only in pitch) for recognition. They used playback to show that even with considerable background noise chicks could still recognize the calls of their own parents in the dense, noisy colonies.

There are fewer studies on songbirds (see reviews by Beecher 1982, 1990), but the calls of various species of swallow provide an interesting story. Bank swallows are colonial (Beecher et al. 1981) and the young are fed in their burrow for the first fifteen days. After fledging they leave and need to be located by their parents who still feed them. The young birds develop an individually distinct two-note call which their parents can recognise, both naturally and in playback experiments. Stoddard & Beecher (1983) found a similar system in the colonial cliff swallow (Fig. 6.8), but not in the closely related barn swallow which does not live in colonies (Medvin & Beecher 1986). Leonard et al. (1997) found no evidence for parent-offspring recognition in the tree swallow which is also non-colonial. Taken together, these studies suggest a relationship between coloniality and the development of individual recognition between parents and offspring in swallows.

Another case where such recognition in a songbird has been detected is when fledglings leave the nest and brood division occurs, the male parent feeding one group and the female the other. Draganoiu et al. (2006) studied the black redstart and found that each parent only responded to playback of begging calls from the chicks it normally fed. The begging calls of chicks are an important signal that stimulates the parents to keep feeding their young. Parents are more likely to feed young which call

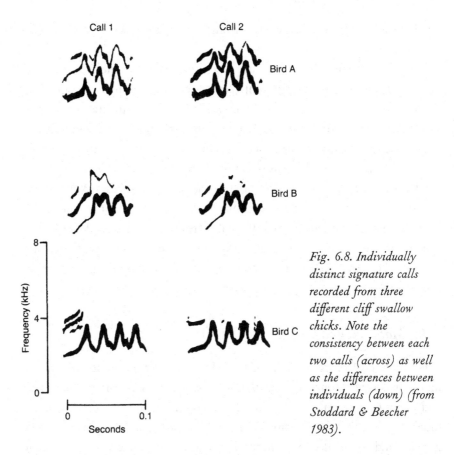

Fig. 6.8. Individually distinct signature calls recorded from three different cliff swallow chicks. Note the consistency between each two calls (across) as well as the differences between individuals (down) (from Stoddard & Beecher 1983).

more, as shown when nestling tree swallows were placed next to a speaker playing back calls at higher rates (Leonard & Horn 2001).

An interesting case of how this can be subverted by a brood parasite has been discovered in studies on the European cuckoo (Davies *et al.* 1998, Kilner *et al.* 1999). European cuckoos lay their eggs in the nests of much smaller species such as reed warblers. The single cuckoo chick soon disposes of the reed warbler brood, but how will it persuade the parents to bring enough food to grow ten times their size? A typical brood of four reed warbler chicks can open their several mouths to beg and stimulate the parents to feed them all but, although the cuckoo has a larger gape, it only has one. However, the reed warbler chicks also make begging calls and sonagraphic analysis has revealed that the single cuckoo makes as many begging calls as a whole brood of reed warblers and presumably stimulates the host parents to feed it much more. Playback experiments at nests were then used to confirm that recordings of a single cuckoo chick elicited as

much feeding as playback of recordings from a whole brood of reed warbler chicks.

It might be assumed that the early begging calls of the cuckoo chick are unlearned, particularly as different races of cuckoo are host-specific, but this turns out not to be the case. Cuckoos in England parasitise dunnocks as well as reed warblers, and Madden & Davies (2006) observed that the cuckoo begging calls are also host-specific. However, when they cross-fostered newly hatched cuckoo chicks between these two species, the chicks developed begging calls more like the new host species. When they tested the cross-fostered calls with playback experiments at the nest, the host parents responded most to the calls produced in nests of their own species. It seems that the young cuckoos soon learn to shape their calls towards the specific structure which obtains them most food from their host parents.

6.4.2 Mate recognition

The problem of recognising your mate from among thousands of similar individuals is also found in seabird colonies. Seabirds are often long-lived and many pair for life, suggesting that recognition of individuals is very much a part of their behaviour. White (1971) studied a colony of northern gannets and found that each individual had a distinctive call used when landing in the colony. When these were played back to sitting birds, they only orientated to the particular call of their mate. The manx shearwater nests in burrows in large colonies on remote islands, and compounds the difficult problem of finding it by returning at night. Each year, males return to the island first, occupying the same or a nearby burrow. When a female returns, she calls from the air and the males call from their burrows in return. Brooke (1978) suspected that the distinctive male call was used by females to locate their mates and return to the right burrow. He tested this by playing back male and female calls to individuals of both sexes incubating in the burrows. Sure enough, he found that whereas females responded selectively to the call of their mate, males did not.

Penguins also live in large colonies and many pair for life. Some, like the emperor penguin, move their solitary egg or chick from place to place, making location even more difficult for a returning mate. Jouventin *et al.* (1979) confirmed that emperor penguins responded selectively to playback of the distinctive call of their mate by leaving the colony and approaching

the speaker. This was confirmed more recently by Clark *et al.* (2006) who studied the magellanic penguin. Speirs and Davis (1991) studied the adelie penguin, and found that both males and females could discriminate playback of their mates from controls. Unlike Clark *et al.* (2006), they also found a more subtle effect: males (not females) also discriminated between playback of neighbours as opposed to total strangers. Males spend more time on their territories than females, and so have more time to learn their neighbours' characteristics. This is the only time a difference in response to neighbours and strangers has been detected in a colonial seabird but, as we shall see in the next section, it is a behaviour well known in territorial songbirds.

There has been far less work on mate recognition in songbirds, but there are several examples where females appear to recognize and respond preferentially to the songs of their mates. These include the zebra finch (Miller 1979b) and the dunnock (Wiley *et al.* 1991). Lind *et al.* (1996) used playback of song to female great tits in nest boxes during incubation. The females would only emerge when called out by the song of their own mate, and ignored the control songs of other males.

6.4.3 Neighbours and strangers

When songbirds occupy and defend their territories, there is also a great deal of singing, as we have seen in Chapter 4. Given the well known abilities of birds to learn their songs (Chapter 3), it seems quite likely that territorial males may also become familiar with the songs of their immediate neighbours. As playback experiments on a variety of species have shown, they will certainly habituate to the constant repetition of a particular song type (e.g. Krebs 1976). However a different kind of experiment is needed to show that males are able to positively recognise the songs of their neighbours.

Weedon & Falls (1959) were the first to develop a new kind of playback experiment, which provided direct evidence that males could identify territorial neighbours and their positions by learning their songs. They originally worked on ovenbirds, but refined the technique on white-throated sparrows (Brooks & Falls 1975a,b, Falls & Brooks 1975). The first step is to demonstrate that males are familiar with the songs of a neighbour by playing back the songs from the normal position at the appropriate territorial boundary. A significantly weaker response is

obtained when compared with playback of the songs of a complete stranger from the same position (Fig. 6.9). The male is clearly used to the songs of his neighbour, and can discriminate these from the songs of the new male. In the second step, the same songs are now played from the opposite side of the territory. Now the resident male responds equally strongly to both (Fig. 6.9): he is not used to hearing his neighbour's song from a different location. The experiment suggests that the resident male recognises the position of a particular song, and can therefore identify the singer.

The above interpretation of this classic neighbour-stranger experiment has resulted in its widespread use as an experimental technique to demonstrate individual recognition by song. Although there can be problems with both design and interpretation (see discussion and review by Stoddard 1996) the basic design has proved to be robust. The differential response to neighbour and stranger is sometimes referred to as the 'dear enemy' effect (Fisher 1954), and its adaptive significance has been discussed in detail by Ydenberg et al. (1988). There are thought to be several

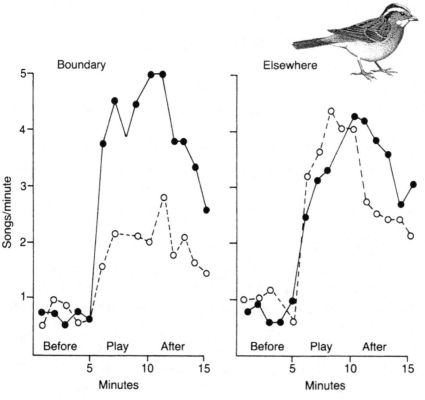

Fig. 6.9. Responses of male white-throated sparrows (singing rate) to playback of songs of neighbours (open circles) and strangers (filled circles). The speaker is either positioned on the usual boundary with the neighbour (boundary), or elsewhere in the territory (elsewhere) (from Falls & Brooks 1975).

advantages to the territory holder who can discriminate between his neighbour and a stranger. The neighbour is a resident in his own territory, who is unlikely to pose a serious threat to established territory boundaries, which have been previously contested and learned. It is a waste of time and energy to respond to an established neighbour every time he sings from his usual song post. A song from a stranger is a very different proposition. A new male could mean a real threat to the resident and his territory, and it is well worth investing in a strong response to see the potential invader off. Similarly, if the neighbour moves his position, it might mean a renewed attempt to acquire more space, and again is worth a strong response.

According to a review by Falls (1992) 22 species had then been studied in this way, and in 21 cases the dear enemy effect was detected, a pattern later confirmed by Stoddard (1996). It had been thought that species with large repertoires of song types (and large numbers of songs to learn) would not show the effect so clearly. However studies on a variety of species with large repertoires such as song sparrows (Stoddard et al. 1990), European robins (Brindley 1991) and banded wrens (Molles & Vehren-camp 2001) suggest that this may not be the case, as did the review by Weary et al. (1992). There is another group of birds in which it was suspected that learning to recognize the songs of neighbours might not occur. The sub-oscines sing quite complex songs but do not learn them (see Chapter 3), so would they be able to learn to respond to the songs of their neighbours? Only two species have been studied so far and the results are equivocal. Wiley (2005) studied the Acadian flycatcher and found no evidence that they could discriminate, yet Lovell & Lein (2004, 2005) obtained clear, positive results from their experiments on the alder flycatcher.

McGregor & Avery (1986) have shown that in the great tit, whereas song learning for singing takes place during a restricted period (see Chapter 3), males continue to learn to recognise new neighbours' songs throughout life. Even if males can recognise song types, it may be that there are other more general vocal characteristics that permit individuals to be recognised by the listener. Weary et al. (1990) have shown that there are indeed such general differences between individual great tits, and Weary & Krebs (1992) demonstrated that males can discriminate between them. Weary and Krebs used an operant conditioning experiment, which trained great tit males to discriminate between songs from the repertoires of two individuals. They then tested the males on unfamiliar songs (which

they had never heard) from the same two individuals. The male great tits were still able to discriminate between the two males, suggesting that in the wild they would be perfectly capable of identifying males, without having to learn all their song types. This would certainly help to explain how species with large repertoires can still identify neighbours from strangers.

Godard (1991) has used the neighbour-stranger design in a particularly elegant experiment, which demonstrates that male songbirds can even remember their neighbours from previous years. Hooded warblers migrate from their winter quarters in Central America to breeding territories in North Carolina. In 1988, Godard demonstrated the classic neighbour-stranger effect on 17 territorial males. In 1989, seven of the males returned to the same territories, and to these she repeated the experiment using songs recorded from their 1988 neighbours. The results were very clear, the males were able to remember the positions of their previous neighbours, and responded accordingly (Fig. 6.10). Remembering individuals and their locations in the long-term is impressive, but short-term memory has also been investigated by Godard (1993). She predicted that a resident should increase his response to a neighbour if that neighbour had recently showed signs of extending his territorial boundary. Incursions were simulated experimentally by playback of neighbour songs well inside the territories of residents. When playback returned to the usual territory boundary, the resident males now showed increased responses to their familiar but expansionist neighbours. Hyman & Hughes (2006) also found that territorial males were able to remember and distinguish between rivals who posed varying degrees of threat. They first used playback to assay the level of aggression in male song sparrows, then later played back songs from both aggressive and non-aggressive neighbours. Residents responded much more to playback of their aggressive neighbours, presumably having recognized them as posing more of a threat to their territorial boundary.

6.4.4 Eavesdropping and communication networks

Such flexibility suggests that territorial males are well aware of the local status quo, and responsive to even subtle changes in the territorial activities of their neighbours. As we have seen in Chapter 4, there is also evidence that males can use the degradation of song as it transmits through

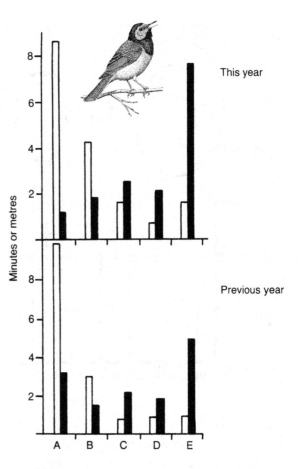

Fig. 6.10. Male hooded warblers recognise and respond to the songs of this year's and their previous year's neighbours at their usual boundary (white bar, low response), or elsewhere (black bar, strong response). Note similar response patterns in both years. The responses to playback shown here are (A) closest distance to speaker, (B) latency to approach, (C) latency to song, (D) time spent near the speaker during and (E) after playback (from Godard 1991).

the habitat as a cue to estimate the range of their rivals, and again modify their responses in an appropriate way. Many songbirds live in neighbourhoods of relatively stable territories. As well as using songs to identify and range their rivals, McGregor (1993) has pointed out that in such communication networks there is considerable scope for low cost information gathering, and what has become known as 'eavesdropping'. With the realization that dyadic communication tells us only a limited amount, and the development of new techniques to study more complex situations (see Chapter 1), there has been a recent move towards studying of communication networks (reviewed in McGregor 2005). Eavesdropping refers to the common situation in animal communication where a signal intended primarily for one receiver is also intercepted by a third party. Such a situation must be extremely common in networks, but McGregor & Dabelsteen (1996) take the definition further and propose that eavesdropping only occurs when the third party 'gains information from an

Fig. 6.11. (a) Design for an interactive playback experiment to reveal eavesdropping among territorial male great tits. The subject first hears a simulated interaction through two speakers next to his territory for two minutes. Fifteen minutes later one of the interactants is simulated to intrude by playback through a third speaker, and the behaviour of the subject recorded for five minutes. (b) The number of songs of the subject in response to a simulated intrusion by one of the two birds it had heard interacting earlier as discussed in the text (from Peake et al. 2001).

interaction that could not be gained from the signal alone.' Thus, eavesdropping differs subtly, but importantly, from merely intercepting a signal. We will now try and clarify this with examples which show how and whether we can detect eavesdropping in territorial males.

The obvious way to test whether eavesdropping has occurred is to allow a third party to observe and listen to an interaction between two other individuals, then see whether the behaviour of the third party changes in some way that is relevant to the interaction. This can be done by observation, but by using an experimental approach more control can be achieved particularly by the use of interactive playback. In recent years there have been several attempts to do this with a variety of designs and results (see reviews by Peake 2005, Searcy & Nowicki 2005).

Naguib & Todt (1997) studied eavesdropping in the nightingale, a species well known for long and complex interactions during which males sing against each other. They achieved a high level of experimental control by replacing singing male dyads with two speakers, and then recording the behaviour of listening third parties. We have already discussed how

overlapping songs signals increased aggressive motivation, and in this experiment they were able to simulate bouts of overlapping or non-overlapping dyadic interactions. Their experimental birds responded more strongly (by approaching closer and singing more) to whichever speaker played songs that overlapped the other. This suggests that the listening males had indeed eavesdropped by monitoring the interactions and showed this by responding most to the simulated intruder who posed the greater potential threat.

Peake *et al.* (2001) used a more sophisticated playback design in their experiments with great tits (Fig. 6.11). A territorial male first hears a dyadic interaction played through two speakers located outside his territory. This interaction involved one of the males overlapping the other. Fifteen minutes later, one of the simulated males is made to intrude by playing back its song through a third speaker located inside the territory. In this case the test males responded by singing back less when the speaker played back the male who had been heard to overlap his rival. When the male had been overlapped they sang back at the same rate as to controls, who either alternated or sang at random (Fig. 6.11). These results are somewhat different and not directly comparable with those from the nightingale experiments, and show that difficulties of interpretation still remain with playback experiments, even with more modern designs. However, both sets of experiments suggest that territorial males do eavesdrop on their neighbours and even remember the outcome of interactions between them. By being able to do this they may well gain some advantage by knowing more about their rival's location and and the level of threat he poses. But we must not forget that another class of birds also spend a great deal of time listening to male song, the females of the species, and in the next chapter we will turn our attention to them.

SEXUAL SELECTION AND FEMALE CHOICE

*Sedge warbler
song flight*

The sweet strains poured forth by the males during the
season of love are certainly admired by the females
Charles Darwin *The Descent of Man and Selection in Relation to Sex*

7.1 Introduction

So far, we have dealt almost exclusively with males. Indeed, some definitions of song proclaim that it is a male behaviour, and we have already spent a great deal of time exploring the complex relationships between male, territory and song. In earlier writings (e.g. Armstrong 1973) the term 'territorial song' is often used and, although some space is given over to the effects of song upon females, they were often relegated to a minor role. One reason for this is, quite simply, that elusive, retiring females have been far more difficult to study than bold, displaying males. As we have already seen, territorial males respond readily to playback, and are also relatively easy to record and catch. However, in more recent times, the study of sexual selection, and female choice in particular, has undergone an explosion of interest among ethologists. At the same time, new techniques have enabled us to investigate the sexual preferences of captive females in the laboratory, and even to follow their behaviour in the field using radio transmitters. DNA fingerprinting can also tell us which males sired their offspring, as opposed to which male they were paired with. As a result, female behaviour, including responses to male song, has been the focus of considerable research in the last decade. The various ingenious techniques used by biologists investigating female choice, in both laboratory and field, will form the basis of this chapter.

7.2 Sexual selection

Charles Darwin (1871) not only originated the theory of sexual selection, but suggested that it was the driving force behind the evolution of the ornate plumage and elaborate songs used by male birds during courtship. He recognised that male competition for females occurred in two different ways. First, males could compete among themselves, the victors acquiring either the females or a breeding territory. This would lead to strong intrasexual selection for traits, such as size or special weapons, which increased success in male-male contests. Secondly, males might influence female choice directly, by producing more elaborate or conspicuous courtship displays. There would then be strong intersexual selection for increasingly effective signals for female attraction and stimulation.

Darwin's theory is appealing to evolutionary biologists who study bird song for several reasons. Like Darwin, we find it difficult to arrive at functional interpretations of bird song in terms of conventional natural selection. Bird songs occur in a bewildering variety of forms, often extremely elaborate, and are as puzzling to explain as the plumage of a bird of paradise or a peacock's tail. One obvious approach is to test predictions which can be made from Darwin's theory. For example, if he is right, then the sexually selected trait should only be found in one sex. As we have already pointed out, bird songs are predominantly found in the male, at least in temperate zone species, a situation paralleled in the under-lying brain structure (see Chapter 2). Other vocalizations such as calls are shared more equally between the sexes. Another prediction which can be tested, stems from the variety of mating systems found in birds. In a polygynous mating system, the increased variance in male mating success should lead to more intense intersexual selection pressure on the evolution of song complexity, and we will follow up this prediction later in the chapter.

Ever since Darwin first proposed his theory, intersexual selection by female choice has been the most controversial area and most difficult in which to obtain convincing evidence. However, in more recent times that has been a great upsurge in interest and a great deal of theoretical and empirical work, culminating in the publication of the major review by Andersson (1994). It is fair to say that the study of bird song played an important part in this revival (Searcy & Andersson 1986, Searcy & Yasukawa 1996). It is now recognized that, not only do we need to investigate whether male song attracts females and increases male repro-ductive success, but we now need to focus upon what benefits the females obtain by choosing singing males. As Searcy & Yasukawa (1996) point out, female benefits can be divided into two categories, direct and indirect, leading to direct and indirect selection on female prefer-ences for song. Direct benefits are those that affect the female's own fitness, such as a strong male to help her feed the young, or a large territory with plenty of food. Indirect benefits come later, as they are those that affect the fitness of the female's offspring. They are the genetic benefits obtained from choosing a high quality male, such as 'good genes' for viability or resistance to disease. The female herself does not benefit directly, but indirectly through the later reproductive success of her superior offspring.

Our main working hypothesis, that song structure may be shaped by the forces of intra- and intersexual selection (Catchpole 1982, 1987, 2000), generates a number of predictions, which can be tested using a variety of techniques (see also Chapter 8). These can involve observations, correlations and experiments, both in the field and in the laboratory. In most of the cases we have selected, field observations and laboratory experiments can be integrated to provide a more powerful test. Evidence for male-male intrasexual territorial functions of song has already been described in Chapter 6, and we will now examine the case for male-female intersexual functions. To do this we will start by asking a very simple question: is there any evidence that females are attracted by male song?

7.3 Female attraction

The most important and direct evidence that male song actually functions in female attraction has been lacking until quite recently. The literature contains persuasive but indirect evidence, such as cessation or reduction of song after pairing (Chapter 5). The breakthrough came from Eriksson & Wallin (1986) who studied a population of pied and collared flycatchers. Flycatchers compete to breed in nest boxes, and Eriksson and Wallin fitted automatic spring nets to each one, so that any prospecting females who perched by a box entrance would be caught. Each box had a male decoy dummy placed outside, but half also had a loudspeaker broadcasting a recording of male song. Nine out of the ten females caught were attracted to the 'singing' dummies, as opposed to the non-singing control boxes (Fig. 7.1). A similar design was used by Mountjoy & Lemon (1991), who found that female European starlings were also attracted to nest boxes playing recorded songs. Johnson & Searcy (1996) found that female house wrens competed aggressively to enter nest boxes containing speakers playing male song compared to silent controls, and even built nests within them. All these experiments demonstrate a female preference for song over no song, but further experiments are necessary to demonstrate preferences for particular song types or song complexity.

Wiley et al. (1991) studied the responses of mated female dunnocks to playback of song in their own territories. They removed the resident males and found that females responded to playback, but only in their fertile period. During this time, they would approach the songs of their own male in preference to those of his neighbour. When the alpha male was removed

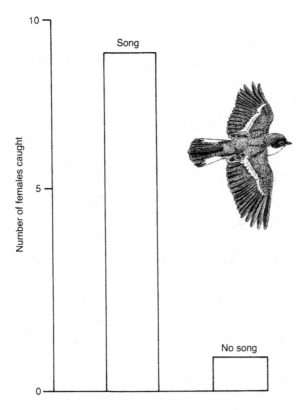

Fig. 7.1. More female flycatchers were caught at nest boxes with decoys and song playback (song) than at control boxes with just decoys (no song) (from data in Eriksson & Wallin 1986).

from polyandrous females, again they responded to his song rather than to those of their neighbours. These experiments suggest that females are capable of recognising and responding to their mates' songs, and do so particularly when seeking copulations.

Male sage grouse do not sing, but during their elaborate strut displays on the lek they produce a number of whistling and popping sounds (Gibson & Bradbury 1985). Gibson (1989) carried out a playback experiment on the lek, using tapes of these sounds to augment the natural displays. More females were attracted to the lek on days when playback was used.

These field experiments provide important direct evidence that male song (or its equivalent in sage grouse) does function in female attraction. Another experimental approach is to use the muting technique already described in Chapter 5, only this time study the effects upon female attraction and pairing. McDonald (1989) did just such an experiment on the seaside sparrow, and found that males muted early in the season failed to attract females, unlike sham-operated and normal controls. When they

recovered their voices and sang once more, the experimental males were eventually able to attract females. Paired males muted later in the season lost their mates, but when they recovered their singing ability were able to attract new females.

Theories of sexual selection based upon female choice rely upon the assumption that females actively choose their mates, rather than just passive attraction to the nearest male stimulus. Active choice must involve sampling several males, and rejecting some before a final choice is made. Observing elusive female birds in this sort of detail is clearly a tall order, but field studies on pied flycatchers by Dale *et al.* (1990, 1992) have shown that females visit up to nine singing males before eventually choosing a mate.

A study on great reed warblers by Bensch & Hasselquist (1992), used miniature radio transmitters to follow individual females as they prospected for mates in a Swedish reed bed (Fig. 7.2). The females visited an average of six male territories before eventually returning to settle with a particular male, and they took between one and three days to make their choice. Catchpole (1983) had previously shown that unmated male great reed warblers sing longer and more complicated songs when advertising for females, and shorter, simpler songs when paired. The females in the Swedish study visited males singing short and long songs, but all of the males eventually chosen sang long songs. The link between male song complexity and active female choice extends even further in this species. Catchpole *et al.* (1986) showed that captive females displayed more to playback of more complex song, and in a parallel field study males with more complex songs attracted larger numbers of females and produced more young (Catchpole 1986).

7.4 Female eavesdropping

We have already discussed (Chapter 6) how males may eavesdrop on each other and appear to gain considerable advantage by intercepting and acting upon songs directed primarily to other individuals. As mate choice is such an important decision for females, we should not be surprised to find that females are also eavesdropping on male-male interactions in order to gain important additional information (McGregor 2005).

Otter *et al.* (1999) used playback to stage interactions between territorial male great tits and simulated intruders. They wondered whether

Fig. 7.2. The search pattern of a radio-tracked unpaired female great reed warbler in a reed bed over two days. The track (dots indicate known radio contacts) passed near or through all six territories containing singing males (open male signs) and the female spent considerable time in several territories before returning to pair with her selected male (filled male sign) (from Bensch & Hasselquist 1992).

eavesdropping females might monitor the relative performance of their males in relation to neighbouring males, and perhaps try to obtain extra-pair matings with any superior neighbours. The intruder playback was made to either escalate by overlapping the owner's songs, or to de-escalate by alternating with them. The listening females were then followed to see how often they left their breeding territory to seek out neighbouring males. If a female left her breeding territory, she was more likely to intrude on a territory where the neighbouring male had performed well during his playback test and appeared to have won his contest more easily. In spite of this, Otter *et al.* (2001) found no evidence from DNA profiling that their manipulations affected female choice for extra-pair paternity.

However, Mennill *et al.* (2002, 2003) repeated this basic design on another species, the black-capped chickadee. They were able to classify males as high or low ranking on the basis of interactions in winter flocks. The results showed that low ranking males that did well in the experiments against intruders lost paternity to the same extent as controls. In the few cases when females did have young fathered by low ranking males, these males had recently won a contest. The high ranking males that did poorly against playback showed a much higher loss of paternity compared to controls. Taken together, these studies offer some evidence that females

may well eavesdrop on male-male interactions and adjust their behaviour accordingly when prospecting for extra-pair mates in neighbouring territories.

7.5 Female stimulation

Although female attraction is clearly of great importance, there is also a line of experimental evidence which suggests that male song has a direct effect upon female reproductive physiology and behaviour. Brockway (1965) showed that playback of vocalizations alone could stimulate both ovarian development and egg laying in captive female budgerigars. Hinde & Steele (1976) studied captive canaries, where nest building behaviour was known to be oestrogen dependent. Females were exposed to playback of song from their own species, from budgerigars, or to no song at all. Those exposed to canary song built far more actively than the controls, suggesting that male song may well act to stimulate their reproductive physiology and behaviour during the breeding cycle. Kroodsma (1976) used a similar experiment to show that playback of more complicated canary songs increased nest building activity when compared to playback of much simpler songs. In a colony of European starlings, Wright & Cuthill (1992) found that males who sang more were paired to females who laid earlier.

However useful physiological responses and nest building behaviour are, biologists studying bird song needed a more immediate method of measuring female choice in the laboratory. King & West (1977) discovered a sensitive behavioural assay, which has been of considerable use in quantifying the responses of females to male song. In many species, a female signals that she is ready to mate with a male by crouching, quivering her wings, and raising her tail to reveal the cloaca. This copulation solicitation display is an obvious invitation to the male to copulate, and has come to be regarded as a valid index of female choice. King and West first used this technique in their studies on the brown-headed cowbird. They found that when they exposed captive females to playback of male song in the laboratory the females responded with sexual displays. These first experiments suggested that male song does have a direct effect upon female sexual behaviour. But by quantifying the displays as a behavioural bioassay, King and West had developed a new experimental technique which could be used to 'ask' females which kinds of songs they really preferred.

Unfortunately, cowbirds are unusual in many respects, and most female songbirds do not display so readily in captivity to recorded song. However, Searcy & Marler (1981) found that priming females with low doses of female sex hormone made them much more receptive. Oestradiol was packed into lengths of silastic tubing, which permits the slow release of a measured dose into the bloodstream. The implants were placed just beneath the skin, and the birds isolated in sound-proof chambers for several days. The first playback experiments clearly demonstrated that female song sparrows would only display to recordings of their own species song, and not to control recordings from closely related swamp sparrows. These important experiments confirmed that male song does indeed stimulate the female into sexual display and the way was now open to ask females much more interesting questions about the details of the song structures they prefer.

We have already seen that there is some evidence that females are attracted and stimulated by male song. Trivers (1972) suggested that as females invest more than males in offspring, females should also be more selective and choosy than males. Searcy & Brenowitz (1988) decided to test this idea using song mimicry of red-winged blackbirds by mockingbirds. Brenowitz (1982b) had previously found that territorial male red-wings responded equally to playback of model or mimic song, even though there are subtle differences in fine structure between them. Searcy & Brenowitz (1988) then used the hormone implant technique and repeated the experiment on captive female red-wings. The females were not fooled by mimic song, and clearly discriminated between it and the model song (Fig. 7.3). This experiment suggests that females are indeed more selective than males in their response to song, and there are very good reasons why this would be favoured during evolution. A male who makes a mistake, and responds to a similar song, may only pay for it by wasted energy chasing off a member of the wrong species. If the female makes the same mistake, she may mate with a member of the wrong species. The cost for her is much too high, as she may forfeit her reproductive success in an infertile hybrid mating. The evolution of male song primarily as a mating signal may have resulted in the female response becoming more finely tuned and much more selective.

Although the hormone implant technique has been a useful tool to investigate female choice in the laboratory, it has two disadvantages. First, implanting is an invasive procedure, and secondly, it introduces an

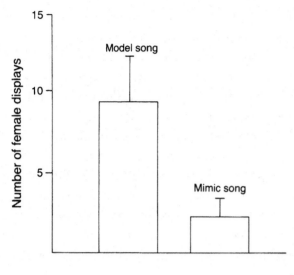

Fig. 7.3. Female red-winged blackbirds displayed more to their species song (model) than to an imitation by a mockingbird (mimic) (from Searcy & Brenowitz 1988).

artificial dose of hormone into the blood system. Work on captive canaries in France (Vallet & Kreutzer 1995, Vallet *et al.* 1998) has shown that, as with cowbirds, females may respond with copulation solicitation display (CSD) to playback of male song alone and no implants are needed. The CSD response is only elicited by playback of certain 'sexy syllables' in male canary song. These have a number of special characteristics (Fig. 7.4), such as a wide frequency range and a high repetition rate which gives canaries that familiar trilling sound. They also have a complex double structure, two sounds produced one from each side of the syrinx (Suthers *et al.* 2004). This suggests that complex sexy syllables may perhaps be more costly and certainly more difficult for the male to produce. Hence they may well be a good indicator of male quality and this is why females respond so strongly to them.

Several recent studies have used playback of canary songs, manipulated to add or remove sexy syllables, in order to investigate their effects upon female reproductive physiology. Sexual selection theory (Andersson 1994) predicts that females may invest more in offspring when mated to more attractive males, and may even manipulate the sex ratio in favour of male offspring. Gil *et al.* (2004) found that exposing female canaries to regular playback of songs containing sexy syllables caused them to deposit more testosterone in the yolk of their eggs, compared to controls exposed to songs with no sexy syllables. This was confirmed in a similar experiment by Tanvez *et al.* (2004). Marshall *et al.* (2005) also repeated this basic design, but this time they found no increase in maternal deposition of testosterone

Fig. 7.4. Sexy syllables in the canary have a special structure, which makes them attractive to females. Sexy syllables have a wide frequency range, a double structure and a high repetition rate when compared with non-sexy syllables. (From Leitner & Catchpole 2002.)

into the egg yolk. However, they did find that females exposed to sexy songs had higher levels of circulating testosterone in their blood. Leitner *et al.* (2006) followed this up by looking to see whether playback of sexy song also affected egg size or the sex ratio of young produced. They found that sexy song playback induced females to lay larger eggs, but no evidence that they also produced more male offspring. On balance, it does seem that hearing more attractive, sexy songs has considerable influence on reproductive physiology and maternal investment in female canaries.

7.6 Song output

If females are selecting males, and using song as an indicator of male or territory quality, then one obvious cue could be the amount of song the male produces. There are several possible ways in which a male might vary song output. Some males (continuous singers) hardly seem to pause for breath, whereas others (discontinuous singers) have long pauses between songs. To increase song output, some birds lengthen their songs, others reduce the inter-song intervals. The latter is the most common method, and is referred to as increasing song rate. Singing may be a costly activity, both in terms of energy expended and also in terms of the risk of exposure to predators. Therefore the amount of song a male produces might well be related to phenotypic or genotypic quality. One way in which this can be demonstrated is by time or energy budget studies on the diurnal rhythm of song production. An advertising male also has to feed and survive, often in a harsh environment, and several studies (e.g. Reid, 1987) have demonstrated that males must make a trade-off. The more time

a male spends feeding, the less time he can afford to allocate to singing (see also Chapter 5). One consequence of this is that a male with a better quality territory, or one who feeds more efficiently, can afford to sing more. Therefore singing rate may indeed be a valid, condition-dependent index of male or territory quality.

For some time there was no direct evidence that female birds were responsive to variations in male song output. However, Wasserman & Cigliano (1991) obtained firm evidence by using the hormone implant technique on captive female white-throated sparrows. They manipulated song playback in both ways, by lengthening songs and by reducing inter-song intervals. In both cases they found that females displayed more to the same songs when song output was increased. Houtman (1992) gave captive female zebra finches a chance to express a preference for different males, as measured by the time they spent in front of their cages. She then let them breed by randomly assigning them partners. When the females commenced breeding, she gave them the chance to mate with other males which they had earlier expressed a preference for. One third of these females solicited and obtained extra-pair copulations (EPCs) with the more 'attractive' male. Analysis of various aspects of male quality revealed that the most important factor which correlated with male attractiveness was song rate. Houtman also found that attractive males with high song rates produced sons with high song rates and fledged heavier offspring, supporting a 'good genes' model of sexual selection. More recently, Nolan & Hill (2004) presented captive female house finches with playback of songs given at fast or slow rates and found consistent preferences for those given at higher rates.

There are a considerable number of field studies which suggest that females might base their choice of male upon his ability to sing more. Payne & Payne (1977) studied a population of village indigobirds in Africa. The males competed for females at dispersed leks located in special trees. Males used their songs to ward off rival males, and females visited several males before mating. The females mated with relatively few males, and one male accounted for over half of all the matings on the lek. The successful males were those which sang more songs per hour than their less vocal neighbours.

Several studies on European passerines have suggested a link between male singing rate, reproductive success and one particular aspect of territory quality – food. Radesater et al. (1987) found that song rates varied consistently between individual willow warbler males. Males who

Fig. 7.5. Male pied flycatchers with extra food and increased song rate ($\bar{x} = 6.9$ per min) attracted females before paired controls with lower song rates ($\bar{x} = 3.2$ per min) (from data in Alatalo et al. 1990).

obtained the best territories spent less time searching for food, and could afford to allocate more time to singing. These males attracted females before their rivals, who spent more time foraging and less time singing. To test the idea that male song rate reliably reflects territory quality, Radesater & Jakobsson (1989) monitored the singing rate of their male willow warblers. The males were then removed, and the singing rates of their replacements were also monitored. A strong correlation was obtained between the singing rates of different males in the same territories.

Alatalo *et al.* (1990) investigated the relationship between food supply, singing rate and male reproductive success more directly. They studied a population of pied flycatchers in deciduous woodland in Sweden. By manipulating nest boxes, males were assigned to random sites. Half received additional daily feeding with mealworms, and half were left as controls. The males with the extra food sang at twice the rate of the controls, who spent much of their time searching for natural food. In eleven out of thirteen cases, the experimental male with the increased singing rate paired before its control (Fig. 7.5).

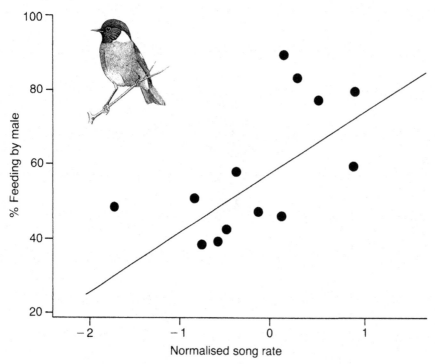

Fig. 7.6. The relationship between singing rate and feeding young in male stonechats (from Greig-Smith 1982).

Whilst there is now some evidence that male singing rate may be related to territory quality, there is less evidence to link singing rate directly with male quality. The one exception is the study by Greig-Smith (1982b) on the stonechat, where two intriguing correlations were obtained between singing rate and male behaviour. Males with high singing rates were found to participate more in nest defence, and helped more in feeding the young (Fig. 7.6). Both of these activities effect reproductive success, and so a female stonechat who selects a strongly singing male, may increase her own fitness by picking a partner who is able to invest more in parental care. But Greig-Smith found little evidence to suggest that males who sing more actually increase their fitness by leaving behind more offspring. However, the study on village indigobirds by Payne & Payne (1977) established that males with higher song rates did achieve more matings. Finally, Møller *et al.* (1998) studied song rate and paternity in a population of the barn swallow using genetic analysis of relatedness based on microsatellites. They found that males with a higher song rate also produced more offspring. One problem with song rate is whether it really is a consistent, reliable, signal, or whether it is more likely to fluctuate even in the short-term with fluctuations in the environment or in male condition. Nevertheless, there is now a line of

evidence which suggests that a male who sings more can influence female choice and may then go on to increase his reproductive success.

7.7 Repertoire size

Another cue upon which females might base their choice is the complexity of male song, and we have already seen how males utilize repertoires when signalling to each other in territorial disputes. Female preferences for the elaborate repertoires of males have been examined by two main techniques, observations in the field and laboratory experiments. Howard (1974) pioneered the field approach, by showing for the first time that male birds in a wild population who sang more elaborate songs obtained mates before their rivals with simpler songs. He studied one of the most elaborate of all avian songsters – the northern mockingbird. Pairing date might well be an important measure of reproductive success, as several studies have shown that young produced earlier in the season are more viable.

Later studies have attempted to obtain more direct measures of reproductive success, such as the number of young produced. Another problem with field studies is that they are confounded by other variables of both male and territory quality. In the laboratory, male and territory quality variables can be eliminated, and the only variable of interest, song complexity, can be manipulated by using playback of natural or even synthetic recordings. Female preference is measured using sexual display as an index, and the hormone implant technique described previously. While both field and laboratory techniques have their limitations, used together they provide a more powerful test of whether or not repertoire size can affect female choice. For this reason, we will only consider species where both techniques have been applied, and those where there is some evidence that females gain benefits from selecting the males of their choice. Finally, we will review studies that have used the comparative approach, and see whether any overall patterns or trends emerge in the evolution of repertoire size in relation to mating systems.

7.7.1 The song sparrow

Having previously demonstrated that captive female song sparrows would only display to their own species song, Searcy & Marler (1984) then tried

tapes containing equal numbers of songs of the same song type, or four distinct song types from the repertoire of one individual. They found that females displayed significantly more to the song repertoire tape, than to the single song tapes. Male song sparrows normally cycle through their repertoire in a particular way. The simplest option is to present the different song types one after the other, with immediate variety (ABCD, ABCD, etc.). In fact males sing with eventual variety, repeating one song type in a bout before switching (AAAA, BBBB, etc.). When exposed to the same repertoire presented in these two different ways, female song sparrows displayed more to the tapes where songs were presented with eventual variety, just as they are sung in nature.

Song sparrows in the wild have a larger repertoire than four, ranging from five to thirteen song types. Searcy (1984) went on to demonstrate that his captive females responded more strongly to eight than to four (Fig. 7.7), and more to sixteen than eight. He then attempted to see whether such clear results in the laboratory were reflected in the field. The results were negative: he found no correlation between male repertoire size and pairing date. A later study (Searcy et al. 1985) also found no correlations between repertoire size and any aspects of male or territory quality. However, Hiebert et al. (1989) studied a different population of song sparrows on Mandarte Island, British Columbia. They found that repertoire size was strongly correlated with several measures of reproductive success. Male song sparrows with larger repertoires obtained bigger territories, held them longer, and had greater annual and lifetime reproductive success (Fig. 7.7). A later study (Reid et al. 2004) on this population has investigated how far repertoire size can predict initial mating success. They found that young males with larger repertoires were more likely to obtain a mate, and that young females laid earlier when mated to males with larger repertoires. What females gain from choosing males with larger repertoires is not always clear, but as well as obtaining a male with a larger territory she may also breed earlier. Reid et al. (2005) have now shown that females may also gain a mate who is less inbred and has superior immunity to parasites, an important component of the Hamilton–Zuk hypothesis (see p. 195). Taken together, the song sparrow studies provide considerable evidence that repertoire size in song sparrows influences female choice in the laboratory, increases male reproductive success in the field, and also suggest that females may obtain both direct and indirect benefits from their choice.

Fig. 7.7. Repertoire size in song sparrows. Above – the relationship between repertoire size and lifetime reproductive success in a population of male song sparrows (from Hiebert et al. 1989). Below, the relationship between male repertoire size and sexual displays in captive females (from Searcy 1984).

7.7.2 The great tit

We have already seen how repertoire size in the great tit enhances territorial defence. Baker *et al.* (1986) demonstrated that hormone-implanted females displayed more to playback of larger (3–5) repertoires rather than to smaller ones (1–2). McGregor *et al.* (1981) had earlier investigated the relationship between the same repertoire range and reproductive success over several years in a natural population. Although males with larger repertoires did not pair earlier, they were more likely to survive and breed in later years. They also occupied higher quality territories and produced heavier offspring which were more likely to survive and reproduce. In a separate study, Lambrechts & Dhondt (1986) confirmed that male great tits with larger repertoires survived longer and produced more viable young. It therefore seems likely that females who choose males with larger repertoires obtain not only direct benefits from a larger territory, but also the indirect benefits of good genes for offspring viability.

7.7.3 The sedge warbler

The sedge warbler is a species well known for producing some of the longest and most complicated of all bird songs, and has now been studied

in laboratory and field for many years (reviewed by Catchpole 2000). Songs are not produced as stereotyped song types, but instead consist of long, variable sequences of syllable types. The number of different syllable types is used as a measure of repertoire size in species such as this. Catchpole (1980) was able to demonstrate that males in a marked population with larger repertoires paired earlier than their rivals with more simple songs (Fig. 7.8). This was confirmed in a later study (Buchanan & Catchpole 1997) when the same positive correlation between repertoire size and pairing date was found in three consecutive years, although song flighting and territory size also had some influence. To overcome the problems with confounding male and territory quality variables, a series of hormone implant experiments were also carried out on captive females (Catchpole *et al.* 1984).

In the first playback experiment, three recordings from the 1980 population were used. The three males selected were the ones which paired first and last, as well as an intermediate one to cover the whole of the natural range. In essence, this experiment gives the females a chance to choose again, but this time under more controlled conditions. The female response was clear, they displayed most to the largest repertoire, next to the intermediate size, and least of all to the smallest repertoire. To control for individual variation and hold singing rate constant, new tapes were prepared using a range of synthetic songs constructed from only one male. This resulted in seven 'different males', each singing at the same rate and in the same way, but with a variety of repertoire sizes ranging from two to fifty syllable types. A strong positive correlation was obtained between increasing repertoire size and female response (Fig. 7.8), adding support to the first more natural experiment and the field studies.

As for later measures of reproductive success, Bell *et al.* (1997) found that polygynous sedge warbler males had larger repertoires and produced more offspring than monogamous males. There has also been a DNA-based study on the relationship between repertoire size and the production of extra-pair young in the sedge warbler. Initially, Buchanan & Catchpole (2000a) used multilocus DNA profiling to simply compare the characteristics of cuckolded males who had lost paternity with those that had not, but found no differences in repertoire size. In a recent study on paternity using microsatellites, Marshall *et al* (2007) found that, when females selected an extra-pair male, he had a lower repertoire size that their social

Fig. 7.8. Repertoire size in sedge warblers. Above, the relationship between repertoire size and pairing date in a population of male sedge warblers (from Catchpole 1980). Below, the relationship between male repertoire size and sexual displays in captive females (from Catchpole et al. 1984).

mate. This suggests that, although females may select their initial social mate on the basis of repertoire size, other factors are then involved in the later selection of an extra-pair mate. As the sedge warbler male ceases singing after initial pairing, this may well be the case here.

In the case of the sedge warbler there has been considerable recent progress in identifying the various benefits females may obtain from selecting a male with a large repertoire. In their field study, Buchanan & Catchpole (1997) found that, in all three years, males with large repertoires also had larger territories. They also found that males with larger repertoires gave more parental care to offspring, as measured by male

provisioning rate at the nest (Buchanan & Catchpole (2000b). These studies suggest that female sedge warblers that choose males with larger repertoires certainly gain direct benefits.

Buchanan *et al.* (1999) found that sedge warbler males infected with blood parasites such as malaria had small repertoires, whereas those free of infection had larger repertoires. This suggests that females choosing males with large repertoires may also be obtaining 'good genes' for parasite resistance (Hamilton & Zuk 1982). In the same population, Marshall *et al.* (2003) then found a strong correlation between repertoire size and genetic diversity, and also obtained evidence that females prefer to mate with males who maximize their offspring diversity. In two separate studies, Birkhead *et al.* (1997) and Nicholson *et al.* (2007) obtained correlations between repertoire size and age, suggesting that females who choose males with large repertoires may also obtain good genes for offspring viability. Males who paired successfully also tended to have larger repertoires and a larger HVC in the song control pathway of their brains (Airey *et al.* 2000a). There is also evidence from isolation experiments on sedge warblers (Leitner *et al.* 2002) that HVC and song structure are under strong genetic control.

Taken together, it does seem that female sedge warblers selecting males with large repertoires could obtain a variety of indirect genetic benefits. How song, and repertoire size in particular, may have evolved as a reliable signal through developmental stress acting on both song and brain, is discussed later in this Chapter.

7.7.4 The great reed warbler

The great reed warbler is a polygynous species with great potential variation in male reproductive success. In the laboratory, Catchpole *et al.* (1986) found that recordings from a polygynous male with a large syllable repertoire obtained more response from hormone implanted females, than the same number of songs recorded from a monogamous male with a smaller repertoire. A field study in Germany (Catchpole 1986) then showed that males with larger syllable repertoires attracted more females and produced more young. Repertoire size was also correlated with some measures of habitat quality, such as the amount of high quality reed within a territory. In a later study on the same population, Forstmeier & Leisler (2004) found no correlations

between repertoire size and breeding success. They suggest that the difference may be due to a smaller, declining population in more recent times. In the earlier German study, Catchpole (1986) found no relationship between repertoire size and age, a finding confirmed in the later study (Forstmeier & Leisler 2004). However, in the Swedish study they did detect an increase in repertoire size with age (Forstmeier *et al.* 2006). It does seem that different selection pressures may operate on different populations of the same species.

The study on great reed warblers in Sweden (Hasselquist *et al.* 1996) extended for 12 years, and has thrown up some particularly interesting results. For example, they found that some males in the population also increase their reproductive success by obtaining extra-pair copulations. DNA fingerprinting revealed that the males responsible for these extra-pair fertilizations all had larger repertoire sizes than the neighbouring males whom they had successfully cuckolded. Another striking finding was a correlation between male repertoire size and offspring viability. Males with larger repertoires fathered young who were more likely to return and breed in future years. This finding from a long-term study is particularly important, as it suggests that females pairing with males who have large repertoires do benefit by obtaining good genes for more viable offspring.

7.7.5 The red-winged blackbird

Yasukawa *et al.* (1980) investigated female choice in a population of red-winged blackbirds. Red-wing males are highly polygynous, and males with larger song type repertoires obtained larger harems of females. This correlation was indirect, as older, experienced males who obtained better territories, also had larger repertoires. In the laboratory, Searcy (1988) then demonstrated that hormone-implanted female red-wings displayed more to repertoires of four than to single song types. He pointed out that extra-pair copulations are very common in red-wings, and suggested that males which are able to switch song types more frequently, might well obtain more EPCs.

Searcy & Yasukawa (1990) developed an ingenious experiment for investigating the effect of females upon the singing behaviour of males. They presented territorial red-wing males with a female dummy, mounted in the copulation solicitation display posture. The males responded by

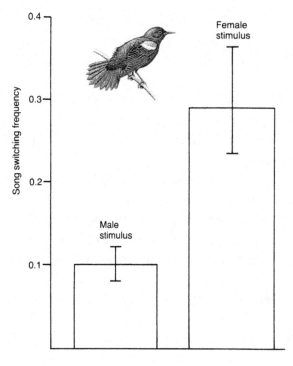

Fig. 7.9. Male red-winged blackbirds switched song types three times more rapidly when presented with a dummy of a displaying female, compared with a male (from Searcy & Yasukawa 1990).

increasing the switching frequency between song types, and by singing more song types than in control periods (Fig. 7.9), or when presented with a male stimulus. It seems as though the male red-wings attempt to demonstrate their full repertoire as quickly as possible whenever a female appears.

7.7.6 The starling

We have already seen that female European starlings can be attracted to nest boxes playing recordings of male song (Mountjoy & Lemon 1991), and that females lay earlier when paired to males who sing more (Wright & Cuthill 1992). Mountjoy & Lemon (1996) extended their experiments with nest boxes further to control for the quality of the box and their small territory within the aviary. By removing nest boxes after the first settlement, and by controlling for nest box preference statistically, they were able to confirm that males with larger repertoires attracted mates earlier. Gentner & Hulse (2000) also used the choice of nest boxes, but in an operant system where males were reinforced by song playback from a choice of boxes. They demonstrated a clear choice for boxes which played

back longer song bouts. Although they did not test for the effects of repertoire size directly, in the starling longer bouts are more complex and bout length and repertoire size are highly intercorrelated (Eens 1997). Duffy & Ball (2002) found positive correlations between bout length and both cell-mediated and humoral immunity, and suggest that starling song may act as an indicator of immunity to prospecting females.

Eens *et al.* (1991a) studied a colony of starlings in Belgium, and found that males had large repertoires (23–67) of song types delivered in long, complex bouts. Males with larger repertoires attracted females earlier, obtained more females, and produced more young. Choice experiments were then carried out in large outdoor aviaries, containing six standard nest boxes and three resident males. Again, females selected the males with the largest repertoires.

In another experiment, Eens *et al.* (1993) showed that male starlings produced more song, and more song types, when females (as compared to males) were introduced into the aviary. They pointed out that male star-lings usually sing prior to both intra- and extra-pair copulations with females. In a major review of song studies on the starling, Eens (1997) confirmed an overall correlation between repertoire size and age, suggest-ing the females choosing males with larger repertoires may well obtain good genes for offspring viability. Repertoire size also correlates with male condition and competitive ability. He concluded that the complex, war-bling songs of starlings function mainly in female choice and are a reliable indicator of male quality.

7.7.7 The comparative approach

Taken together, the integrated field and laboratory studies described above constitute a convincing case that repertoire size in several species can influence female choice, that males with larger repertoires can increase their reproductive success, and that females may gain a variety of benefits by choosing to mate with them. Quite a different approach is to test sexual selection hypotheses using the comparative method. One hypothesis here is that in polygynous mating systems increased variance in male mating success leads to more intense intersexual selection pressure on the evolu-tion of song complexity. The prediction most frequently tested is that species which are polygynous will have larger repertoires than those which are monogamous.

Kroodsma (1977) studied nine species of North American wren, and was able to confirm the prediction: the more polygynous species did have larger repertoires. However, two studies on European genera showed exactly the opposite trend. Catchpole (1980), working on *Acrocephalus* warblers, and Catchpole & McGregor (1985), using *Emberiza* buntings, found that in both cases the polygynous species had smaller repertoires. The explanation seems to be that, in these cases, polygynous males are defending large resource-based territories to which several females are attracted. The female is usually left to feed the young alone, and so must base her choice upon territory quality, rather than male quality. The male has therefore developed a song largely for territorial defence, rather than direct female attraction. Conversely, in monogamous species, the males defend smaller territories with few resources, and both sexes are needed to feed the young. The female therefore bases her choice upon a high quality male (rather than a high quality territory), and one aspect of this may be song structure, leading eventually to increased repertoire size.

These studies were on quite small sample sizes, but have shown that sexual selection on repertoire size can result in quite different patterns in different groups. Irwin (1990) studied 17 species of New World icterine blackbird, and found no overall association between repertoire size and mating system, but again different patterns occurred in different groups. Shutler & Weatherhead (1990) studied a larger sample of 56 species of paruline wood warbler, and also reported no overall relationship between repertoire size and mating system.

The most ambitious comparative study is that by Read & Weary (1992), who investigated a database of 165 passerine species. They used several different methods to search for associations at different taxonomic levels while controlling for the confounding effects of phylogeny. Syllable repertoires (but not song repertoires) were found to be higher in polygynous species, whereas song repertoires were larger in species which provided more parental care. Both song and syllable repertoire size were higher in species which migrate, confirming a suggestion by Catchpole (1980) that such species have less time for mate selection, leading to more intense sexual selection on song complexity. Read and Weary also found that their associations were often different at different taxonomic levels. For example their correlation between syllable repertoire size and polygyny only holds for species and not higher taxa.

In their pivotal paper, Hamilton & Zuk (1982) used song structure and parasite prevalence in a comparative study which suggested that song complexity had evolved as a signal of parasite resistance. However, a later analysis by Read & Weary (1990) contradicted this view. Møller *et al.* (2000) have suggested that in such comparative studies, parasitism scores should be based upon measures of immune function rather than just prevalence. They used the relative size of the spleen, important in immune defence, and in 38 species of birds found a positive correlation between repertoire size and spleen size. Garamszegi *et al.* (2003) used a similar sample size and measured the immune response directly and found a positive correlation between repertoire size and the cell-mediated immune response.

Searcy (1992a) maintains that another prediction from sexual selection can be tested by a comparative study. He postulates that, across species, the strength of the female response should be positively correlated with an increase in repertoire size. Because the same response measure has been used in different hormone implant studies, and the same playback design, Searcy was able to confirm his prediction (Fig. 7.10). Another possibility to consider is that the female response merely reflects a general sensory bias which the male trait subsequently exploits. The sensory exploitation hypothesis (Ryan *et al.* 1990) has suggested that sensory tuning in the auditory system of female frogs pre-dates the evolution of male mating calls. Evidence for this is that females in one species are sensitive to a male trait which is only found in another species. Supporting evidence that this may have happened in the evolution of song repertoires in birds is equivocal. Searcy & Marler (1984) studied field sparrows and white-throated sparrows, which both lack repertoires, and found no female preference for four as opposed to one song type. However common grackles also have only one song type, and Searcy (1992) found that females showed a clear preference for repertoires of four.

Comparative studies suffer from a number of problems, as discussed by Read & Weary (1992). For example, if confined to one particular group, then generalizations cannot be made to birds in general. Yet smaller studies by one investigator do not suffer from 'the comparability problem' encountered by Read & Weary (1992) when attempting to make valid comparisons between heterogeneous data collected from many different studies. This is particularly true of data on bird song complexity and mating systems, where widely differing criteria may be used in estimating repertoire sizes and deciding whether a species is monogamous or

Fig. 7.10. The relationship between repertoire size and the strength of female sexual response in six species of songbird (white-throated sparrows, field sparrows and common grackles have median repertoires of one, swamp sparrows three, red-winged blackbirds four, and song sparrows nine) (from Searcy 1992).

polygynous. The latter has become increasingly difficult as DNA finger-printing studies continue to blur the boundaries of conventional mating system classification (Birkhead & Møller 1992). The difficulties of obtaining accurate, comparable estimates of repertoire size are well known, but in addition comparative studies often obtain data from secondary sources. For example, the most recent by Garamszegi et al. (2003) used a mixture of estimates from earlier comparative studies and measurements of sonagrams taken from handbooks. Although there have been considerable advances in the techniques now used in comparative studies (Harvey & Pagel 1991, Felsenstein 2004), in some ways they raise more questions than they answer. But perhaps this should be their main role. Whilst comparative studies must clearly be treated with some caution, there is no doubt that they reveal interesting relationships which should then tested by further observations, field studies and experiments.

7.8 Reliability and honesty

The reliability and honesty of signaling is a central question in the study of animal communication, and has recently been reviewed in some depth by

Searcy & Nowicki (2005). They suggest that a signal is reliable if some characteristic of it correlates with some attribute of the signaler that the receiver will benefit from knowing. We have already seen in this Chapter that there are many correlations between various aspects of song and male or territory quality. In modern sexual selection theory (Andersson 1994) there is general agreement that the honesty of signaling is maintained by costs incurred by the signaler. Thus, weaker, inferior males are prevented from cheating as they cannot afford to pay the extra costs. But what exactly are the costs of singing in birds? According to Nowicki & Searcy (2005), there are three main categories to consider: production, maintenance and developmental costs. As we will see, there is now considerable debate about the nature of these costs (e.g. Gil & Gahr 2002), and also where and when they are paid.

7.8.1 Costs and performance

The most obvious cost is that incurred by simply increasing the amount or rate of singing. We have already seen that there is a trade-off between the amount of time spent singing and foraging (Reid 1987). The more time spent foraging the less time is left for singing, and there are several studies which have shown that singing increases after food supplementation of territories (e.g. Alatalo *et al.* 1990). High male song output may well be a reliable signal to a prospecting female that food is plentiful on a particular territory.

But how much does it cost a male to produce his songs? To answer this we need to measure the metabolic cost of singing, as opposed to the basic resting metabolic rate. The standard way to do this is to measure the oxygen consumption of a singing bird in the laboratory. When Eberhardt (1994) first did this on a Carolina wren, he found that oxygen consumption increased by a factor of 3.9 over the resting metabolic rate (RMR). This suggested a substantial metabolic cost of singing. However, later studies have found somewhat lower costs. Oberweger & Goller (2001) studied several different species and found factors of between 2 and 2.6, and Ward *et al.* (2003, 2004) reported a similar range between 2.1 and 2.7. These values were obtained by comparing the metabolic rates of resting or sleeping birds with those singing. If the comparison is made between the birds standing while awake compared to singing, the values drop even

more, towards 1 rather than 2 (Ward *et al.* 2004). It therefore seems that the metabolic costs of singing are rather less than previously thought, although birds close to their energy limits might still find it difficult to afford the extra costs.

There is also a recent body of work which suggests that the quality of the song itself may also be an indicator of male quality. Vocal performance refers to the ability of males to produce songs which are at the very limits of physical and physiological constraints. These songs may or may not be more expensive, but they may also be more difficult to produce, and so provide another test of male quality.

There are two examples of this, which we have already discussed in earlier sections, and the first is the canary. A series of studies (Vallet & Kreutzer 1995, Vallet *et al.* 1998) have shown that female canaries respond much more when certain phrases are included in the playback of male songs. These 'sexy syllables' have particular characteristics such as wide frequency range and rapid repetition in long trills. They also have a special two-note structure and recent research shows that these two notes are produced separately from the two sides of the syrinx (Suthers *et al.* 2004). This performance may well be not only more costly but more difficult to produce. Female canaries have now been shown to respond preferentially to sexy syllables by producing more sex hormones in their blood (Marshall *et al.* 2005), depositing more hormones in their eggs (Gil *et al.* 2004) and also laying larger eggs (Leitner *et al.* 2006). As we have seen in Chapter 2, playback of sexy syllables even affects the brain directly by stimulating production of immediate early genes such as ZENK (Leitner & Catchpole 2005).

Work on another species, the swamp sparrow, has also shown how females pay particular attention to song performance (see also Chapter 2). Like canaries, swamp sparrows produce fast trills with a wide frequency range. Podos (1996, 1997) has shown that to achieve this performance they must open and close the bill widely but also quickly. However, there is a limit to how fast this can be done and a trade-off exists between bandwidth and repetition rate. Some birds perform better than others and are close to the upper limit of performance. Ballentine *et al.* (2004) tested a range of songs and found that during playback, females responded preferentially to songs close to the performance limit. It does seem that, in both these species, females prefer males whose singing performance is the best they can achieve.

7.8.2 Song and developmental stress

The costs of producing a more complex song by increasing repertoire size are much less obvious than by increasing the amount of song, but recent research is starting to focus on the possible costs of learning and storing these extra syllables in the brain. As we have seen in Chapter 2, there is a complex neural pathway in the brain which controls song production and learning. The idea that the size of this pathway might increase in order to store more, or more complex songs, was first emphasized by Nottebohm et al. (1982). In Chapter 2 we also reviewed the evidence which suggests that, both within and between species, there are positive correlations between repertoire size and the main song control nucleus HVC. The question which then arises is: are there significant costs involved in the production and maintenance of the extra neurons required for further song learning? It is fair to say that in recent years this has become one of the biggest questions in relation to the evolution of repertoire size by sexual selection. Whilst it is reasonable to assume that there are some metabolic costs in singing, would it really cost much more just to add a few syllable types or another song type to the repertoire?

The debate about the metabolic costs of singing continues (Gil & Gahr 2002), but there is now a hypothesis which attempts a more plausible explanation of how and when the costs of song complexity might be paid. Nowicki et al. (1998a, 2002) pointed out that, in a young bird, the development of both brain and behaviour are particularly vulnerable to nutritional shortages which may have lasting effects. We already know that the song control pathway develops over a period of several weeks and that this is critical if normal song learning and production is to occur. In the 'nutritional stress' hypothesis, Nowicki et al. (1998a, 2002) suggested any disruption at this stage could have lasting effects which manifest themselves in the final structure of the adult song. The key point is that, any costs are paid during development, not during song production itself. Only young which are of high quality and can pay the extra costs to compensate for any stress will have the necessary neural hardware to produce a more complex song. Those of lower quality cannot afford to pay the costs and will signal their inferior quality as adults by producing a more simple song. Buchanan et al. (2003) have suggested that environmental stresses other than nutrition might have similar effects and that the hypothesis should now be considered as a more general 'developmental stress hypothesis.'

There are now several studies which have tested the developmental stress hypothesis in the laboratory. Nowicki *et al.* (2002) subjected nestling song sparrows to nutritional stress and found that as adults they showed poor copying fidelity of model songs. They also found that the volume of HVC and RA were significantly smaller in the food-stressed males. Buchanan *et al.* (2003) exposed young starlings to nutritional stress by removing their food for several hours each day. As adults the young birds sang shorter song bouts and had smaller repertoires than controls. In an experiment on zebra finches, Buchanan *et al.* (2004) subjected nestlings to two experimental regimes, either food diluted by mixing it with husk, or a daily dose of the stress hormone corticosterone given in a food pellet. In both cases, the experimental males produced shorter, simpler songs than controls (see also Spencer *et al.* 2003). This study also investigated the effects upon the underlying song control pathway. In the experimental birds, both groups showed a significant reduction in the volume of the

Fig. 7.11. The effects of experimental malaria infection during development upon both repertoire size (a) and the size of the song control nucleus HVC (b) in adult male canaries (from Spencer et al. *2005a).*

main song control nucleus HVC, but not in others also measured. Spencer *et al.* (2005a) tested the idea that other environmental factors such as infection might also be a cause of developmental stress. They infected nestling canaries with an avian malaria and monitored their infection and subsequent recovery. Later they found that adult males had smaller song repertoires, and again HVC was significantly reduced in volume compared to other brain nuclei (Fig. 7.11).

Developmental stress may well result in the modification of adult song, but are the experimental changes we have produced enough to convince discriminating females that such males should be avoided? Spencer *et al.* (2005b) have recently tested this in the zebra finch using an active choice apparatus. This gives females the chance to show their song preferences by selecting perches which then play back a particular song. The songs used were the ones produced in the experiments just described above (Buchanan *et al.* 2004). The results were very clear: females preferred to perch where they heard normal control songs and tended to avoid the perches where stressed songs were played. MacDonald *et al.* (2006) found that nutritional stress selectively reduced the size of HVC in both male and female song sparrows, and it now remains to be seen whether stressed females are impaired during female choice for song.

In summary, it seems that developmental stress can affect both song and brain development resulting in the production of inferior song structures that females are more likely to avoid. Whilst the developmental stress hypothesis does not explain all our questions about why male songs have become so complex, it is certainly a step forward in explaining how song could evolve through sexual selection as a reliable indicator of male quality. There may also be other factors which help to explain the evolution of song complexity, and these will be discussed in the next chapter.

THEMES AND VARIATIONS

*Chorusing group of
plain-tailed wrens*

The merry lark hir mattins sings aloft
The thrush replyes, the mavis descant playes
The ouzel shrills, the ruddock warbles soft
 Edmund Spenser *Epithalamion*

8.1 Introduction

In some birds, such as the zebra finch, each male has only a single, simple and stereotyped song which he repeats monotonously over and over again. In others, song is a virtuoso performance, with a variety that makes the finest operatic aria seem positively dull. The mockingbird in North America, the superb lyrebird in Australia and the nightingale in Europe are among the best examples here. The majority of birds lie somewhere between these extremes with repertoires ranging from a few song types to several hundred. In this chapter we will survey this diversity; we will discuss how birds use their repertoires and what functions repertoires are thought to serve. At first sight the diversity seems daunting: we will end by suggesting how it may be understood by pulling it all together within an evolutionary framework.

8.2 Repertoire sizes

While it is clear that species differ enormously in the variety of their song, repertoire sizes are not at all easy to define and estimate in a way that is useful for comparisons between species (Krebs & Kroodsma 1980, Kroodsma 1982). In some species, songs are simple and discrete, and there is little if any sharing of elements between song types. But in other cases different song types may be made up by reassorting elements. The variety of song organisation itself makes the measurement of repertoires in a way that allows comparison between species a considerable practical problem.

Although no single measure of repertoire size is entirely satisfactory, the most usual one is to look at the number of song types that each individual possesses. Table 8.1 gives typical song repertoire sizes for a range of different species in which whole songs are relatively stereotyped and discrete and shows what diversity there can be even amongst close relatives. In the thrushes, for example, the European redwing has a single type, the blackbird around 50 and the song thrush 200. In these birds, as in all those chosen for Table 8.1, we are talking about distinct songs, with very little if any sharing of elements between them. However, some birds build up what sounds like a formidable repertoire from a smaller variety of elements assorted in all sorts of different ways. Catchpole (1976) has suggested that a male sedge warbler may never repeat exactly the same

Table 8.1 The variety of song repertoire sizes in songbirds

Species	Repertoire size	Reference
Ovenbird	1	Falls (1978)
White-crowned sparrow	1	Baptista (1975)
Rufous-collared sparrow	1	King (1972)
European redwing	1	Bjerke & Bjerke (1981)
Splendid sunbird	1	Grimes (1974)
Chaffinch	1–6	Slater (1981)
Great tit	2–8	Krebs *et al.* (1978)
Dark eyed junco	3–7	Williams & MacRoberts (1977)
Western meadowlark	3–12	Horn & Falls (1988a)
Hermit thrush	6–12	Rivers & Kroodsma (2000)
Song sparrow	7–11	Beecher *et al.* (2000b)
Cardinal	8–12	Lemon (1974)
Banded wren	15–24	Molles & Vehrencamp (1999)
Starling	15–70	Eens (1997)
Red-eyed vireo	12–117	Borror (1981)
European blackbird	34–63	Rasmussen & Dabelsteen (2002)
Marsh wren	33–162	Kroodsma & Verner (1987)
Eastern meadowlark	36–84	Lemon *et al.* (2000)
Yellow-breasted chat	46–81	Dussourd & Ritchison (2003)
Mockingbird	53–150	Howard (1974)
Nightingale	160–231	Kipper *et al.* (2004)
Song thrush	138–219	Ince & Slater (1985)
Five-striped sparrow	159–237	Groschupf & Mills (1982)
Brown thrasher	1500+	Kroodsma & Parker (1977)

sequence of elements twice during the course of his life. This is because each song is extremely long, and the 50 or so elements that a bird possesses follow each other in highly varied orders. At one level (that of the song) his repertoire size is enormous, but at another (that of the element) it is much more modest.

If a bird has only a few song types, working out its repertoire size is not particularly difficult, especially if he sings each type in turn in an orderly fashion as some birds do. But, where repertoires are very large, it is almost impossible to count their size accurately, and one must resort to making an estimate.

Two different methods have been used to estimate large repertoires. The usual one is to plot the number of songs recorded against the number

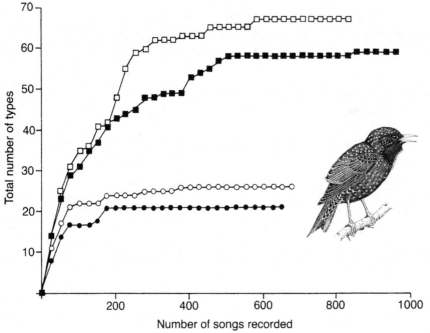

Fig. 8.1. A cumulative plot showing how the number of new song types found rises as an increasing number of songs are analysed. The results shown are from four male starlings. Initially each song recorded is new, but eventually the plot approaches an asymptote at which every song type recorded has already been recorded earlier (after Eens et al. 1991b).

of different types found (Wildenthal 1965). As more and more songs are recorded, fewer and fewer new ones are found (see Fig. 8.1). Eventually, when all of a bird's types have been discovered, the plot becomes horizontal. This asymptote is the bird's full repertoire size. Even before it is reached, however, the form of the curve approaching it can be used to extrapolate and so estimate where the ceiling will be. If new songs turn up randomly as recording continues, the shape of the curve is an exponential one and the formula describing it is:

$$n = N(1 - e^{-T/N}).$$

In this equation e is the exponential constant, T is the number of songs that has been recorded (the figure along the x axis in Fig. 8.1) and n is the number of different song types found in that sample (the equivalent figure on the y axis). N is the total number of song types the bird possesses and it can be estimated from the sample by fitting the other values into the equation.

An alternative technique, which was suggested in our first edition and subsequently picked up by a number of authors (e.g. Garamszegi et al. 2002, 2005), is to employ the capture–recapture technique used by population biologists. If 100 fish are marked and put back in their pond, and 10 out of 100 caught in a later trawl are found to have the mark, then we can estimate

the fish population of the pond at $100 \times 100/10 = 1000$. In the same way, if 200 songs are recorded and 10 out of the second 100 are repeats of types recorded in the first 100, the repertoire size can be estimated at 1000.

These ways of estimating repertoire size have their drawbacks. Most notably, they assume that the song output of the bird is not structured in any way (just as it is assumed that the fish in the pond become totally mixed up between samples). The samples must be random. As the next section shows, this is far from so for many of the birds studied to date. The exact way in which a bird sings may lead the estimate to be too high or too low. For instance, if it repeats the same song many times before switching to another (shows 'eventual variety'), a short sample may suggest that the bird has only one song type, whereas if it sings with 'immediate variety', only singing each type once before switching to the next and cycling through its repertoire, a short sample may include no repeats, suggesting that its repertoire size is infinite.

For both these techniques, therefore, a good long sample is required and care must be taken to ensure that differences in song organisation do not throw out the estimates too much. The precise technique employed is often best adapted to the singing style of the particular species under study, a point made clearly by Garamszegi *et al.* (2005) who, in something of a *tour de force*, examined how capture–recapture methods might be best applied to 18 different species. Of course, if a bird does not often sing the same song type twice, this may be either because it has a huge repertoire of distinct types or because it is improvising so that its repertoire is continuously changing. The latter may occur in some species, as suggested for the Sardinian warbler by Luschi (1993) (see Fig. 8.2). Some other species change the songs they sing more slowly, from one part of the season to another or from one year to the next. Where they appear to generate new songs much more rapidly, this may be because a limited repertoire of elements is being reassorted between songs. In such cases, it is more appropriate to examine repertoire sizes at the level of elements rather than at that of songs. Jumbled and rambling though many bird songs may appear to be, examination of sonagrams reveals that they are in reality usually far from so; exactly the same elements occur again and again, albeit perhaps in very different orders.

Where repertoire size is really vast, looking right through a long sample of the bird's output and searching for repetitions may be impractical. The brown thrasher is the best example here, as this species is the current

Fig. 8.2. Two successive song bouts as sung by a male Sardinian warbler. Forty songs were produced and these were composed from 53 different elements, the black dots indicating the exact elements that occurred in each song. New elements continued to appear throughout the series, and a variety of those common in the first bout were rare or absent in the second (from Slater & Lachlan, 2003, after Luschi 1993).

record holder for song repertoire size (see Table 8.1). As each bird has over 1000 song types, estimating its repertoire size is not an easy matter. Boughey & Thompson (1981) did use Wildenthal's method, but the labour involved was obviously enormous. They recorded nearly 2000 songs from each of two birds and discovered 1147 and 1116 song types, respectively. Application of the formula gave estimated repertoire sizes of 1967 and 1553. However, these figures may well have been underestimates as new song types were still occurring at a rapid rate towards the end of their samples. What Kroodsma & Parker (1977) did to solve this problem was to take every 100th song in a sample of 4654 and see if it recurred anywhere else in the whole sample. There were 45 different song types amongst the 46 examined, and those types accounted for 116 of the songs in the whole sample. In other words they were repeated an average of 2.6 times each. The repertoire of the bird could therefore be estimated as 4654/2.6 = 1805.

8.3 The organisation of repertoires

As with many other aspects of singing in birds, there is huge variation between species in how repertoires are used. There have been many studies of how song output is organised, and song has proved an especially useful subject for studies of the patterning and sequencing of behaviour. Models of song organisation can become quite complex but tend to fall into one of two different types. First are the competitive models, in which each

song type is seen as competing with others for access to a single output pathway. Each type may be thought of as governed by a control unit, with facilitating or inhibitory interrelations with other such units (Todt 1975a, Whitney 1981). The alternative approach is to look at sequential relationships between song types. At one extreme these might be deterministic (so that A always follows B) and at another they might be random (so that the identity of the previous song does not influence which will be the next one). Usually, however, song sequences are somewhere between these extremes with the probability of a particular song type depending on one or more of those that precede it (Chatfield & Lemon 1970). If it depends only on that immediately before, song organisation can be described as a first-order Markov chain (Lemon & Chatfield 1971, Martin 1990); if dependence stretches further back, a higher order Markov chain may provide a better description (Dobson & Lemon 1979).

These two types of model are not incompatible: they are simply alternative ways of describing song organisation. The sorts of processes they represent, such as facilitation, inhibition and competition, may underlie the production of song sequences. With time, the rather clear structure of song-repertoire use in many species seems a good prospect for a link with neurobiology. At present we must be content with describing the organisation and modelling how it may be produced. A few case histories will give some impression of the variety of processes likely to be involved.

8.3.1 The chaffinch

The way in which chaffinches use their songs has been studied particularly by Hinde (1958) and Slater (1983b). Most chaffinches have two or three song types, though some have only one and others may have up to six. Each song type is fixed in form and consists of exactly the same sequence of syllable types every time it is produced, though syllables in the trill part of the song may be repeated a varied number of times, and there are also differences between individuals in this and various other measures of timing and duration (Riebel & Slater 2003, Leitão et al. 2004). Two songs within the repertoire of a single bird can be very similar, although with practice they can usually be distinguished by ear.

The chaffinch is typical of a species with a small repertoire that sings with 'eventual variety': a male with more than one song type will sing

a whole series of one type before switching to another. Furthermore, if a male chaffinch has three or four types, he will usually sing a sequence of each in turn before returning to the first again. Thus, a typical sequence from a bird with three song types might be:

... AAABBBCCCCCCCCCAAAAAABBCCC ...

But the order A, followed by B, followed by C is not fixed. On another day the same bird might well sing for long periods in the alternative order:

... AAACCCCCCBBBAAAAAAACCCCCCCCCBBBA ...

Note that the songs are not necessarily sung with equal frequency: in the sequences shown above, B is rarer than the other two types. It is quite common in many species for one song to be a much larger part of the bird's output than another. Indeed, a song type may be so rare that its presence in the bird's repertoire is only discovered after many hours of recording. The duration of bouts is more a matter of time than of number of songs: a bird that is singing fast tends to produce more songs of one type before switching to another (Riebel & Slater 1999b).

While all these features may not be true of other species, it is certainly common for birds with small repertoires to sing with eventual, rather than immediate, variety. Examples of other species that have been found to do so are the great tit (Krebs 1976), western meadowlark (Falls & Krebs 1975), dark-eyed junco (Williams & MacRoberts 1977) and northern cardinal (Lemon & Chatfield 1971).

Slater (1983b) argued that the organisation of chaffinch song can only be completely understood if it is assumed both that there are sequential relationships between song types and that the different song types compete with each other for the output. While the sequences found were, for the most part, quite straightforward if both these processes were taken into account, they were more complicated in some individuals, where a given song type appeared to occur in two different contexts. Rather than proposing that this arose because of complex sequential relationships, Slater suggested that it might be because one song type had been learnt twice and so occupied two positions in the sequence. The bird might, therefore, have a repertoire of four song types, but appear only to have three because two of the types were identical. On this assumption, the temporal organisation of song in such birds could be understood in the same way as that of others without requiring a more complex model.

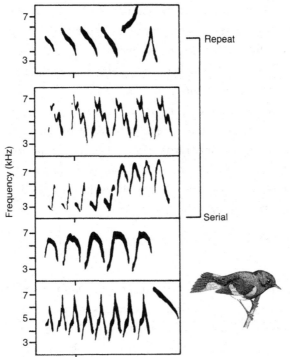

Fig. 8.3. An example of an American redstart song repertoire. The male has five songs, one of them sung in the 'repeat' mode, the others used during 'serial' singing (from Lemon et al. 1987).

8.3.2 The American redstart

The American redstart, which nests on the eastern side of Canada and the United States, also has a small repertoire of song types, but it uses them in a very different way from the chaffinch (Lemon *et al.* 1985). The number of song types that a male has can vary from one to eight, with a mean of 4.3. In birds with three or more songs, one song type, which Lemon and co-workers call its 'repeat song', is sung in long bouts (... AAAAA ...), while the others, its 'serial songs', are sung with immediate variety (... BCDECBCE ...). Fig. 8.3 shows the typical repertoire of a single bird. Birds vary in how predictable the sequences are: some cycle through their serial repertoire in a more or less determined sequence, whereas in others it is less easy to predict from the preceding songs which will come next. Birds with repertoires of six or more serial song types may use these in groups. For example, one bird with six types studied by Lemon *et al.* (1993) sang these in two groups of three, most often switching between songs within a group and much less often switching to the other group.

The Paruline warblers, a large North American group of which the redstart is one, will be discussed further in Section 8.7, as many of them

have more than one type of song used in different contexts. Repeat singing is thought to function between the sexes and is particularly prevalent at the time of mate acquisition, while serial singing seems to act mainly as a signal between males. In the case of the redstart, there is experimental evidence of this from playback experiments by Weary *et al.* (1992, 1994a). In addition to differences in their context, the repeat and serial songs of redstarts differ in development. Most males retain the same repeat song from one year to the next, whereas many birds add to or delete from their serial song repertoire between years (Lemon *et al.* 1994). This links with the function of this repertoire in male–male interaction, as the changes that occur lead to greater sharing between neighbours.

8.3.3 The nightingale

The common nightingale is famous amongst European birds for its virtuosity and apparently endless variety. Despite this, each song type that a male has occurs in identical form whenever it is sung, except for the minor point that a particular element may be repeated a variable number of times. The appearance of huge variety is partly because each male has a very large repertoire, which may include over 200 song types, and partly because it is unusual for the same song to be sung twice close together.

Studies of nightingales in the laboratory suggest that their songs fall into small groups, or packages, the songs in which tend to be sung close together (Hultsch & Todt 1989a,b), and these packages in turn tend to be clustered into higher level groupings in a hierarchical manner (Todt & Hultsch 1998). Thus a bird may start off ... ABCDEF ... and then, perhaps half an hour later, it might sing ... BFDE ... There is nearly always immediate variety: the same song is not usually sung twice in a row. But the order within a grouping is not necessarily identical each time it appears, nor is the sequence as the bird moves from one package to another. Every song is not sung each time the bird cycles through its repertoire. Thus, in the example above, song types A and C have been omitted the second time round and will not occur again until the next time that package crops up. In a bird with 200 types, it may be common for songs to show a 'recurrence interval' of around 70 types, rather over half the songs being omitted during each passage through the repertoire. Birds often share song types with others in their neighbourhood, and may also share associations between song types, but individuals do differ

substantially in the organisation of their singing, while the same bird will follow much the same pattern in successive years (Kipper *et al.* 2004). Given these characteristics, it is not surprising that the nightingale has its reputation as one of Europe's most varied singers.

8.3.4 The starling

Starlings are well known for their long and complex songs and for the mimicry of other species that occurs within them. But, what at first may seem a random outpouring is in fact highly organised (Eens *et al.* 1989, 1991b). A bout of singing lasts an average of 25 seconds, and consists of a sequence of phrases or song types with very short intervals (mostly less than 0.1 seconds) between them. Each phrase tends to be repeated several times before the bird moves on to the next. Repertoire size varies from around 15 to 70 (Eens 1997) (see Fig. 8.1).

The order of song types within a bout of singing is quite stereotyped, and the types fall into four broad categories, which also tend to occur at particular points in the sequence (see Eens 1997 for a detailed account). The bout normally starts with a number of whistles, each male having a repertoire of 2–12 of these. These are then followed by a series of variable and complex phrases including cases of mimicry: Hindmarsh (1984) found each male to have 9–31 different imitations in his repertoire. The next part of the bout consists of rattle song types. These phrases include a rapid succession of clicks, sounding like a rattle, and each male has a repertoire of 2–14 of them. Finally, most song bouts which are not interrupted earlier end with some loud high-frequency song types. A male may have up to six of these.

Again, as with the nightingale, starling song may seem endlessly varied to the ear, but closer analysis reveals that each male has a limited repertoire of song types and that these are ordered according to quite well-specified rules.

8.3.5 The sedge warbler

The sedge warbler is a good example of a species in which elements can be reassorted to make many different song types (Catchpole 1976, 2000, Buchanan & Catchpole 1997). Here successive songs consist of different combinations of elements and elements are recombined continuously, so

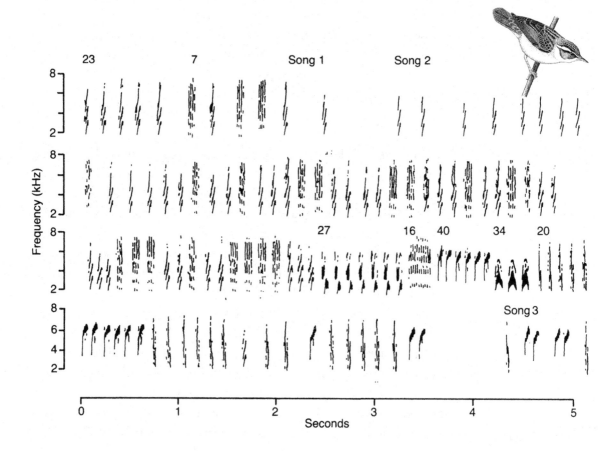

that there is no fixed repertoire of song types as in a chaffinch or nightingale.

Each bird has a repertoire of up to 75 different element types. The song of a perched bird is typically 20 seconds in length and can contain up to 300 elements (see Fig. 8.4). It starts with a long section in which short series of two elements alternate with each other. There is then an abrupt switch to a louder, more rapid and complex central section in which 5–10 new elements are suddenly introduced in quick succession. In the last part of the song, the patterning is similar to that at the start, except that the two elements used are selected from among those that occurred in the central section. These same two elements are typically those that are employed at the start of the next song. Because it is long and has this varied patterning, the song of the sedge warbler is extremely complex. But this huge variety is built up, once again, from a relatively small repertoire of subunits put together according to fairly specific rules.

Fig. 8.4. A complete song recorded from a male sedge warbler, together with the end of the song that preceded it and the start of that which followed. Syllable types are numbered when they first appear. Syllables 23 and 7 alternate at the end of song 1 and the start of song 2. Several syllables first appear in the middle of song 2 and two of these, 40 and 20, alternate at the end of that song and the start of song 3 (from Catchpole 1976).

In common with many other open-country birds, sedge warblers also show song flights. Songs produced during these are similar to perched songs but are even more complex as more element types are introduced into the central section.

The five main examples used in this section can only give something of the flavour of repertoire use among birds. None of these species is typical of a huge array of others; almost any other species that could have been described would differ in some ways from all of them. But they do indicate that song patterning is rich in its variety.

8.4 Duets and choruses

A special, and particularly fascinating, case of repertoire use is where more than one individual combines to sing. Most such cases are where the two members of a pair produce a 'duet', but sometimes males may sing in synchrony with each other, as in the display of pairs of males in the long-tailed manakin (Trainer *et al.*, 2002), and sometimes groups of socially living birds may all join in a chorus.

8.4.1 Duets

Duetting is much commoner and easier to analyse than choruses involving more than two birds. Some duets have phenomenal precision of timing. Indeed, while bouts may overlap, the sounds themselves may not do so, the birds fitting their sounds together so precisely that it is hard to believe that more than one individual is involved. This form of duetting, in which male and female use different notes and sing alternately, is known as antiphonal singing (Hooker & Hooker 1969). It was extensively studied by Thorpe (1972), particularly in African shrikes. A good example of it is in the white-browed robin-chat (Todt & Hultsch 1982), in which series of elements from the two birds alternate, with very little overlap between them yet astonishingly brief latencies. In the Eastern whipbird (see Fig. 9.4), the male produces a long thin whistle followed by a loud crack, like that of a whip. This is followed on around 30% of occasions by the more varied response of its partner, with a very short latency (Rogers 2005, Mennill & Rogers 2006).

Duetting is uncommon in the temperate north and common in the tropics. Although there are around 700 breeding bird species in both North

America and Panama, Farabaugh (1982) found that there are 101 that duet in the latter area and only 23 in the former. Holding of year-round territories is also common amongst duetting birds, even considering only those in North America, where it is not a common feature generally as it is in the tropics. This in turn is associated with birds that form long-term monogamous pairbonds, suggesting that duetting may have an important role in pair-bonding.

One other association often claimed is that between duetting and sexual monomorphism, and Malacarne et al. (1991) did find such an effect among European non-passerines. However, in her examination of Panamanian species, Farabaugh (1982) found 63% to be monomorphic overall, while amongst duetting species 65% were monomorphic. She therefore argued against such an association, but she did say that it is possible that mono-morphy may be commoner than expected among species with very precise duets. Her failure to find the association may, therefore, stem from the very broad definition of duetting that she took. Her analysis may also have been biassed by including many related species from a few groups: there is an almost total lack of monomorphism amongst the many ant-thrushes (Formicariidae) in Panama that duet and all of them are included as separate data points in her analysis. Nevertheless, the associations that Farabaugh did find are strong and clear. The common occurrence of antiphonal singing in the tropics is especially striking to someone from the temperate north, who is likely never to have encountered it before going there. It seems to be associated with the relative constancy of the tropical environment.

But why do they do it? The evidence was thoroughly reviewed by Hall (2004), who lists a variety of hypotheses (see also von Helversen 1980). Some of these have attracted rather little support. For example, Thorpe (1972) considered maintenance of contact in dense vegetation as a likely function in the shrikes that he studied. However, Wickler (1976) reported that duetting only starts in several species when the pair are in visual contact and is, therefore, unlikely to serve to maintain contact between them. In a comparative study, Malacarne et al. (1991) failed to find a correlation with dense habitat, and in some species duetting is actually less likely if the birds are further apart (e.g. Hall & Magrath 2000). Paternity guarding by males is another idea that has attracted little sup-port. If it was the case one would expect duetting to be most frequent during the female's fertile period, and also that duets would be female led,

the male joining in to label her as mated. Neither is generally true. In the bay wren, where the duet cycles are female led, males do not answer more when their females are fertile than at other times (Levin quoted by Hall 2004). In the buff-breasted wren the proportion of songs of one sex answered by the other did not differ between breeding stages (Gill *et al.* 2005), again arguing against paternity guarding being an important function.

There are several more likely ideas. Perhaps duetting helps to maintain reproductive synchrony (Dilger 1953), an idea in line with its greater occurrence in the tropics where fewer environmental cues are available (Slater & Mann 2004). Firm evidence on this is, however, lacking and, given the occurrence of duetting throughout the year and the fact that changes through the breeding cycle are not usually great, the effect would have to be a somewhat subtle one. However, Sonnenschein & Reyer (1983) argued that different duet types perform different functions in slate-coloured boubous and that one of them, particularly prevalent at the start of breeding attempts, might help to bring the pair into synchrony.

The signalling of commitment between partners has also been suggested as a reason for duetting (Wickler 1980). For example, if achieving a high degree of song coordination between partners takes time and investment, then desertion will become costly because of the need to invest again with a new partner, thus tying members in to the pair. However, the most detailed studies suggest that it does not take time to achieve high coordination (Levin 1996a, Hall 2004). Despite this, pairs may combine particular phrases in unique ways and the consistency of these combinations may improve with time (Marshall-Ball *et al.*, 2006). So other aspects of duetting may be a sign of bonding. Against this, Todt & Hultsch (1982) showed that duetting need not be essential to pair bonding, as the pair-bond in robin-chats was maintained even when their duetting was disrupted (albeit in a captive study where there was no opportunity for the birds to reassort). Nevertheless, it seems likely that in many species, as well as acting as a signal to neighbours and intruders, duetting does function as a signal between the members of a pair. Where the two both have repertoires, it is usual for particular phrases to be combined with each other non-randomly (e.g. Mann *et al.* 2003, Hall 2006), to form what Logue (2006) refers to as a 'code'. Different pairs often combine the same phrases in different ways, making the code a rather private one most likely to

function between the members of the pair rather than in interactions with others.

By far the most commonly proposed functions for duetting are mutual mate guarding and joint territorial defence. Though some data are counter to it (Mulder *et al.* 2003, Gill *et al.* 2005, Hall & Magrath 2000, Logue & Gammon 2004) there is some good evidence for duetting functioning in mate guarding: by joining in a duet, the two members of a pair advertise their partner's mated status and may therefore deter intruders of their own sex. In many species, playback of solo song is most likely to elicit a response from the same sex as that of the playback (Marshall-Ball *et al.* 2006, Seddon *et al.* 2002). In plain wrens, (Marshall-Ball & Slater 2004) members of neighbouring pairs also tend to match phrases with each other, male with male, female with female, but not whole duets. These results are consistent with mate guarding as they suggest that intruders of the same sex are more of a threat. However, they could also arise if, as has been suggested by Levin (1996b) for example, territorial defence is sex-specific, males defending against males, females against females. Better evidence comes from the study of tropical boubous by Grafe & Bitz (2004), who found birds more likely to join with their partners to form duets in response to playback of solo song by their own sex, as one would predict from mate guarding. Females in this species sometimes join in with male solo playback, and their partners will then often 'jam' the song of the rival. A similar jamming effect by females has been described in the moho (Slater *et al.*, 2002), with the suggestion that it might guard against polygyny. Perhaps the best evidence for mate guarding comes in cases where birds respond more aggressively to solo song of their own sex than to duets (e.g. Grafe & Bitz 2004, Seddon & Tobias 2006, Rogers *et al.*, 2006). This result is the opposite of what one would expect in territorial defence, as two birds would be a greater threat to the territory than one.

The evidence for duets functioning in joint territorial defence is stronger than that for any other hypothesis, though the majority of studies argue that this particular function is not exclusive but one among several, mate guarding in one form or another being the second most favoured (Grafe *et al.*, 2004, Levin 1996a,b, Marshall-Ball *et al.* 2006, Seddon 2002, Sonnenschein & Reyer 1983). Consistent with duets functioning in territorial defence, playback of duets tends to lead to aggressive approach by both members of a pair. Playing the song of one sex from one speaker and that of the other from another can lead to different responses by the two,

for example the male being more likely to approach the male playback (Mennill 2006) or the female the female (Logue & Gammon 2004). However, driving out rivals of the same sex could be interpreted as sex specific territorial defence or as mate guarding. In a two speaker experiment on magpie-larks, Rogers *et al.* (2004) found both pair members tended to approach the same speaker, usually that from which the initiator of the duet was broadcast. This is more convincing as evidence of joint territory defence. Even more so is the experiment by Molles & Waas (2006) on the kokako, a rare New Zealand endemic with an extraordinary slow motion duet. In this species the whole performance can be produced either by a single individual or as a duet by a pair. Playback of the whole song from a single speaker elicited a less strong response from pairs than splitting the phrases between speakers to simulate the intrusion of a pair. This points to pairs being more threatening than single individuals, as one would predict of joint territorial defence but not of mate guarding.

Despite the large amount of recent work on duetting, its functional significance still remains a matter of controversy. Its prevalence in the tropics correlates with the higher incidence of year round territoriality, long term pair bonds and prolonged breeding seasons in those regions. These characteristics are also common in Australia, and duetting is shown by many species there even though the continent stretches well south of the tropics (Robinson 1949). Year round territoriality and long breeding seasons may have led to sex role convergence and hence to mutual mate guarding and to joint territory defence, the two reasons for duetting that have the most current support. In addition, however, it almost certainly has functions within a pair. Neither by observation nor experiment it is easy to tease apart the different functions that have been suggested: the confusing variety of evidence we have at present suggests that duetting often serves several functions within a species and that the mix of functions also varies between species.

8.4.2 Choruses

Duetting is a feature of a pair, chorusing one of a group. The best known example of a chorusing species is the Australian magpie, in which groups produce a remarkable cacophony, varying from quiet warbling to loud 'carolling'. These birds are communal breeders, and both males and females join in the chorus. The sounds produced are highly varied, and

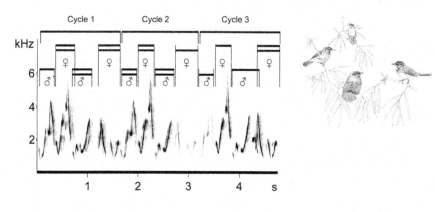

Fig. 8.5. Sonagram of a song sequence produced by a group of plain-tailed wrens. Male and female contributions are labelled above. Two males and two females are involved in seven of the twelve phrases as shown by the paired lines; one bird of each sex is closer (upper line) and the other more distant (lower line). Where two birds are singing they sing the same phrase in synchrony, the slight time difference between them leading to an 'echo' visible on the sonagram (after Mann et al. 2005).

many of them are specific to the individual, though there is more sharing within a group than between groups. There is also more sharing among birds of one sex than between the sexes, and females have larger and more complex repertoires than males (Brown *et al.* 1988, Farabaugh *et al.* 1988, Brown & Farabaugh 1991).

Another interesting case is that of the white-browed sparrow-weaver studied by Voigt *et al.* (2006). This species breeds cooperatively and, as well as the dominant male having a solo song and the dominant pair a duet, the group as a whole sings choruses. The syllables used by subordinates are the same as those of the pair in their duet, and two or more individuals may sing the same syllable at the same time. Most syllables are shared between the sexes and may, both in duets and in choruses, be sung in synchronised unison or antiphonally to form a sequence of syllables to which two or more birds contribute.

The chorus of the white-browed sparrow-weaver is wonderfully coordinated compared with that of the Australian magpie. Rather more complex still is that of the plain-tailed wren (Mann *et al.*, 2005). These birds live in groups, though whether they breed cooperatively is not yet known. The males and females have different repertoires and, like many other wrens, sing in a cycle during which the two sexes alternate (Fig. 8.5). In this case, however, the cycle has four components (ABCDABC . . .) rather than two (ABABA . . .), males singing A and C, females B and D. In fact these phrases are linked so that A C and B D can each be thought of as a single phrase in the middle of which is a gap for the contribution of the other sex. Each sex has a repertoire of 15–20 phrases and, where more than one bird of that sex is singing simultaneously, they almost invariably choose the same phrase.

As with duetting, these choruses may help to coordinate group activities, but their primary function seems to be in territorial defence. Playback to plain-tailed wrens leads a group to surround the speaker singing their chorus in response, something highly likely to be intimidating to intruders. In line with this, Seddon & Tobias (2003) found that sub-desert mesite groups approached playback of communal singing more when number of simulated intruders was smaller and, in particular, when the number of males in their own group was greater than in that from which the playback had been recorded.

8.5 Matched countersinging

Duets and choruses occur within a pair or group. Another form of coordinated singing is the matching that often occurs between territorial neighbours. It is striking that, when two such males sing against each other, they will often sing the same song type if both have it in their repertoires. Armstrong (1973) suggested that this was 'naming an opponent in song duels', repertoires enabling birds to interact with specific individuals. Bertram (1970) wrote of matched countersinging as involving 'flinging an insult back at the rival'. It may be, as this suggests, that hearing a song closely matching its own is a particularly effective deterrent for neighbours or other intruders.

Examples of song-type matching are numerous (e.g. cardinal, Lemon 1968; western meadowlark, Falls 1985; tufted titmouse, Schroeder & Wiley 1983b; song sparrow, Kramer & Lemon 1983, Stoddard *et al.* 1992; great tit, Krebs *et al.* 1981a) but there are also quite a number of counterexamples. Corn buntings do not match (McGregor 1986). Wood thrushes avoid matching: they are more likely to produce a particular type if they hear a song similar to it than if they hear an identical one (Whitney & Miller 1983, Whitney 1990). In yellowhammers, Hansen (1981) found that males that shared a song type tended, in five out of six cases, to avoid singing it in bouts that overlapped in time with each other. In the bobolink, a species in which all individuals have two song types, males tend to respond to playback with the opposite type to that being played, thus giving a result that is functionally identical to song matching (Capp 1992).

One of the most thorough studies of singing interactions between birds was that by Falls and his collaborators on western meadowlarks. In playback experiments, the extent of matching was found to depend on the exact

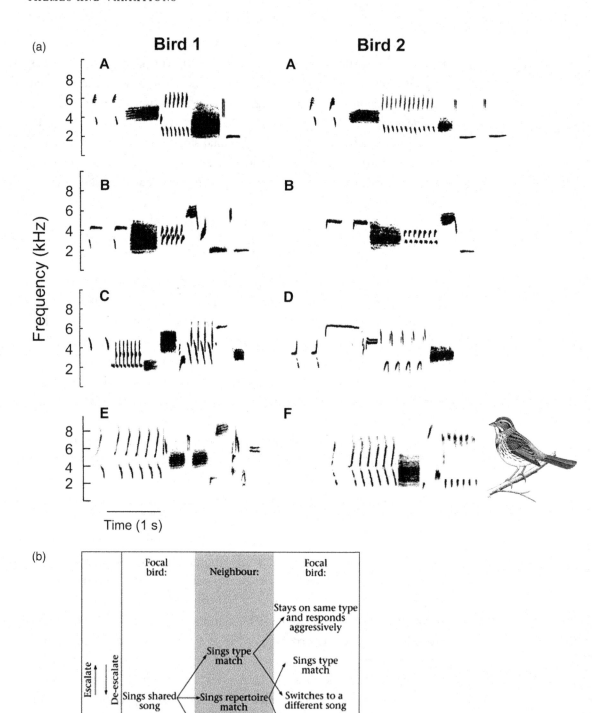

Fig. 8.6. Song matching in song sparrows. (a) Four song types as they might be sung by two adult male song sparrows. If the first bird sings song A, the second might song match *by also singing A, might* repertoire match *by singing B, another song they both share, or might sing D, an* unshared *song. A further possibility is a* partial match, *so if the first bird sings E the second may reply with F, the two songs sharing the initial trill but not later elements (sonagrams from Beecher et al., 2000a, Burt et al. 2002a). (b) How the responses of two neighbouring song sparrows to each other can lead to escalation or de-escalation of a contest between them: matching is an aggressive signal and singing an unshared song an unaggressive one, with repertoire matching intermediate in threat (after Beecher & Campbell 2005).*

source of the song. Birds matched their own songs strongly but showed less response to the same type as sung by a stranger, and they did not match that of a neighbour above chance (Falls 1985). In the playback of songs recorded from strangers, their matching depended on the precise similarity: they showed no tendency to match songs that were close to their own but not the same (Falls *et al.* 1988). Given their lack of response to the playback of neighbour songs, it is not surprising that neighbours did not match during interactions (Horn & Falls 1988b). However, their singing did influence each other: the switching of song types tends to be synchronous, and they were more likely to sing in response to a neighbour if that neighbour was using a shared type (Horn & Falls 1988c). Neighbours did not share at more than chance levels in this species (Horn & Falls 1988a).

The most extensive studies on song matching have been those in recent years on song sparrows and these paint a rather different picture from the meadowlark one, made more complicated by the fact that populations within this species also differ. In Washington State, males tend to share around 2 of their 6–10 song types with each neighbour (Beecher *et al.* 2000a), while in Pennsylvania neighbours share very little (Hughes *et al.* 1998). But the story in the two places may not be as different as this makes it appear. Despite not sharing whole songs, birds in Pennsylvania can match neighbours with parts of their songs and do so to playback at above chance levels. Where they do so they also approach the speaker more closely, pointing to this being an aggressive response (Anderson *et al.* 2005). By contrast, the most aggressive response of Washington birds is to match whole songs, which their level of sharing allows them to do (Beecher *et al.* 2000a); if they cannot match precisely, they tend to use the closest song type that they have (Burt *et al.* 2002a). Fig. 8.6 illustrates these different levels of matching. A rather less aggressive response is to 'repertoire match': not to use the particular song type the neighbour is singing, but another of those the two birds share (Beecher *et al.* 1996), and this is more common late in the season when boundaries are more settled (Beecher *et al.* 2000a). The least aggressive response of all, which tends to de-escalate contests, is to use a song type that is not shared (Beecher & Campbell 2005). All in all, therefore, the possession of repertoires, some songs in which are shared in whole or in part with each neighbour, allows a sophisticated communication system embodying various levels of threat. Sharing, rather than repertoire size as such, seems the most important

thing, as the former rather than the latter best predicts lifetime territory tenure (Beecher *et al.* 2000b).

At first sight, the picture on the role of repertoires in male–male interactions is a confusing one, with the level of matching varying greatly even between populations. However, as more species are studied it may emerge that repertoire matching and partial matching are more widespread phenomena that help to pull together what appear at first sight to be strong species differences, Perhaps another reason why no clear pattern emerges is because species vary in the extent to which song is a signal to neighbours rather than to strangers. As well as having a part in the interactions between birds already established on their territories, it may have an important part as an advertisement of ownership and a signal of repulsion to potential intruders looking to set up territories anew.

If it is advantageous for males to be able to countersing with neighbours using the same song type, we might predict that neighbours should share song types more than expected by chance. This is not very easy to establish because the very fact that song varies geographically means that birds close together tend to be more similar than those further apart. However, one can examine whether birds share songs more with their immediate neighbours than with birds one or two territories away. The evidence here is highly varied. In some species, neighbours do indeed share more than one might expect (Schroeder & Wiley 1983b, Dufty 1985, Morton 1987). However, in others there is no tendency towards sharing (Beletsky 1982, Horn & Falls 1988b) and there is evidence for several species that birds share more with non-neighbours than with neighbours (Grant & Grant 1979, Borror 1981, 1987, Bradley 1981). Grant (1984) found that Darwin's large cactus finches had greater breeding success if their neighbour had a different song type than if he had the same one, so in this species there may be an advantage in their failure to share.

In contrast to this last finding, if there is an advantage to matched countersinging we would expect that birds which share song types with their neighbours would do better than those that do not, as Beecher *et al.* (2000b) found for territory tenure in song sparrows; in the same species, Wilson et al. (2000) found both male survival and territory tenure to increase with the fraction of song types shared with neighbours. Good evidence for this has also come from the indigo buntings studied by Payne and co-workers (Payne 1982, Payne *et al.* 1988a): these extensive studies

have found that males which shared did better in mating, nesting and fledging young.

Another possible advantage might come from sharing songs with the previous occupant of the territory, an idea also put forward by Payne (1981a). The individual might then benefit from the learnt avoidance of this song type that neighbours would show. Indeed, even if the newcomer had had no opportunity to learn the song of the previous territory holder, the larger his song repertoire the more likely it would be to include this song or one like it (Slater 1981). So an advantage of repertoires might be that they would be more likely to include such an aversive song. The evidence on this idea is mixed. While Payne & Groschupf (1984) found that village indigobirds did tend to match the songs of their predecessor, McGregor & Krebs (1984a) could find no evidence of it in great tits.

It is clear from these examples that small repertoires of song types often have a role in male–male interactions, as indicated by the widespread occurrence of matched countersinging. However, their exact role certainly differs between species, and the fact that neighbours often do not share song types argues against matched countersinging, at least at the level of the whole song type, always being an important factor. While longer, more complex songs and larger repertoires, as we shall see below, are more likely to be linked to female attraction, small ones have most probably evolved because of the part they play in the relationship between neighbouring males. This may account for the apparent paradox among those North American warblers that show serial and repeat singing. Here the more complex serial singing behaviour occurs when males are singing to each other, but the repertoire of songs used is a small one and fits in well with those of species where males countersing with one another. It is another question to decide why, in this particular case, the song sung to females is so short, simple and stereotyped.

We can conclude that the extent to which birds share song types with their neighbours varies enormously, but there is an increasing amount of evidence, particularly from studies of song sparrows, that interaction with neighbours using repertoires plays an important role in their relationships. Given the additional role of temporal relations between singers, such as the threat that overlapping presents (see Chapter 6), singing with small repertoires clearly provides rich possibilities.

8.6 Versatility, habituation and exhaustion

Two theories about song repertoires have been put forward which suggest that it benefits birds to be highly versatile in their singing, either because this reduces habituation on the part of listeners or because it enables the singer to avoid becoming exhausted. We will consider each of these ideas in turn.

8.6.1 Habituation

Hartshorne (1956, 1973) first argued that birds with highly varied songs tend to sing with only short gaps between them ('continuously'), whereas those with simple ones usually sing with longer gaps ('discontinuously'). A measure of continuity is 'performance time', the amount of time actually occupied by song when a bird is singing. To take an extreme example which illustrates Hartshorne's relationship, in the song thrush song occupies about 60% of the time (species 7 in. Fig. 8.7). This is a species with a large repertoire of song types, which sings with immediate variety. On the other hand, the European redwing (species 2 in Fig. 8.7), with its one song type, sings with long gaps, so that song only occupies about 30% of the time (Ince & Slater 1985).

There is a good deal of controversy about whether Hartshorne's correlation exists or not. It has been criticised by Dobson & Lemon (1975) and by Weary & Lemon (1988, 1990), but defended in turn by Kroodsma (1978b, 1990a). Dobson & Lemon (1975) looked more formally than Hartshorne had done, over a range of North American species, and failed to find any such relationship. However, they took repertoire size as their measure of versatility, whereas this differs from what Hartshorne meant (Kroodsma 1978b). A bird with a large repertoire may still sing the same song many times in a row and so not be versatile at all. Another problem, common to many analyses of this sort, was in the species that they included. Their correlations were biassed by the large number of vireos in their calculations. If a group of species like this all behave in the same way this is probably because their common ancestor did so, and to regard each of them as an independent data point in a correlation is not justified. As the patterning of vireo song is unusual, this problem might indeed have biassed their results against finding Hartshorne's correlation.

In a subsequent analysis, examining only North American warblers, Weary & Lemon (1988) again failed to find a relationship. A possible reason

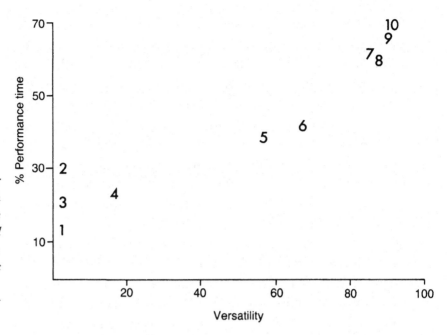

Fig. 8.7. A plot of percentage performance time against versatility for ten species of thrush, each labelled by a number. The measure of versatility used is based on the proportion of songs in a sequence that are different from each other (from Ince & Slater 1985).

here is that the 19 warbler species they studied do not show a huge variety of singing patterns: for example repertoires range only from one to five and performance time from 9% to 28%. They are looking for relationships at a very fine level of detail compared with those amongst the more varied singers examined by other studies (e.g. Kroodsma 1978b, Ince & Slater 1985).

Results on this topic do not allow a clear conclusion. This is perhaps not surprising, for such studies require a great deal of judgement and objective comparisons are difficult to achieve. Versatility and continuity can be scored in a variety of ways. In looking for correlations it may be critically important which species are included or not. But, even if a relationship is found, as amongst the thrushes, what does it mean? Why is there this difference between species? Hartshorne himself interpreted it in terms of what he called the 'monotony threshold principle'. In this he argued that species with more monotonous songs break up their performance more so that listeners do not habituate to it. While this is an interesting idea, it does have problems. As pointed out by Krebs (1977a), it leaves open the question of why birds should habituate: if it is to the listener's disadvantage to habituate, why does it not avoid doing so regardless of the patterning of song? It is necessary to propose either that habituation has some advantage to the listener, which the pattern of song is selected to overcome, or that habituation is a fundamental property of living things which selection cannot override.

That habituation may occur in some species because it is beneficial was suggested by Krebs (1977a) in his Beau Geste hypothesis (see Chapter 6). As part of this idea, he proposed that birds might actually use habituation as a means of assessing crowdedness. If there is a bird singing only one song type in the area, they will rapidly habituate and so approach, but they will be less likely to do so with four birds each singing its own song. What the Beau Geste idea does is to suggest that one bird singing four song types effectively mimics the latter situation. Attractive though this idea is, there are a number of theoretical obstacles to it, and practical attempts to test predictions from the hypothesis have had mixed fortunes, as our discussion in Chapter 6 showed.

The idea that habituation to a monotonous signal is inevitable was suggested by Hartshorne himself and by Kroodsma (1982), though it seems more an argument of last resort than one with clear justification. Some studies have been carried out with the specific aim of testing this idea. In a study on song sparrows, Lemon et al. (1981a) played back tapes in which either the continuity or versatility of singing had been manipulated. Their results were mixed and could certainly not be simply explained in terms of differences in habituation. A more clear cut result was obtained by Simpson (1984). She carried out playback to male Carolina wrens from outside their territory boundaries, using either single songs on the tape or repertoires of different sizes. She found no decline in response when only single songs were played, nor did birds respond differently to repertoires of different sizes compared with single songs. Thus, she obtained no evidence of habituation regardless of whether or not repertoires were played. As the Carolina wren normally repeats the same song a number of times before switching to another, Simpson suggested that individuals may have evolved resistance to habituation when played simple signals like this. Her results do not support the idea that repertoires function to reduce habituation. However, Searcy et al. (1994) found that repeated presentation of the same song type to male red-winged blackbirds did lead to habituation of their aggressive display, which then reappeared when song types were switched. They argue that repertoires in this species consist of dissimilar song types because this maintains the responsiveness of listeners despite habituation. Once again the evidence is therefore mixed.

Another drawback of the monotony threshold principle is that it only proposes a correlation. It does not suggest why some species have complex

and continuous songs while others have simple ones with longer gaps between them. We either need a further theory to account for the particular level of variety found in different species, or an alternative one that can account for this as well. One such alternative is based on the idea that song serves two main functions, mate attraction and rival repulsion, and that the relative importance of these differs between species: this is a theory to which we will return at the end of this chapter.

8.6.2 Exhaustion

If switching between song types does not lead to less habituation on the part of listeners, perhaps it causes less exhaustion in the singer. For example, a long series of the same song type might lead to fatigue in the particular muscles involved in its production so that the song became more difficult to generate. Thus, the song itself would become more sluggish, or longer gaps might be required between utterances to allow recovery. Switching to another song type, involving a different pattern of contractions on the part of the syringeal muscles, would then allow a higher singing rate to be maintained.

Are there measures of performance that show such a decline as the same song type is repeated many times? Lambrechts & Dhondt (1987, 1988), who proposed the anti-exhaustion hypothesis, suggest that there are. They describe a phenomenon they refer to as 'drift'. Male great tits tend to show longer intervals between songs as a bout of one song type continues, and in some cases the songs become shorter, so that their 'percentage performance time' declines (see Fig. 8.8). They revert to shorter intervals and/or longer songs after switching to another type. The lengths of the pauses within songs were found to drift as well as those between them (Lambrechts & Dhondt 1987).

Drift is not a universal phenomenon in bouts of great tit song. Lambrechts & Dhondt (1988) found it in 31 out of the 52 bouts they studied. In coal tits, Adhikerana (1992) could detect it in only 6 out of 68 bouts, and found increasing percentage performance time in 10. He concluded that there was no evidence for drift having a significant effect in this species. However, we have to be cautious because the hypothesis was proposed as an evolutionary explanation for the existence of song repertoires. If repertoires are effective in this way, drift may no longer be detectable, or only so when birds sing at high rates or with long bouts of

Fig 8.8. Changes in the percentage performance time of a male great tit singing during the dawn chorus. After 21 songs of type A the bird switches, then sings 39 of type B, followed by 34 of C. Each point represents a mean based on five songs (but four or six as indicated at the end of the three bouts shown). The plot shows that the performance time diminishes as the bout continues, then rises again after the bird switches song types (after Lambrechts & Dhondt 1988).

a particular song type. An experimental approach is preferable. Adhikerana & Slater (1993) found that coal tits challenged with tape-recorded songs presented at an accelerating rate actually switched types less frequently than when the rate was constant or decreasing. This is the opposite of what the anti-exhaustion hypothesis might predict. In chaffinches, too, Riebel & Slater (2003) failed to find drift towards the end of song bouts, as did Brumm & Slater (2006a). Such results are not particularly clinching, however, as they might just indicate that switching occurs before exhaustion sets in. If true, this makes the exhaustion hypothesis very difficult to test. One result that does favour it is that chaffinch songs with rapid trills, which are presumably more difficult to perform, occur in shorter bouts than those with slow trills (Brumm & Slater 2006a).

Weary et al. (1988) argued that the results of Lambrechts & Dhondt could as well be caused by motivational decline as a bout continues as by neuromuscular exhaustion. If so, they suggest, playback of the appropriate song late in the bout should boost singing rate again by raising motivation, whereas it would have no effect if neuromuscular exhaustion was responsible for the decline. Their results do show such a boost in singing rate and they therefore suggest that declining motivation is the more likely explanation for drift. However, Lambrechts (1988) points out that their birds were not singing at a particularly high rate in the first place, as opposed to

the birds originally studied by Lambrechts & Dhondt which were singing flat out, with performance time at the start of bouts being up to 80% (see Fig. 8.7). Exhaustion might not be expected where birds were singing relatively slowly. Lambrechts did not find recovery in response to playback in two birds he tested at dawn when song rate is much higher.

An attempt was made to resolve this issue with a joint experiment by Weary *et al.* (1991) in which males were tested both at dawn and later in the day to see whether playback would boost song output under both circumstances. The birds did show higher performance time at dawn and their singing rate declined in both tests, but it also showed an equivalent recovery in response to playback in both of them. While the authors did not appear to be fully in agreement about the interpretation of the results, they are more in line with the idea that decline in song rate stems from a reversible change in motivation, rather than from exhaustion, which can only be countered by switching song type.

Implicit in the anti-exhaustion idea is that it benefits males to sing at the highest possible rate. Indeed Lambrechts & Dhondt (1987) suggest that singing rate and song length are measures of male quality. However, repertoires make it rather difficult to test for this in species like the great tit. McGregor (1988) avoided this difficulty by studying song length in the chiffchaff, a species with one simple song type which varies considerably in length. He tested the response of birds to playback, reasoning that if song length indicates quality they should sing particularly long songs when stimulated in this way. The contrary was the case: song length decreased in response to playback.

Current evidence casts doubt on Lambrechts and Dhondt's theory. However, there is certainly a need for more studies on this topic, preferably on a wider range of species, as the idea is an attractive one. But, for the moment, the ideas that repertoires may have arisen because they counter either exhaustion on the part of the singer or habituation on the part of listeners have not received widespread support.

8.7 Songs with different functions

In section 8.3.2 we described the singing behaviour of the American redstart and some other North American warblers where there are two sorts of singing behaviour, 'serial' and 'repeat' in the terminology of MacNally & Lemon (1985). The prevalence of the two types differs with time of day,

stage of the breeding cycle and behavioural context, and this suggests that they function in different ways, repeat singing being a male–female signal and serial singing a male–male one.

This is quite a striking example because it involves different patterning and organisation of singing rather than different song types. But there are now many examples, largely from among North American and European warblers, where birds with small repertoires have been found to have songs of two different types. In North American warblers, the two types of song are sometimes referred to as Types I and II (Beebee 2002) or A and B songs (Staicer 1989) but in many species as accented and unaccented songs, because of a distinctive ending found in the accented ones (Fig. 8.9) (e.g. Morse 1966, Lein 1972, Kroodsma *et al.* 1989, Byers 1996b). The unaccented song, like serial singing, appears to function primarily between males. It is produced mainly at dawn and dusk (Highsmith 1989, Morse 1989) and in contexts of territorial defence (Morse 1970). By comparison, the accented song is produced more when the female is present and has been associated with courtship and pair-bond maintenance in several studies. As the unaccented song is the only one given in response to playback in black-throated green warblers, another explanation is that it is produced in response to stronger stimulation (Morrison & Hardy 1983). However, this is an explanation in terms of cause and does not really compete with the notion that the two songs function differently.

The most complex repertoire in this category is that of the chestnut-sided warbler studied most recently by Byers (1995, 1996a,b), in which each male has several types of each category of song. Accented songs are few in number and highly stereotyped, both within and between individuals, with distant sites having much the same mix of types. They are sung largely by unmated males interacting with females. The unaccented songs are more numerous and variable, between birds and between populations, and they are used mainly in male–male interactions.

For several species, the two sorts of song that have been described differ primarily in length. In the five-striped sparrow, there is a long and variable song thought to act as a female attractant, and a short and stereotyped one used in male–male interactions (Groschupf 1985). The great reed warbler has a short song which resembles its long song except for being prematurely cut off (Catchpole 1983) (Fig. 8.10). Again, the contexts in which they are produced suggest that short songs act between males and long ones as a signal to females. Playback experiments show that males and

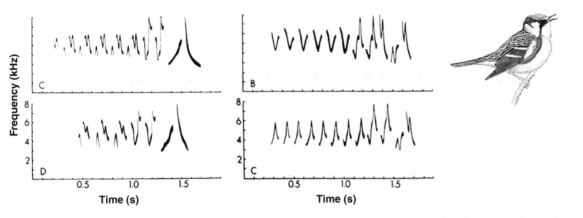

Fig. 8.9. Two accented (left) and two unaccented (right) songs as sung by male chestnut-sided warblers (after Lein 1978).

females respond differently to them (Catchpole *et al.* 1986). The European redwing has a simple song and a long and more varied warbling part which follows it but becomes shorter or is dropped after egg-laying: this is again consistent with the idea that longer and more varied songs attract and stimulate females (Lampe & Espmark 1987). The songs of mated pied flycatchers are also shorter than those of unmated ones (Espmark & Lampe 1993). In the European wood warbler, each male has two very different song types, although it is not known whether these differ in function. However, song is reduced and both of them become shorter after pairing. Longer songs at this stage only persist far from the nest and may function to attract a second female (Temrin 1986).

As we saw in Section 8.3.4, starling songs are highly varied, but each male has phrases falling into certain categories. It has been argued by Adret-Hausberger & Jenkins (1988) that the more complex warbling songs function in female attraction, while the simpler whistles are a male-male signal. However, on the basis of a quantitative experimental study, Eens *et al.* (1993) found the opposite: male starlings whistle more when presented with females than with males. They argue that different types of song in starlings cannot be categorised according to function and that, while the primary role of song in this species is in communication between the sexes, it also plays a part in male-male interactions (see also Eens *et al.* 1990, 1991a).

Further examples of different songs acting in different ways come from the aquatic warbler (Catchpole & Leisler 1989), willow warbler (Jarvi *et al.* 1980), three North American titmouse species (Gaddis 1983, Schroeder & Wiley 1983a, Johnson 1987), yellow-headed blackbird (Cosens & Falls 1984b), bobolink (Wittenberger 1983, but see Capp & Searcy 1991) and Cuban grassquit (Baptista 1978). The phenomenon has primarily been

Fig. 8.10. An example where different song types convey different messages: a long song and two short songs as sung by a great reed warbler (after Catchpole 1983).

described among species with small repertoire sizes, and generally seems to take two possible forms. One is where a repertoire of songs is used in male-male interactions, as in many other species, but there is an additional simple song of which the main role is mate attraction. The other is where song is more complex early in the season at the time of mate attraction, and becomes shorter and simpler when the male is mated. But the reader is no doubt accustomed by now to the idea that any attempt to generalise about bird song must be qualified with exceptions! Yellow warblers use one of their song categories in interactions with males, but females solicit equally to it and to the other form (Beebee 2004).

8.8 The puzzle of song complexity

At first sight, the diversity of modes of singing amongst birds is so great that it defies explanation. Many species, particularly the non-passerines, have relatively simple calls or songs. Clearly, they can get by without undue complexity, so why have some songbirds evolved such elaborate signals? Looking at repertoire sizes alone (Table 8.1) there are also enormous differences between songbird species: how can we attempt to explain such diversity? One starting point in setting out to answer these questions is to consider what functions bird songs appear to serve.

8.8.1 The functions of song

We have already covered various different theories for the function of song in this and earlier chapters and will not repeat the arguments here, but will

refer back to relevant sections. In Chapter 5 we established that, at least in the temperate north, it is usually only males that sing and that there is a broad correlation between song output and the breeding cycle (Section 5.2). But this simple correlation does not imply a single role for song, as males are involved in two main activities at this time: obtaining a territory and attracting a female. We, therefore, come up against the difficulty of teasing apart these two possible functions of male song. In a few species, such as the sedge warbler, song production ceases dramatically after pairing, suggesting that mate attraction is particularly important. In others, such as great tits or starlings, song production continues and may be related to territorial defence or attracting other females. Even in such species, removal of the female after pairing results in a resurgence of song, suggesting indirectly that female attraction may still be an important function later in the season.

The evidence reviewed in Chapter 5 is indirect and circumstantial, but it suggests the rather simple hypothesis that song has the two main functions of mate attraction and territorial defence. There may be other possible functions, but most of these hinge on the main idea that song communicates with two important classes of individuals: males and females. Other features of singing behaviour dealt with in Chapter 5, such as female song (Section 5.4), the dawn chorus (Section 5.5) and competing for a time to sing (Section 5.6) can also be interpreted within this broad perspective.

But what real evidence is there that song has these two functions? Observations confirm that males use song in territorial disputes, but Chapter 6 dealt with more direct evidence of its territorial role based on two different types of experimental test. One of these deprives the bird of its ability to sing, the other replaces the bird by a system of loudspeakers; in both of them territorial boundaries are monitored before and after treatment. There have been several muting studies to date (Section 6.2.1) and in general they show that muted males suffer more frequent invasions, have to fight harder to maintain their territories and may even lose them to rival males. In some studies, males that recovered their songs were also found to recover their territories. Speaker replacement experiments attempt to isolate the effects of song itself, by removing the real male from the experimental design (Section 6.2.2). These also show consistent results: territorial intrusions are delayed when compared to control territories with no song or with another sound stimulus. Recent interactive playback experiments also confirm the importance of more subtle aggressive signalling, such as overlapping a rival's song. The evidence is,

therefore, consistent with the hypothesis that male song does function in territorial defence.

Chapter 7 reviewed the evidence for female attraction and stimulation. Working with females in the field is more of a problem, and consequently there is rather less direct evidence to implicate male song in female attraction (Section 7.3). The main evidence comes from several experiments in which more females were found to be attracted to nest boxes at which recorded songs were played, compared with control ones with no songs. There is also one other field experiment in which females within their territories were found to be attracted to playback of their mates' songs. However, most work has concentrated on captive females and their responses in the laboratory (Section 7.4). A number of studies show that male song has a stimulating effect upon their reproductive physiology. However, the real breakthrough in measuring the responses of captive females to playback of male song was the hormone-implant technique developed by Searcy & Marler (1981). Using this technique, it has been possible to show that females of several species will only perform sexual displays to playback of their own species song. Taken together, these field and laboratory studies provide good evidence that females are attracted and responsive to playback of male song.

Experimental evidence supports the view that male song may function in both female attraction and territorial defence. However, in a review of song function, Kroodsma & Byers (1991) concluded that 'Male bird song may attract and stimulate females and may repel rival males, but the evidence that song is essential for accomplishing these "main functions" is not abundant'. However, in a more recent extensive review, Collins (2004) concluded that 'song is clearly important for both competition and attracting a mate'. Although it is certainly correct that more evidence is desirable, in our view there is reasonable support for the general idea that bird song subserves two functions, at least as a working hypothesis. Furthermore, at this stage, no alternative hypotheses have been proposed.

8.8.2 Evolution and song complexity

We now return to the main problem: why is song so complex? If the primary functions of song are in mate attraction and territorial defence, then perhaps individuals that develop more complex songs gain some selective advantage and increase their fitness. It is not easy to answer such

evolutionary questions. One problem is that the important events may well have happened in the past. While we can examine present function (current utility), we can never be sure about evolutionary origins. But are more complex songs currently advantageous, and what selective advantage might they confer? In Chapter 7 we argued that important selective pressures might operate through the two different forms of sexual selection: intrasexual selection acting on song in its role in male–male competition, and intersexual selection acting on song through female choice. In terms of current utility, these are related to the male–male territorial function of song and that in male–female attraction, respectively.

Considering possible intrasexual functions first, we will concentrate upon repertoire size as an index of song complexity. Chapter 6 dealt with the evidence that males with larger repertoires might gain an advantage in territorial defence (Section 6.2). Two speaker replacement experiments to test the Beau Geste hypothesis both found that repertoires of song types were more effective than single-song types as a first line of territorial defence, but more such tests are clearly needed. There are many subtle ways in which repertoires may function in territorial behaviour, such as in reducing habituation and in matched countersinging, as discussed earlier in this chapter.

Turning to intersexual selection on song complexity through female choice, most of the evidence comes from studies on repertoire size, and this time there is rather more of it. In Chapter 7 we considered studies on six species, selected because they combined field and laboratory evidence (Section 7.6). The five laboratory studies that used the hormone-implant technique all found that females displayed more to larger repertoires, although Kroodsma (1989b, Kroodsma & Byers 1991) has criticised these experiments for using too few different playback tapes. However, the field evidence points in the same direction, and in all six species correlations have been obtained between repertoire size and various measures of reproductive success.

When one draws together all the observational and experimental evidence we have considered in this book, there is considerable support for the suggestion that song has two main functions, and for the idea that intra- and intersexual selection have been important evolutionary forces in shaping it. If this is so, then we might expect to find two different kinds of song: those with mainly a territorial function and those which function primarily in mate attraction (Catchpole 1980, 1982; Slater 1981). As we

saw in the previous section (8.7), there are some species that do appear to have shorter, simpler songs for territorial defence and longer, more complex ones for mate attraction, but in many other species song of a single structure appears to function in both ways. It seems possible that length and complexity are the result of runaway intersexual selection, whereas intrasexual selection tends to produce shorter, simpler songs. Slater (1981) has pointed out that these selective forces might also lead rather simply to the continuous or discontinuous singers stressed in Hartshorne's theorising (Section 8.6). If songs are addressed to rival males in territorial disputes, then gaps during which the singer could listen for replies would seem to be essential. As we have seen, males use replies to assess both the position (Section 6.4) and distance (Section 4.4) of neighbouring males. By comparison, if songs are addressed to females, then we might expect a continuous signal to be used, as females listen for the songs of males but do not sing themselves. There is some evidence (Section 7.5) that listening females respond more to higher song production and song rate, as well as to larger repertoire sizes. On these arguments, one would expect a correlation between versatility and continuity, such as Hartshorne suggested, but for a very different reason from the anti-habituation mechanism he proposed.

The comparative method has also been used to examine how sexual selection may have shaped song complexity, but the results so far have been mixed (Section 7.6.7). Correlations between mating systems and repertoires have been found, but in some groups repertoires are larger in monogamous species, and in other groups they are larger in polgynous species. The most extensive and recent comparative survey by Read & Weary (1992) found that such associations varied with taxonomic level. They concluded that, in general, greater song complexity is correlated with those factors likely to produce more intense inter- or intrasexual selection. In some species, for reasons which still remain a puzzle, intersexual selection and female choice may have led to the evolution of a more complex song than in others. In some cases, such as the *Acrocephalus* warblers (Catchpole 1980, 1987), attempting to tease the two functions and selective pressures apart has offered us a brief glimpse of what the evolutionary history might have been. In most species, however, the two functions are so inextricably linked together that songs of the same structure serve both of them. A more general comparative examination then leaves us to puzzle over the resulting richness and variety that evolution has created.

Perhaps it is naive to attempt a general, overall explanation for the evolution of song diversity. Even though we have stressed a duality of main function and argued that sexual selection is strongly implicated, we should not relegate other evolutionary pressures entirely to the sidelines. If there is one broad theme in this book, it is that many different evolutionary constraints and pressures must be considered if one is to understand the form and function of song. In our view, there is now considerable support for the 'dual function hypothesis', and no alternative functional hypothesis has yet been proposed. In our last chapter we will consider other important parts of the evolutionary story: how songs change within space and through time.

CHAPTER NINE

VARIATION IN TIME AND SPACE

*Montezuma
oropendola
displaying*

The voice I hear this passing night was heard
In ancient days by emperor and clown
John Keats *Ode to a Nightingale*

9.1 Introduction

It is well known that bird song often varies from place to place. At an extreme, someone expert on a particular species can tell roughly where they are by listening to the songs round about them, rather as one can sometimes tell where people come from by their accents. But this is unusual for songs. While the word 'dialect' is frequently used to describe differences in song between places, as we shall see many different patterns exist, and it is not very helpful to clump them all under such a blanket heading.

In this chapter we will consider some of these patterns, looking both within and between populations. We will also describe what happens to the songs of populations that become isolated from each other, such as those that occur on islands. We will ask whether the differences that have been described have any functional significance. Lastly, we will move into the fourth dimension to look at cases where songs have been found to change with time. Such change, as with that in space, is most obvious and rapid in birds that learn their songs, and much of our discussion will be about these. But whether songs are learnt or not, one form of change with time that undoubtedly involves genetics is that during the course of evolution, which has led to the wonderful variety of songs between species that one observes today. It is fitting to end this chapter, and the book, with a discussion of how the variety of song fits in to this evolutionary framework.

9.2 Variation within a population

Where individual birds have only a single song type, it is common for this to be similar in form among groups of neighbours. With larger repertoires, birds may also share several of their songs. At an extreme, whole repertoires are shared. This is the case in English populations of the corn bunting where each male has two or three song types, and groups of birds on adjacent territories have exactly the same ones (see Fig. 9.1) (McGregor 1980, McGregor & Thompson 1988). An even more striking example is in the village indigobird, in which each male's repertoire is some 20 song types, all of which are shared with neighbours in displaying groups (Payne 1985). In these species, birds appear to learn

Fig. 9.1. Dialects in a population of corn buntings on the Berkshire Downs, England, in 1979. Each symbol represents a singing male. Each dialect consists of two song types and most males sang only one dialect (symbolised as ▲, ● and ■), although the four with combined symbols sang both the dialects these represent (from McGregor 1980).

their whole repertoires from each other as a package, just as if they were single song types.

Such complete sharing of song repertoires between neighbours is unusual and, not surprisingly, is commonest in species where each male has only a single song. But birds with single song types do not necessarily share them with neighbours. In the Kentucky warbler, males have only one type, but there are a variety of different types in the population, and there is no tendency for neighbours to share more than birds that are further apart (Tsipoura & Morton 1988). In snow buntings on Svalbard, Espmark (1995) found that the range of element types varied between localities but that the majority of them (70%) were unique to particular individuals, again making neighbour sharing unusual. Species with repertoires range from complete sharing, through those where neighbours share many or most of their songs (e.g. house finch, Mundinger 1975, Bitterbaum & Baptista 1979; tufted titmouse, Schroeder & Wiley 1983b; thrush nightingale, Greissmann & Naguib 2002) to ones where the amount of sharing between neighbours is no greater than that between individuals a few territories apart (e.g. western meadowlark, Horn & Falls 1988a; chaffinch, Slater & Ince 1982) (see Fig. 9.2). Finally, there are some species that have been found to share less with neighbours than with more distant individuals (white-eyed vireo, Bradley 1981, Borror 1987; European

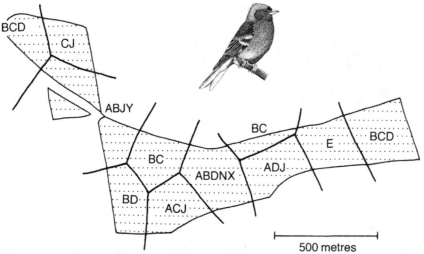

Fig. 9.2. The song types, each represented by a letter, sung by 11 chaffinches in a wood in Orkney, Scotland, in 1977. Repertoire size varies from one to five song types. Some song types are unique to particular individuals; some are sung by many of the birds.

blackbird, Wolffgramm 1979; Darwin's finches, Grant & Grant 1979; chaffinch, Lachlan & Slater 2003).

There is, therefore, wide variety in the amount of sharing between neighbouring birds. The commonest pattern is probably where males on adjacent territories share at least some of their song types with each other, allowing matched countersinging, as discussed in Chapter 8. This may arise where young birds, settling on their territories for the first time, learn songs from their neighbours, or where they arrive having memorised many songs and selectively use those that they share. But it could also occur if birds seek out breeding sites where those around them sing in the same way, an idea explored further in Section 9.5.2. Where sharing with neighbours is no greater than that with non-neighbours, or even less than it, it is most likely that birds learn their songs before arriving in the breeding area and then either take no account of neighbour songs when settling or actively avoid birds that sing similarly to themselves. The advantage of avoiding similar songs is not clear but, if matching has a deterrent effect during territorial encounters, as discussed in Section 8.4, it may also be aversive to birds seeking territories.

The exact pattern of song sharing can vary quite markedly even within a species, and a major factor here is likely to be how static the population is. Nelson *et al.* (2001) found a high degree of neighbour sharing in two sedentary white-crowned sparrow populations, while neighbours and non-neighbours shared equally in four migratory populations. Song sparrows from migratory populations also share less with neighbours than those

from sedentary ones (Peters *et al.* 2000), and populations differ in the extent to which they learn whole songs rather than recombining sections (Hughes *et al.* 1998). On the west coast, where the learning of whole songs seems to be the norm, males will typically share 20–40% of their song types with each neighbour (Wilson *et al.* 2000, Beecher *et al.* 2000b), a pattern consistent with birds attempting to find territories close to the area where they learnt their songs. A similar picture emerges for chaffinches, though here young birds may disperse rather further as they share at only chance level with their neighbours (Slater & Ince 1982) and have most songs in common with birds around 500m away (Lachlan & Slater 2003).

9.3 Geographical variation

Song can thus vary a lot or rather little between individuals within an area. When we start to compare between areas, we also tend to find differences. The distinction is sometimes made here between microgeographic and macrogeographic variation (Mundinger 1982). The former is the variation that occurs between neighbouring groups of birds which might, at least potentially, interact or interbreed with each other. Dialect areas with sharp boundaries between them, which are discussed in the next section, are the best known cases here. Macrogeographic variation, however, refers to differences that are found between populations many kilometres apart, individuals from which are unlikely to meet each other.

Differences between areas are most obvious in species where each bird sings only one or a few song types, and where neighbour sharing is high, in other words where there is little variation within an area. In such cases, differences may be striking to the ear, as in species singing just one song type, such as the European redwing (e.g. Bjerke & Bjerke 1981, Espmark *et al.* 1989) or white-crowned sparrow (e.g. Baptista 1975, Baker & Cunningham 1985), two species for which the microgeographic variation has been particularly extensively studied.

With larger repertoires, and more variation between individuals within a population, differences between them may be more difficult to establish. In the chaffinch, there may be many song types within a wood, some of which will also be found in other woods close by. But, while all songs share the same basic structure (a trill followed by a flourish), the exact sequence of elements that we label as 'a song type' will not be found in two places many kilometres apart (Slater *et al.* 1984). In the song of the

chaffinch there is no single feature that all males in an area share, which is distinct from those in other areas and could, therefore, be labelled as a dialect characteristic. The nearest approach to it is perhaps the 'kit' element which sounds like a great-spotted woodpecker call and often follows the flourish of chaffinches in parts of continental Europe, but only very exceptionally in Britain (Thielcke 1969). However, it is not produced by all birds in an area where it occurs (Mundry et al. 1994) and cannot therefore, strictly speaking, be termed a dialect feature.

An alternative to looking at whether neighbours share more or less than birds further apart, and one that is in many ways preferable, is to look at changes with distance. In great tits, the rate of song sharing between males is greatest where birds are less than 100 m apart, roughly the width of one territory, but declines rather steeply thereafter (McGregor & Krebs 1982a). The closer together birds are the more likely they are to share. In chaffinches in New Zealand the sharing of whole song types, which involve sequences of several different syllables, declines sharply with distance, while shorter sequences are shared more and individual syllable types to a greater extent still (Lynch et al. 1989). A similar pattern has been found in the savannah sparrow in southern California. Here songs are highly stereotyped and fall into four distinct sections, each of which can vary between localities. Individual sections of a particular form tend to be much more widespread than complete sequences in which all four sections are the same (Burnell 1998).

Thus, although songs often vary considerably from place to place, the phrases or syllables that comprise them may vary much less. This is even more striking at the level of the individual element. All swamp sparrows construct their songs from a limited range of element types (Marler & Pickert 1984) (see Fig. 9.3), and the same is true of indigo buntings (Payne et al. 1981; Baker & Boylan 1995). A more striking case is that of the common rosefinch (Martens & Kessler 2000). This species has elements of only five different types which are strung together in sequences of 3–9 to form particular song types. Many different song types can be formed as a result but, purely by chance, the same song type can be found in locations far distant from one another. One further extreme, is the case of the eastern whipbird in Australia, a duetting species in which there is clear regional variation in the female contribution, but that of the male is highly stereotyped, though all males may produce a whip-crack of rising or falling frequency (see Fig. 9.4). Mennill & Rogers (2006) suggest that the

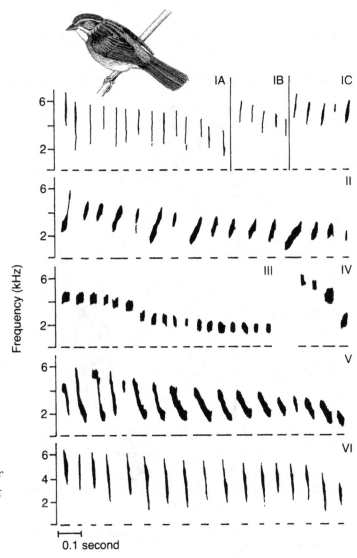

Fig. 9.3. The element types found in a sample of swamp sparrow songs from Ulster County, New York. The sample of 1307 types found could be split into the 96 types shown, which in turn fell into six broad categories (from Marler & Pickert 1984).

exceptionally broad sweep of the whip-crack may be particularly hard for males to produce, and be a cue on which female choice is based, forcing individuals to conform to a uniform pattern.

Even in species where the element repertoire is extensive, it is common for similar elements to occur in places far apart. Indeed, it is perhaps obvious that variation is not unlimited and that only certain element types are capable of being produced by a particular species. But the limitations that do exist are much tighter in some than in others. As far as whole songs are concerned it is usual for a different array of song types to be found in

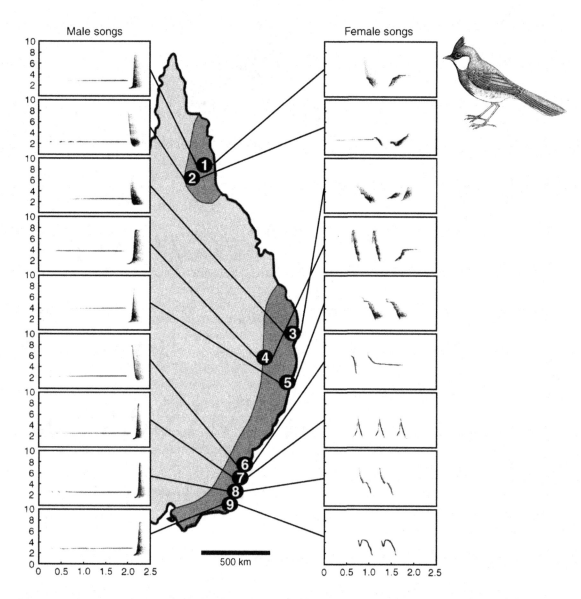

Male songs Female songs

500 km

populations some distance apart, even if they are made up of a restricted range of syllables. Just as birds often share song types more with neighbours than with individuals a few territories away, so sharing tends to fall away with increasing distance until there is usually none at all (e.g. Morton 1987, Brown *et al*. 1988, Farabaugh *et al*. 1988). While the exact array of song types may vary from one place to another, this does not mean that there are necessarily obvious differences between populations: some cases have been described where variation in song within populations is as great as that between them (e.g. vesper sparrow, Ritchison 1981). In the

Fig. 9.4. Examples of male (left) and female (right) songs from nine populations of eastern whipbirds up and down the east coast of Australia. Dark grey areas show the range of two races of this species. (From Mennill & Rogers 2006.)

chaffinch, comparison between the song types of two widely separated British populations found a significant difference between them on only 2 out of 20 measures of frequency and timing (Slater *et al*. 1984). Even this may be an exaggeration of the real difference between the sites, as song types within an area tend to share features, so that they are not truly independent statistical samples.

It is all very well for us, either by ear or from sonagrams, to argue that there are differences in song between localities, but are these differences detected by the birds themselves? The results of playback experiments suggest that they are. In most such tests, it has been found that males respond more to songs from their own area than those recorded some distance away (e.g. corn bunting, McGregor 1983; white-crowned sparrow, Milligan & Verner 1971, Tomback *et al*. 1983). Remarkably, this is even the case in song sparrows, despite their rather larger repertoire of song types (Searcy *et al*. 1997). The difference here is not affected by substituting elements from different areas between the songs used in playback, indicating that the effect is due to the structure of the song above the level of the element (Searcy *et al*. 2003). Searcy *et al*. (2002) have tested both males in the field and the copulation soliciting response of captive females in response to songs from varying distances. While females prefer songs from close to (<34 km), males fail to respond only to songs from very distant sites (540 km). They suggest that accurate production of local song may be a cue to male quality that the females use. A similar argument has been used for local dialects of flight whistles in brown-headed cowbirds (Rothstein & Fleischer 1987).

The tendency of males to respond less to distant songs is interesting as it contrasts with the stronger response usually found to strange birds from within their area than to familiar ones. It suggests that the greatest response is to a moderate discrepancy from the familiar rather than either a small or large one. Where a strong response has been found to birds from a distant location, as in the swamp sparrow (Balaban 1988b), this may be because geographic variation is not as great as in other species.

9.4 Sharp dialect boundaries

The most striking examples of geographical variation in song are where birds have only a single song type or a very small repertoire and there is close sharing between neighbours. In such cases not only may the songs

present differ between localities, but it is also frequently found that the change from place to place is not gradual, rather a sudden switch from one dialect to another occurs at a particular place. There have been several detailed studies of this phenomenon.

9.4.1 The white-crowned sparrow

The best known example of sharp boundaries is in white-crowned sparrows (*Zonotrichia leucophrys*) and particularly the sedentary race *nuttalli* originally studied by Marler & Tamura (1964) and by Baptista (1975). Very few birds in this subspecies sing more than one song type, and the songs of all birds in a particular area are similar to one another. However, travelling over a few kilometres, change does not occur gradually. Instead, the same songs will be heard for some distance and then there is a sudden sharp switch to a different song type, the change often being detectable by ear. Some hundreds of birds may belong to one of these dialect groups (see example in Fig. 9.5).

There have now been many detailed studies of song in this race of the white-crowned sparrow. These indicate that this picture of large dialect groups with sharp borders between them is an oversimplification. All white-crowned sparrow songs have certain features in common. They start with a whistle portion and move on to a trill, which often includes both complex and simple syllables. In some areas there maybe buzzes and whistles during or after the trill, but the fundamental whistle–trill sequence is a species characteristic. It is at a finer level that dialect areas differ: some include a 'buzz' and some do not; some have one form of complex syllable, some another; the structure of the simple syllables in some differs from that in others. But the boundaries defined by each of these changes are not necessarily in the same place. In a study by Baker & Thompson (1985), two dialect areas were defined by differences in the structure of complex syllables, while there was no clear distinction between their introductory (whistle) sections, and within each area several subdialects could be defined by their simple syllables. A simple syllable subdialect might also spread across the border between two complex syllable dialect areas. Baker & Cunningham (1985) argued that these different levels of distinction may differ in their functional significance, while Baker & Thompson (1985) suggested that the formation of dialects in this species may result from local extinctions and colonisation

Fig. 9.5. White-crowned sparrow dialects found by MacDougall-Shackleton & MacDougall-Shackleton (2001) at various sampling sites across the Sierra Nevada in California. Black dots show recording sites with the number of birds recorded at each indicated beside it. Thick lines encircle sites that share the song type shown in the sonagram.

events, which must be frequent in an area where habitat often changes in the wake of fires.

Other subspecies of the white-crowned sparrow do not necessarily show the same pattern as that in *nuttalli*. In one more northerly subspecies, *pugetensis*, the main dialect areas can be very much larger than in *nuttalli* (Heinemann 1981). Baptista (1977) recorded six dialect areas in *pugetensis* on the basis of the terminal trill. Some 16 years later Chilton & Lein (1996), in a study over a more limited area, found two of these to persist but with the borders between them shifted, while one that had been scarce in the earlier study had disappeared and two new ones had arisen. No dialect boundaries were detected in the Alaskan race *gambelii* by DeWolfe et al. (1974), but Chilton et al. (2002) did find song differences there, though only over very long distances. The size of dialect areas correlates with latitude and the extent that the population

migrates, which may in turn relate to differences in experience and in dispersal.

9.4.2 Other dialect species

One of the clearest cases of sharp dialect boundaries is in the corn bunting studied by McGregor (1980). In England each male sings two or three song types, and all those in an area share the same repertoire, although the distribution of song types between males was found to be more complex in a Portugese study by Latruffe et al. (2000). A dialect area comprises some 30 to 100 males. There are sharp borders between these areas, although occasional birds sing a repertoire inappropriate to their location and some birds on borders between areas sing both dialects (see Fig. 9.1). These boundaries between areas remained in the same location over a period of eight years at one site (McGregor & Thompson 1988), but had undergone radical changes, as had the song types involved, over 14 years at another (Holland et al. 1996). The latter authors suggest that land use changes, making areas more or less suitable for this species, may underlie this pattern.

Dialects have also been described in starling whistles (Adret-Hausberger 1983), although there is a limited range of variants in this particular part of the song so that a similar variety of whistles has been found in samples from France, Austria and Australia (Adret-Hausberger 1989). Each bird incorporates several whistles in its song and each of these occurs in a particular dialect area, but the boundaries between areas are not the same for different whistles (Adret-Hausberger & Guttinger 1984), rather like the boundaries for different parts of the white-crowned sparrow song. Drawing on examples such as these, and his own studies of house finches, Mundinger (1982) argued that dialect boundaries in bird song are like those observed for human speech where they vary in location for different features. However, if contours ('isoglosses') are drawn round groups of individuals sharing particular features, borders will often be found where several of these contours fall in the same place (Fig. 9.6). Thus, a number of different characteristics switch from one form to another in the same geographical location suggesting a genuine dialect boundary.

Boundaries seem more simple in several other species that have been studied. Most European redwings sing a single song, though often followed by complex twittering, particularly early in the season (Lampe

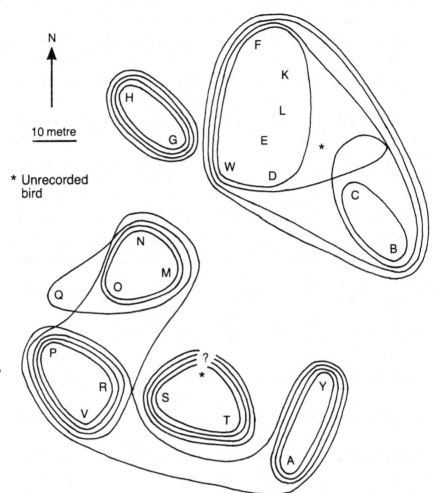

Fig. 9.6. A map showing the positions of singing male little hermit hummingbirds as recorded by Wiley (1971). Each male is represented by a letter and the lines are isoglosses enclosing individuals that share a particular syllable (after Mundinger 1982).

1991). Near Oslo, songs fall into clear dialect areas, with borders between them that were found to remain in the same place over a period of years (Bjerke 1980, Bjerke & Bjerke 1981). However, in the population studied by Espmark (1982) near Trondheim, the differences are not so clear cut, although songs do vary less within a locality than between localities.

Call notes may also vary from place to place in a simple manner, as observed for the rain call of the chaffinch, which has a number of different forms: all birds in one area tend to call in the same way, but a few kilometres away the birds may all adopt another version of the call (Bergmann *et al.* 1988, Baptista 1990, Haartman & Numers 1992). On the macrogeographic scale, the same form of call may crop up in different places far apart: the distribution is a patchwork made up of a limited number of variants.

In brown-headed cowbirds on the eastern slopes of the Sierra Nevada in California, dialects are also found in the flight whistle (Rothstein & Fleischer 1987). The dialect areas are some 10–50 km across and probably contain hundreds of individuals. They are, at least in part, separated geographically from each other by unsuitable habitat. Interestingly, surveys in the 1930s revealed no cowbirds in these areas, so the dialect populations must have been set up since that time. Another case where dialects may be of recent origin is that of the orange-tufted sunbird in a suburb of Tel-Aviv in Israel, where there are two very different dialects which have a sharp border between them, though some birds on the border are 'bilingual' (Leader *et al.* 2000). Two separate populations appear to have come to abut on each other some 40 years ago, with the expansion of two groups of human habitations to meet each other, and the dialects have persisted since then. Though there is a strong difference between the two areas in the frequency of the trill part of the song, this seems unrelated to habitat acoustics, the higher frequency trill being generally much more subject to attenuation (Leader *et al.* 2005).

A picture is emerging of the likely reasons underlying sharp dialect boundaries. Where they remain in the same place over long periods of time, some geographical or topographical barrier seems likely. In other cases habitat changes or colonisation events seem likely causes, so that populations with separate song traditions come together after a period apart. Boundaries may then persist over time, particularly if young males prefer to learn songs like those of their fathers and seek territories close to where they hatched.

9.5 The significance of geographical variation

We have now described some examples of the geographical variation found in bird song. There are clearly a lot of different patterns, both in the distribution of song types or syllables among birds within an area and in contrasts between areas. What, if any, is the significance of these patterns? In answering this question we need to identify both the historical processes that have led to the formation of dialects and other forms of geographical variation, and the reasons why these differences persist. Several different ideas have been put forward on these issues: they hark back to and build upon the brief discussion on the advantages of vocal learning at the end of Chapter 3.

9.5.1 Habitat matching

In Chapter 4 we discussed the possibility that the structure of song might be matched to the transmission characteristics of the habitat, following the idea put forward by Hansen (1979) that song learning might have evolved as a means of matching song to habitat. If a species occurs in a variety of different habitats, geographical variation in its song might stem from this process of acoustic adaptation. We would then expect dialect boundaries to coincide with habitat changes, and similarity to exist between the songs of birds in the same habitat even if they are some distance apart. Some cases of the latter have been described (see Chapter 4), but there is little evidence for such an effect in most species showing strong local dialects. The main exception is the song of the rufous-collared sparrow in Argentina.

The dialect system of the rufous-collared sparrow was originally studied by Nottebohm (1969b) and by King (1972). As in the white-crowned sparrow, each male in this species usually has a single song type, and neighbouring birds tend to share the same type. There is variation from place to place in the initial part of the song (or 'theme') over fairly short distances, but the final trill is relatively uniform over large areas with more or less sharp dialect boundaries between them. The form of the trill varies with habitat, so that it shows rather little change over wide areas of plain but alters quite rapidly on the slopes up mountains. Trill rate differs between woodlands, foothill brushlands and mountain semi-deserts (King 1972, Nottebohm 1975).

Handford (1981) confirmed that the song changes previously found in a transect across a range of mountains were repeated elsewhere in that range, and in more recent work Handford (1988) demonstrated the generality of the trill/habitat association over a very large area, including most of north-western Argentina. As might be expected from the acoustic adaptation hypothesis, trill rate tends to be high in open areas and much slower in woodland and forest, and boundaries generally occur close to habitat changes (Fig. 9.7). However, there are some inconsistencies. Boundaries are often not entirely coincident with vegetation changes (Lougheed et al. 1989) and the very slowest trills occur in open desert scrub (not present in the transect shown in Fig. 9.7) where one would not expect them to confer an advantage (Handford 1988). Handford argues that factors such as which other species sing in the different environments may complicate matters: the habitat through which sound must travel

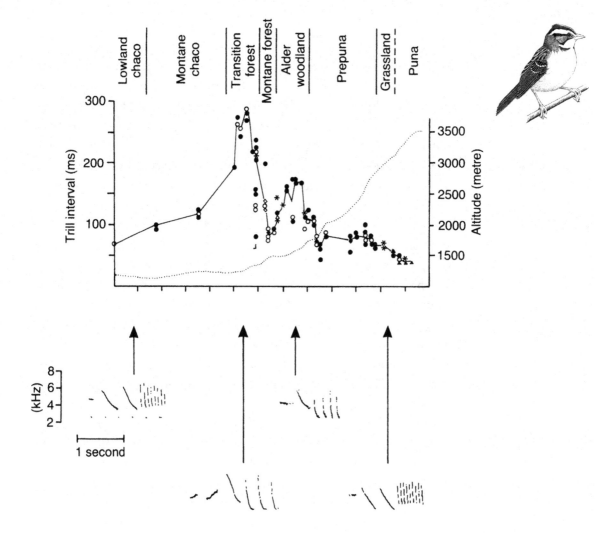

Fig. 9.7. How trill rate changes in rufous collared sparrows along one transect across various original vegetation types and with increasing altitude. The sonagrams give examples of songs from four different vegetation types (after Handford 1988).

consists not just of vegetation but also of the sounds of other species. Tubaro & Segura (1994) similarly found slower trills in open grassland than in woodland: they suggested that the windiness of grassland areas, together with the low density of birds making it necessary for signals to carry further, might be responsible.

Both Nottebohm (1975) and Handford & Lougheed (1991) also found that song is lower in frequency at higher altitudes, in line with the fact that individuals there are larger. However, there are some complicating factors

here too. Habitat also varies with altitude, so that differences in song may stem from its match with habitat rather than its association with body size. Furthermore, it is unlikely that sound frequency is constrained by body size, as the size of the body within this species is not related to that of the syrinx; in line with this, song frequency and body size are not simply related to each other (Handford & Lougheed 1991).

A final intriguing finding made by Handford (1981) is that the trill rate of the song of this species in cultivated areas is similar to that found in the forest habitat that no longer occurs there but used to do so before the spread of agriculture, often over 100 years ago. Indeed, analysis of several features of song by Handford (1988) and Handford & Lougheed (1991), suggested that contemporary vegetation explained much less of the variation in song than did the vegetation originally present. A possible interpretation of this, suggested by Handford, is based on the fact that rufous-collared sparrows are often by far the most abundant species in farmland and occur at great densities there. As a result, neither the sound transmission characteristics of the habitat nor the songs of other species may have had much influence on their song structure since the habitat changed. As Handford points out, this idea does, however, depend on a very high degree of fidelity in the cultural transmission of song if its characteristics are to remain largely unaltered over such a long period of time. While dialects may be retained in this way where they occur over large areas of new habitat, Tubaro et al. (1993) found the dialect appropriate to woodland to have been lost in a small area where it had been replaced by field and parkland around 50 years ago.

The evidence for an association between song and habitat in the rufous-collared sparrow is impressive, but there is so far only limited evidence for similar effects in other species. An exception is the recent study by Nicholls & Goldizen (2006) on the advertisement call of the satin bowerbird in which it was found that call structure at sites up and down Eastern Australia converged in similar habitats and diverged in different ones. In line with the acoustic adaptation hypothesis, calls in more open habitats had higher frequencies and more frequency modulation. Beyond these two species, while there is a good deal of evidence that differences in song between species are affected by habitat (see Chapter 4), there is much less pointing to within species variations. In great tits, song has been found to vary between forest and open woodland (Hunter & Krebs 1979), between noisy and quiet areas of a city (Slabbekoorn & Peet 2003) and between

cities and nearby forests (Slabbekoorn & den Boer-Visser 2006). But variation with habitat is not so obvious in other cases (e.g. chaffinch, Williams & Slater 1993). In the case of noisy environments, rather than changing the form of song nightingales sing more loudly (Brumm 2004) and chaffinches introduce greater redundancy by repeating the same song type more times in a row before switching (Brumm & Slater 2006). These are matters discussed more fully in Chapter 4.

9.5.2 Genetic adaptation

The genetic adaptation hypothesis was also put forward by Nottebohm (1969b, 1972) but has been advocated most forcefully by Baker and his co-workers (see Baker & Cunningham 1985 for a review, together with a variety of commentaries) to account for dialects in white-crowned sparrow song. The suggestion here is that it is beneficial for birds to mate with individuals from the same population as themselves as this avoids the break-up of co-adapted gene complexes and so favours the development of local adaptations. Song dialects might give rise to such a mating pattern if males learn their natal dialect and settle to breed in the same area, and if females prefer to mate with males of their natal dialect. It is not clear that this mechanism would favour the formation of sharp dialect boundaries but, once these had developed, they would tend to persist if birds settled in their own area.

Whether or not white-crowned sparrows show genetic adaptation was for long a very contentious issue. Much of the early evidence was in favour of this view, but later findings called it into question. There are several different issues involved. First, while early experiments suggested that birds learned their songs before dispersal, and would therefore sing their natal dialect, these involved tape recorded song (Marler 1970; Cunningham & Baker 1983). Birds will learn from live tutors much later (Baptista & Petrinovich 1984, 1986, Petrinovich & Baptista 1987), and most wild birds were found not to sing their natal dialect (Baptista & Morton 1988). Second was the issue of whether females settle in their natal area. This was tested by injecting them with testosterone and seeing which dialect they sang: the results were varied, some producing the local song (Tomback & Baker 1984), some not (Baptista & Morton 1982, 1988). Third, as far as their preferences were concerned, Baker and co-workers (Baker et al. 1981, Baker 1982) found captive female white-crowns to

respond more to songs of their own dialect than alien ones. However, such females would certainly have experienced this dialect much more than any other. Chilton *et al.* (1990) examined the preferences of females from the population of the *oriantha* subspecies that they studied in which two different song types are equally common. Females did not prefer to mate with males singing the same type as their fathers, nor did their mates in successive years tend to have the same song type as each other.

If dialects did function in this way, one would expect them to be barriers to dispersal, birds being more likely to settle within their natal area than outside it. Baker & Mewaldt (1978) suggest that this is the case, but their methods were questioned by Petrinovich & Patterson (1981) and Petrinovich *et al.* (1981) also present data suggesting that boundaries do not act in this way. Evidence on other dialect species also argues against boundaries limiting dispersal. In the corn bunting, dispersal across boundaries is common (McGregor *et al.* 1988). Sons also do not tend to sing songs of the same dialect as their fathers, and females do not tend to mate with males of the same dialect as that which their fathers belonged to. Dialect boundaries in this species are probably maintained because birds learn from neighbours after dispersal. The same is likely in European redwings: Espmark *et al.* (1989) found less than 1% of birds to return to nest in their natal dialect area.

Of course, the crucial question is whether genetic differences really exist between dialect areas. If there are no such differences then any evidence on the other questions we have discussed above is of little relevance. Once again, early evidence on white-crowns was mixed, with Baker and his co-workers (Baker 1974, 1975, Baker *et al.* 1982) arguing that there is reduced gene flow between dialect populations, but with others arguing against (Zink & Barrowclough 1984, Hafner & Petersen 1985). However, recent studies help to resolve this issue. Soha *et al.* (2004) looked at genetic and song differences between 11 sites spanning six dialect areas in the *pugetensis* subspecies. When controlling for geographic distance, they found no correlation between genetic and dialect differences. Furthermore, they concluded the same for the original results of Baker and his colleagues on *nuttalli*, using more modern startistical techniques that had not been available to the original authors. In another study, MacDougall-Shackleton & MacDougall-Shackleton (2001) examined 18 *oriantha* sites covering eight dialect regions in the Sierra Nevada. While the effect was not a strong one, they did find a correlation between genetic difference and song dissimilarity, suggesting a slight reduction in gene flow, though certainly not a

barrier, between dialect regions. Correlations can occur for a number of reasons and, even where genetic differences are found between dialect regions, as here, it does not mean that the dialects themselves are responsible for this effect.

Overall, we can conclude that the evidence for dialects leading to genetic differentiation among populations in the white-crowned sparrow is poor. This conclusion also holds for other species. In the rufous-collared sparrow, Lougheed & Handford (1992) found that dialect variation was of no value in predicting genetic variation among populations, nor is variation in mitochondrial DNA related to song variation (Lougheed *et al.* 1993), so that the authors reject the genetic adaptation hypothesis for this species. In the yellow-naped amazon, a parrot studied by Wright (1996), there are clear dialects in the contact call, but a high degree of gene flow between dialect areas (Wright & Wilkinson 2001).

In the swamp sparrow, Balaban (1988a) found sharp discontinuities in both song syllable variation and genetic variation at some dialect borders but not at others. It is possible that song differences may restrict gene flow between adjacent populations, but Balaban is cautious on this score: as pointed out above, observing a correlation between song and gene frequency changes does not necessarily mean that one causes the other. If, for some reason, birds do not cross a particular line, both songs and genetic characteristics will diverge on the two sides of it. Furthermore, while birds may react to dialect differences in territory establishment and in mate choice, as Balaban (1988b) found, this does not mean that song learning and song dialects arose for this reason.

9.5.3 Social adaptation

A further reason why dialects may form and persist is if it benefits individuals to learn the songs of their neighbours as it enables them to interact more effectively. There is a good deal of evidence that birds either learn from neighbours when they settle or select songs for use that they share with neighbours (see Chapter 3) and that song matching occurs during social interactions (see Chapter 8). Young males may benefit from learning the songs of established or dominant males (Payne 1981a), and this mimicry may even be viewed as deceptive if it leads others to treat the mimic as if it were the model. First-year indigo buntings tend to learn from single neighbours and, while the reason for choosing one bird to copy rather than

another is not yet clear, age is one factor involved. Yearlings tend to copy from older males, and in one population first-year birds with bluer, more adult plumage were more likely to be copied than browner ones (Payne & Payne 1993).

An alternative to this 'deceptive convergence' is that of 'honest convergence' put forward by Rothstein & Fleischer (1987). This proposes that possession of the signal appropriate to an area may indicate that a male is an experienced member of the local population, and, thus, be indicative of high status and quality. Such a situation might arise where a signal could not be copied easily or quickly, so that inexperienced yearling males and newcomers to the area were unable to reproduce it until they were older and more experienced. Rothstein & Fleischer (1987) argue that the cowbird flight whistle dialects they studied, having arisen in geographic isolation, may be maintained in this way. The whistles of yearlings are more likely to be foreign to the local dialect, so that possession of the appropriate whistle is an indicator of age, which correlates in turn with male mating success. This hypothesis is similar to one of the predictions of the nutritional stress hypothesis, that accuracy of song learning might be an indicator of male quality (Nowicki *et al.* 1998a).

9.5.4 Historical processes and geographical barriers

A final theory, originally put forward by Andrew (1962) but reiterated by various authors since (e.g. Bitterbaum & Baptista 1979, Wiens 1982), is that dialects in some species may be a functionless byproduct of vocal learning. As we saw in Chapter 3, song learning can be very accurate, but mistakes in learning do occur. In species where there are many song types in an area, these often show greater similarities to each other than they do to those at more distant sites. This suggests that they have been derived by inaccurate copying. The form of song may 'drift' (Lemon 1975), or birds may develop new forms by changes such as adding or omitting sections or splicing together parts of two different song types (e.g. Slater & Ince 1979, Marler 1981). In areas that are distant from each other, the precise changes that occur will be different so that the mix of song types that arises will differ between them, leading to geographical variation. In some cases, birds are thought to base their songs on the average of several individuals that they hear (Thielcke 1987). This will have a conservative influence within an area, so that the songs of all of them will tend to be similar,

though their songs may still change gradually in concert. In other species, individuals may learn each song from a single model so that new variants are more likely to persist: a form of 'particulate' inheritance similar to that of genes (Slater 1986). In either case variation between populations would be expected to arise.

While this theory will explain the fact that song varies geographically, it does not account for the occurrence of sharp dialect boundaries. However, as noted above, these might arise for a number of reasons. The most obvious is where the boundary coincides with a geographical barrier, such as a stretch of unsuitable habitat. An example here is in the winter wrens studied by Catchpole & Rowell (1993), where two populations on either side of a lake were found to have distinct song types. Historical processes, such as local extinctions and founder effects, are another potential reason why sharp boundaries might arise. Jenkins (1978) studied a population of the saddleback, a rare New Zealand passerine, small numbers of which had been introduced to the island where he worked as part of a conservation effort. He found small groups of neighbours to share songs, with sudden switches between them. He suggested that young birds had set up territories around the original founders and learnt their songs from them. As the habitat filled up these song groups spread out until they came to abut on one another, thus forming an abrupt border. Such colonisation may also occur in other species where birds spread to occupy previously uninhabitable areas, such as in corn buntings where there are changes in the use of fields from year to year (Holland et al. 1996), or in white-crowned sparrows, where the regeneration of habitat after brush fires may lead birds to colonise from different directions, so leading populations with different song characteristics to meet (Baker & Thompson 1985). The dialects in cowbird whistles on the Sierra Nevada also formed as the species spread into previously unoccupied areas (Rothstein & Fleischer 1987).

Once formed, dialect boundaries have sometimes been found to alter with time (e.g. Chilton & Lein 1996) but can persist in much the same place for many years (e.g. Seibt et al. 2002). In a computer simulation, Goodfellow & Slater (1986) examined how a border between two dialect areas would change as birds died and were replaced by others learning from one of their neighbours, and this did suggest that such borders would be likely to change rather gradually. These simulations, based on rates of mortality and of song miscopying known to occur in chaffinches, also

suggested that small dialect areas or 'song neighbourhoods', such as those described by Payne (1978a,b, 1987) and Bjorklund (1989), could easily arise. The much larger dialect areas of white-crowned sparrows, European redwings or corn buntings would, however, require much greater fidelity of copying. However, more recent chaffinch studies by Lachlan & Slater (2003) suggest that song miscopying is much less frequent (<5%) than previously thought (15%), with much of the change in songs in an area being due to immigration of birds with song types learnt elsewhere. In such a case, large dialect areas might arise more easily provided males bred close to where they learnt their songs

9.5.5 Songs on isolated islands

A special case of geographical isolation is that of islands which, if far from the nearest mainland, may have isolated populations on them with little if any contact with those elsewhere. While dispersed archipelagoes may become species rich as a result of populations becoming isolated within the island group, individual islands or groups amongst which birds can move easily tend also to have impoverished avifaunas. What tends to happen in such cases is that the songs become simpler but more variable (Marler 1960).

Simplicity of songs on islands or, more generally, in populations isolated by unsuitable habitat (e.g. Naugler & Smith 1991) may result from the lack of need for complexity as a species marker where there are few other species (see Naugler & Ratcliffe 1994). Thus Gill & Murray (1972) found the songs of blue-winged and golden-winged warblers to be less variable where the two species occurred together than in areas where only one was found. Songs have also sometimes been described as more similar in allopatry than in sympatry, the presence of a related species leading to divergence or 'character displacement' (Brown & Lemon 1979, Wallin 1986). For example, blue tits in areas around the Mediterranean where there are no great tits lack the distinctive species-specific trill (Doutrelant et al. 1998, Doutrelant & Lambrechts 2001). However, another theory, put forward by Thielcke (1973), is that islands are often colonised by young birds that have not yet learnt their songs so that the songs they develop are similar to those of birds reared in isolation. This 'withdrawal of learning' hypothesis is not really an alternative: it might account for the origin of simplified songs on islands, while the lack of avifaunal complexity could be

a reason for its maintenance. Whether on not it is young birds that colonise an island, the number of syllables or song types found there may be impoverished because of a founder effect, as argued by Baker (1996, Baker *et al.* 2001) on the basis of studies of singing honeyeaters on an island off western Australia. Thus only a small subset of the species repertoire may have been introduced in the first place. In addition, just as species may often go extinct on islands because of their small populations, so may songs or syllables (see also Lang & Barlow 1997). This is another reason why island songs may be simpler and repertoires smaller.

In this context, the word 'island' need not just refer to patches of land surrounded by water: a small area of suitable habitat in the midst of a sea of land that is unsuitable is logically the same phenomenon. For example, Gammon *et al.* (2005) describe how black-capped chickadee song diverges more among small populations than in ones continuously distributed along rivers. Habitat loss tends to lead to fragmentation and this in turn may have an effect on the songs of birds in the patches that remain. Rogers (2003) describes a similar effect in rufous bristlebirds in south-eastern Australia. Laiolo & Tella (2005) compared the songs of Dupont's larks in areas where the habitat was and was not fragmented. Fragmentation led to more neighbour sharing and less sharing among non-neighbours, the former they argue through more intense competition, the latter through lack of interaction across habitat barriers. High neighbour sharing in these circumstances could equally be attributed to each bird having a lower average number of neighbours from which to learn and with which to interact. Whatever the explanation, the splitting of habitat into islands may affect bird song in a variety of ways, effects which might even be used as markers of environmental deterioration.

9.5.6 Conclusion

The results described in this section suggest that geographical variation and dialects in bird song may arise simply as a consequence of vocal learning. But we should not be too hasty to assume that this simple answer will account for all the variation that is observed. The evidence for habitat matching, in rufous-collared sparrows particularly, is impressive, genetic adaptation remains a possibility in some other species and social adaptation is likely to be a factor in yet others. What we can be fairly sure of is that a single all-embracing theory will prove inadequate.

9.6 Cultural change

If geographical variation results from the accumulation of copying errors as song passes from one bird to another, then we would also expect to find changes with time in the songs present in an area. There is a good deal of evidence that such changes do take place, both from studies of how songs differ from year to year and from longer term comparisons of the songs present in a particular location at different times.

9.6.1 Year to year changes

In some bird species, songs can change quite rapidly so that those present in an area are quite different even between two successive years. This is probably because birds of these species modify their songs in adulthood, and they do so in concert with those round about them so that they remain matched to their neighbours. Thus, in yellow-rumped caciques, Trainer (1989) found 78% of the songs present in one year to be different from those in the previous one. Likewise, in bobolinks, there is very high turnover of song types, with only the commonest ones in the population persisting from one year to the next (Avery & Oring 1977). Sorjonen (1987) found that thrush nightingale song also changes from year to year, so that the song of a male in a given year is more similar to those of his neighbours than to his own song in the previous year. The songs of neighbouring males also become more similar as the song season progresses. The introduction of new songs appeared to be largely because immigrants moved into the area with songs learnt elsewhere.

In a detailed study of village indigobirds, Payne (1985) also looked at year to year changes in song. This species is a brood parasite and small groups of males sing together in display areas: as mentioned earlier, each of them has more than 20 song types, and group members share all of them. Each type changes from one year to the next and may change during the season. Their persistence seems to depend on how common they are: those sung by two or more males tended to be recorded in much the same form in subsequent years, but no song unique to a particular individual ever persisted from one year to the next. Payne suggested that birds may copy the individual with the greatest mating success so that his songs are both shared and persistent.

These examples concern cases where males can modify their songs in adulthood so that change can occur within the lifetime of a single individual. Where song is fixed in adulthood, the rate of change is slower, with new song types arising only through immigration or through inaccurate copying by young birds. The study by Jenkins (1978) on saddlebacks is a particularly good example here. Because he knew all the individuals in his isolated population, he was able to trace the passage of song from one bird to another and so spot changes as they occurred (Fig. 9.8). He thus detected several instances of what he referred to as 'cultural mutations', where a young male setting up his territory for the first time sang a song that was different from that of any other bird on the island.

Fig. 9.8. Cultural change in saddleback songs. Arrows indicate the probable sequence of cultural transmission as judged by the location of each bird. Most transmission is accurate, but the final element was altered when copied from RA-G to W-AR and was omitted when copied from W-AR to GA (after Jenkins 1978).

9.6.2 Sampling occurring some years apart

Moving to cases where songs have been compared at two points some years apart, the original studies were by Payne *et al.* (1981) on indigo buntings and by Ince *et al.* (1980) on chaffinches.

In the indigo bunting, all individuals can be viewed as composing their songs from a relatively limited range of element types: a catalogue of these has been drawn up and it amounts to around 100 different forms which can be recognised with reasonable reliability. Each bird has a single song type, which consists of six to eight of these element types sung in a fixed sequence, one of them repeated two to three times before the bird passes on to the next. Sampling the songs at a site in Michigan during two studies 10–15 years apart, Payne and co-workers looked to see how the sequences of element types present had changed. Some of them had persisted over the period, but there was also evidence of substantial change. Their measure of change was the exponential half-life of a song type, and they estimated

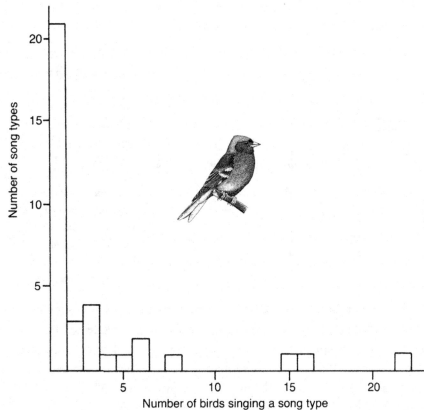

Fig. 9.9. The distribution of song types in a population of 42 chaffinches in Stanmer Great Wood, Sussex, England, in 1978. Of the 36 types present, 21 were sung by single individuals, but the commonest type was sung by 22 birds (after Slater 1986).

that this averaged 3.8 years. As sequencing varies a good deal in this species, song types are not easy to define. The definition Payne and co-workers used was that two songs must include at least three of the same element types in the identical sequence to be classed as the same. Some such definition is needed but, had it been more or less rigorous, the half-life would clearly have been lowered or raised accordingly. The process of change appears to be a random one. While there is some evidence that birds have more breeding success where they share songs with neighbours, whether or not particular songs persist from one year to the next depends mainly on how widespread they are rather than on any benefit they might confer on their possessors (Payne *et al.* 1988a).

The case of the chaffinch is rather easier, as song types in this species are much more discrete and uniform in length as well as being fixed in sequence. Slater *et al.* (1980) studied a population of 42 birds in a wood in England. The average repertoire was 2.9 song types, and there was a total of 36 song types in the wood. The commonest song type was sung by

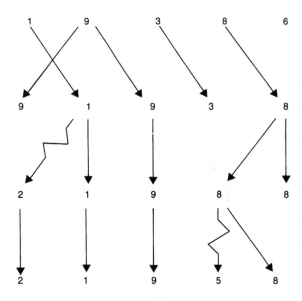

Fig. 9.10. How cultural change takes place through inaccuracies in copying. The numbers represent different song types, with straight arrows indicating accurate copying and crooked ones inaccurate. Some song types are not copied and so become extinct, but variation is maintained because new types arise when miscopying occurs.

more than half the birds in the wood, but there were many songs unique to single individuals (Fig. 9.9). A computer simulation suggested that this distribution did not arise by preferential copying. All it was necessary to propose was that new birds settling to breed either introduced songs from elsewhere or miscopied songs already present on 15% of occasions. A lower error rate would produce more shared songs and fewer unique ones, while higher levels of inaccuracy would lead to much less sharing. Subsequent simulations by Williams & Slater (1990, 1991) suggest that much of the apparent complexity of the song types and their distribution within chaffinch populations can be explained on the basis of accumulating copying errors.

A 15 % rate of introduction of new songs, either through copying errors or introduction by immigrants, also accounts for changes observed with time. Recordings made 18 years previously had found 23 song types in the same wood as that in which Slater *et al.* (1980) recorded their 35 song types. A detailed comparison of these two samples revealed that they only had three types in common, while simulations with a 15 % rate of intro- duction or copying error predicted 3.2 (Ince *et al.* 1980): a remarkably close match! Again it appears that random processes will account adequately for the distribution of chaffinch songs within a population and for the changes in those songs with time. Much of the copying is very accurate, but where errors occur new types appear; equally some song types fail to be copied and become extinct. The distribution and variety of

types in the population arises because the emergence of new types and the extinction of old ones are in balance (see Fig. 9.10).

This picture fits in with a number of other studies of chaffinches. In a study in New Zealand, where this species was introduced in the previous century, Jenkins & Baker (1984) found a good deal of differentiation amongst populations and clear differences from songs recorded in England, from where many of the original colonists had been transported. Given the rate of cultural change it is also not surprising that Thielcke (1988) could find no evidence of chaffinches singing tree pipit elements in an area where many birds that had been taught tree pipit song had been released in the early eighteenth century.

Lynch and his co-workers have carried out detailed studies of cultural change in chaffinch song in both New Zealand (Lynch *et al.* 1989) and the Atlantic Islands (Lynch & Baker 1986, 1993, 1994). Taking a 'meme' as a syllable or group of syllables that passes from one individual to another and employing some of the methodology of population genetics, they have developed what they refer to as a 'population memetics' approach. They have found the diversity of memes to vary between island and mainland populations and argued that this is the result of differences in population density and in avifaunal complexity, in line with our discussion in Section 9.4.3. However, they find no evidence that cultural evolution has been subject to selection.

In some cases, directional change in the form of song has been observed. Gibbs (1990) found that males of Darwin's medium ground finch with less common song types lived longer and left more young. Darwin's finches are amongst the few species where young males often sing the songs of their fathers, and 75% did so in Gibbs' study. The result was that the rarer song types increased in frequency and the commonest one decreased over a period of years (Fig. 9.11). At first sight this result is rather perplexing, as one might expect such changes to lead all song types to become equally common. However, song structure in this species is related to beak morphology (Huber & Podos 2006), and the distribution of beak sizes in the population had been found to change with seed availability following a drought in 1977, just before the period of Gibbs' study (Grant 1986).

Another example of cultural change is that in rufous-collared sparrows sampled on two occasions 13 years apart in a place where two dialects meet and there is a cline in song structure (Kopuchian *et al.* 2004). While the position of the dialect boundary remained the same, songs on both sides of

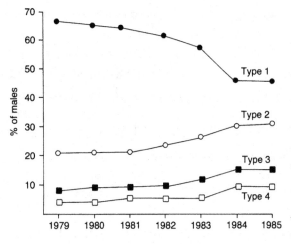

Fig. 9.11. Changes in the frequency of four song types in a population of Darwin's medium ground finch over a period of seven years (after Gibbs 1990).

the boundary differed between the two years: trill intervals had become longer and minimum frequencies lower in both. This might occur if the habitat as a whole had become more closed during this period, but there was no evidence of this. The authors suggest that more subtle changes in habitat or climate may be involved. Such directional changes are rare and the precise reasons for them remain a matter of speculation.

The extent to which song changes over time, is likely to depend on the complexity of the song at the level of analysis being used. If comparison is made between long and complex song sequences all sorts of differences are possible, as where elements drop out or their order changes. But if single syllables or brief phrases are compared change may be less dramatic, especially if the species is one thought to have a limited range of syllable types. Thus Hansen (1999) found little cultural change in the elements comprising yellowhammer song over 15 or more years. Just as island songs are prone to extinction, the extent of change with time may also depend on how small and isolated the population is. Harbison *et al.* (1999) found only slight change in white-crowned sparrow song over 26 years at two sites, but a great deal at two others where the populations were much smaller: they attributed this to the greater importance of extinction and the introduction of new syllables at such sites.

9.7 Evolutionary change

Much of our discussion so far in this chapter has concerned songs that are learnt, as these show much more dramatic changes from place to place

and over time. Unlearnt vocalisations can be very similar over large areas (e.g. Nuechterlein 1981, Trainer & Parsons 2001), but in other cases there can be gradual or subtle change with distance (e.g. Isler *et al.* 2005, Lindell 1998). Where distinct forms of unlearnt song exist in different localities, as in the sub-oscine bright-rumped attila, it even raises the question of whether one is dealing with different species (Leger & Mountjoy 2003).

Whether songs are learnt or not, change over evolutionary time is much more difficult to study because of its slow pace, and the best solution is to rely on the comparative approach to reconstruct how the songs of different species have diverged since their common ancestor. A number of such studies have now been carried out. The simplest cases are those concerning unlearnt vocalisations, as these lack the variety overlaid by learning. For example, by fitting their vocalisations into a molecular phylogeny, McCracken & Sheldon (1997) were able to identify ancestral traits amongst the calls of herons and determine how these had changed during subsequent evolution: some features were clearly related to phylogeny while others were more variable among relatives and linked to the sound transmission characteristics of their habitats. In a detailed study of 17 species of doves in the genus *Streptopelia*, ten Cate and his colleagues have likewise looked at how their coos differ, again using molecular data to assess the role of phylogeny. While these coos are not learnt, the 'perch coo' in particular attracts females and repels rivals so it very much the equivalent of song. Within a species these are somewhat variable among males, and in the Eurasian collared dove it appears that only high quality individuals can produce the sudden jumps in frequency that are intimidating to other males (ten Cate *et al.* 2002). Between species the perch coos of close relatives tend to be similar, in line with the influence of phylogeny. Environmental differences are unlikely to be important, as the species all occupy rather similar habitats. However, another important factor may be whether or not other members of the genus are present in the area. Playback experiments on two African species suggest that they respond more to the calls of other species that do not occur in the same area than ones that do, although this may be because of habituation to the local calls rather than because they are more different (de Kort & ten Cate, 2001). In a comparative study across a wide range of neotropical sub-oscines, also non vocal learners, Seddon (2005) did find that songs of closely related pairs of species that overlapped in range were more

different than those that did not. She also concluded that divergence in song between species was partly due to acoustic adaptation and partly to morphological changes, frequency varying with body size and temporal patterning with bill size.

Turning to songbirds, despite the confusing diversity that learning brings to their songs, some studies have still been able to extract phylogenetic trends. Certain aspects of song may be more constrained by phylogeny than others. For example, in North American wood warblers, van Buskirk (1997) found song structure to vary with habitat, repetition rates for example being lowest in the densest forests, while the structure of the elements comprising the songs was much more dependent on phylogeny. As Podos (1997, 2001) has argued, the capacity to produce certain sounds is limited by morphology, and such constraints are likely to be shared among relatives (see also the results of Seddon discussed above). A particularly detailed study of song evolution has been that on the oropendolas by Price & Lanyon (2002, 2004) (Fig. 9.12). Many song features were found to vary very much along phylogenetic lines, with little evidence for convergence. The greatest song differences appeared to have arisen where sexual size dimorphism had increased, pointing to a role for sexual selection in leading to divergence. However, the exact nature of the divergence differed in different lineages.

A final question, which remains a somewhat controversial one, is what role if any song learning plays in speciation. We have already concluded that there is little evidence for dialect boundaries limiting dispersal. But if populations become separate and song comes to diverge more markedly might there not come a stage when female preferences for local songs drove speciation? The clearest evidence for such an effect is in the very unusual case of the viduid finches that are brood parasites on other finches in various parts of Africa, in most cases each of them specialising in one particular host (Payne *et al.* 1998, 2000). Males learn song features of their host species and females are attracted to that particular song. Cross-fostering young to a new host species leads to different male songs based on that host's, females that prefer those songs and, remarkably, that prefer to lay in the nests of the new host. Thus song learning has played an important role in the rapid speciation that has taken place in this group.

What about more conventional song learning? It has sometimes been argued that the species diversity of the songbirds, amounting as they do to nearly half of all known bird species, might have arisen because they learn

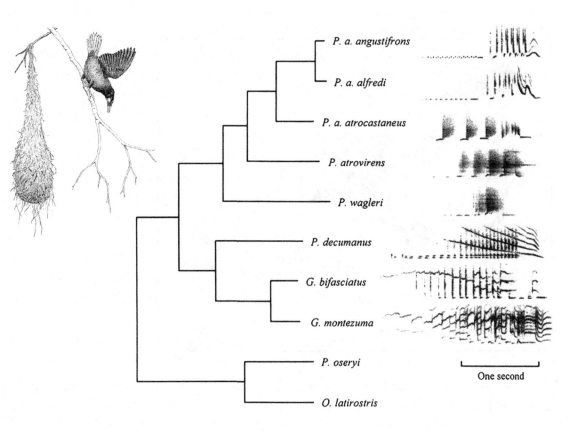

P. a. angustifrons

P. a. alfredi

P. a. atrocastaneus

P. atrovirens

P. wagleri

P. decumanus

G. bifasciatus

G. montezuma

P. oseryi

O. latirostris

One second

Fig. 9.12. Sonagrams of oropendola songs shown alongside a tree based on molecular phylogeny (after Price & Lanyon 2002).

their songs. Baptista & Trail (1992) argued against this idea, partly on the basis of the lack of influence of dialects in assortative mating, partly that some non-learning groups are also very diverse. However, the issue cannot be viewed as totally resolved. Theoretical analysis by Lachlan & Servedio (2004) suggests that song learning would be likely to hide unusual male genotypes that females might otherwise select against, and so accelerate speciation. The role of song in mate choice, and the way in which it varies with morphology and ecology, suggest that there is huge, and largely unexplored, scope for it having a role in evolution and speciation (Slabbekoorn & Smith 2002a). An interesting case is that of the greenish warbler, studied by Irwin (2000, Irwin *et al.* 2001) which shows a ring distribution around the Tibetan plateau (Fig. 9.13). In the south the song is relatively simple, but increases in complexity, though in different ways, as one moves northwards on either side of the plateau. Irwin argues that sexual selection acts more strongly on these more northern populations. North of the plateau the populations from the two sides, with their

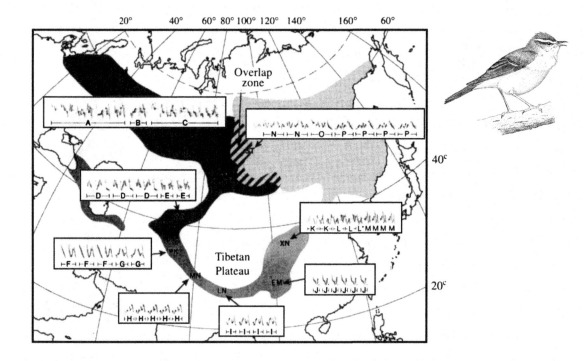

different songs, meet up again but are reproductively isolated. The two forms do not respond to playback of each other's songs.

9.8 Conclusion

We have come a long way. At first sight bird song may seem a subject more for aesthetic appreciation and for the pen of poets than for scientific research. Yet we hope to have illustrated in this book that studies of song in the last few decades have helped to illuminate a wide variety of biological problems. From physical constraints on sound transmission to brain mechanisms involved in song perception and production, from the role of song in mate choice to song learning and its consequences for individuals and populations, research on bird song has been a fruitful source of insights into fundamental biological mechanisms. All the signs are that it will continue to be so, as a lot of unanswered questions remain. In this book, we have aimed to introduce the subject and provide something of a progress report. We hope that we have persuaded the reader that the study of bird song has many important biological themes running through it - but also some fascinating variations, the significance of which we are making good progress towards understanding.

Fig. 9.13. Songs of the greenish warbler recorded at various sites around the Tibetan plateau. Although the two most northerly sites are amongst the closest to each other, and the songs were long in both places, detailed analysis showed songs at these two sites to be more distinct from each other than at any other neighbouring pair (From Irwin & Irwin 2004.)

APPENDIX

COMMON AND SCIENTIFIC NAMES

Accentor, alpine	*Prunella collaris*
Amazon, yellow-naped	*Amazona auropalliata*
Antbird, dusky	*Cercomacra tyrannina*
Attila, bright-rumped	*Attila spadiceus*
Bellbird, bare-throated	*Procnias nudicollis*
Bellbird, three-wattled	*Procnias tricarunculata*
Bishop, red	*Euplectes orix*
Blackbird, chopi	*Gnorimopsar chopi*
Blackbird, European	*Turdus merula*
Blackbird, red-winged	*Agelaius phoeniceus*
Blackbird, yellow-headed	*Xanthocephalus xanthocephalus*
Blackcap	*Sylvia atricapilla*
Bluethroat	*Luscinia svecica*
Bobolink	*Dolichonyx oryzivorus*
Boubou, slate-coloured	*Laniarius funebris*
Boubou, tropical	*Laniarius aethiopicus*
Bowerbird, satin	*Ptilinorhynchus violaceus*
Bristlebird, rufous	*Dasyornis broadbenti*
Budgerigar	*Melopsittacus undulatus*
Bullfinch	*Pyrrhula pyrrhula*
Bunting, corn	*Emberiza calandra*
Bunting, indigo	*Passerina cyanea*
Bunting, lazuli	*Passerina amoena*
Bunting, snow	*Plectophenax nivalis*
Cacique, yellow-rumped	*Cacicus cela*
Canary	*Serinus canaria*
Cardinal, northern	*Cardinalis cardinalis*
Catbird, grey	*Dumetella carolinensis*
Chaffinch, blue	*Fringilla teydea*

Chaffinch, common	*Fringilla coelebs*
Chat, yellow-breasted	*Icteria virens*
Chickadee, black-capped	*Poecile atricapillus*
Chiffchaff	*Phylloscopus collybita*
Cowbird, brown-headed	*Molothrus ater*
Cuckoo, European	*Cuculus canorus*
Dipper, white-throated	*Cinclus cinclus*
Dove, Eurasian collared	*Streptopelia decaocto*
Drongo, greater racket-tailed	*Dicrurus paradiseus*
Dunnock	*Prunella modularis*
Fairy-wren, splendid	*Malurus splendens*
Fairy-wren, superb	*Malurus cyaneus*
Finch, house	*Carpodacus mexicanus*
Finch, large cactus	*Geospiza conirostris*
Finch, medium ground	*Geospiza fortis*
Finch, strawberry	*Amandava amandava*
Finch, zebra	*Taeniopygia guttata*
Flycatcher, Acadian	*Empidonax virescens*
Flycatcher, alder	*Empidonax alnorum*
Flycatcher, collared	*Ficedula albicollis*
Flycatcher, least	*Empidonax minimus*
Flycatcher, ochre-bellied	*Mionectes oleagineus*
Flycatcher, pied	*Ficedula hypoleuca*
Flycatcher, willow	*Empidonax traillii*
Gannet, northern	*Morus bassanus*
Goldfinch, European	*Carduelis carduelis*
Grackle, common	*Quiscalus quiscula*
Grassquit, blue-black	*Volatinia jacarina*
Grassquit, Cuban	*Tiaris canora*
Greenbul, little	*Andropadus virens*
Grosbeak, black-headed	*Pheucticus melanocephalus*
Grouse, sage	*Centrocercus urophasianus*
Guillemot, common	*Uria aalge*
Gull, herring	*Larus argentatus*
Gull, laughing	*Larus atricilla*
Honeyeater, regent	*Zanthomiza phrygia*
Honeyeater, singing	*Meliphaga virescens*
Hummingbird, Anna's	*Calypte anna*
Hummingbird, blue-throated	*Lampornis clemenciae*
Hummingbird, little hermit	*Phaethornis longuemareus*

Indigobird, village	*Vidua chalybeata*
Junco, dark-eyed	*Junco hyemalis*
Kokako	*Callaeas cinerea*
Lark, Dupont's	*Chersophilus duponti*
Lyrebird, Albert's	*Menura alberti*
Lyrebird, superb	*Menura superba*
Magpie, Australian	*Gymnorhina tibicen*
Magpie-lark	*Grallina cyanoleuca*
Meadowlark, eastern	*Sturnella magna*
Meadowlark, western	*Sturnella neglecta*
Merlin	*Falco columbarius*
Mesite, sub-desert	*Monias benschi*
Mockingbird, northern	*Mimus polyglottos*
Moho	*Hypergerus atriceps*
Mynah, Indian hill	*Gracula religiosa*
Nightingale, common	*Luscinia megarhynchos*
Nightingale, thrush	*Luscinia luscinia*
Oropendola, Montezuma	*Psarocolius montezuma*
Ovenbird	*Seiurus aurocapillus*
Parrot, African grey	*Psittacus erithacus*
Parrotlet, spectacled	*Forpus conspicillatus*
Peacock	*Pavo cristatus*
Penguin, adelie	*Psygoscelis adeliae*
Penguin, emperor	*Aptenodytes forsteri*
Penguin, gentoo	*Pygoscelis papua*
Penguin, king	*Aptenodytes patagonicus*
Penguin, magellanic	*Spheniscus megallanicus*
Pewee, Eastern wood	*Contopus virens*
Phoebe, eastern	*Sayornis phoebe*
Prinia, tawny-flanked	*Prinia subflava*
Quail, Japanese	*Coturnix japonica*
Redstart, American	*Setophaga ruticilla*
Redstart, black	*Phoenicurus ochruros*
Redwing, European	*Turdus iliacus*
Robin, European	*Erithacus rubecula*
Robin-chat, white-browed	*Cossypha heuglini*
Rosefinch, common	*Carpodacus erythrinus*
Saddleback	*Philesturnus carunculatus*
Skylark	*Alauda arvensis*
Sparrow, chipping	*Spizella passerina*

Sparrow, field	*Spizella pusilla*
Sparrow, five-striped	*Aimophila quinquestriata*
Sparrow, Harris's	*Zonotrichia querula*
Sparrow, house	*Passer domesticus*
Sparrow, Lincoln's	*Melospiza lincolnii*
Sparrow, rufous-collared	*Zonotrichia capensis*
Sparrow, savannah	*Passerculus sandwichensis*
Sparrow, seaside	*Ammodramus maritimus*
Sparrow, song	*Melospiza melodia*
Sparrow, swamp	*Melospiza georgiana*
Sparrow, vesper	*Melospiza melodia*
Sparrow, white-crowned	*Zonotrichia leucophrys*
Sparrow, white-throated	*Zonotrichia albicollis*
Sparrowhawk, Eurasian	*Accipiter nisus*
Sparrow-weaver, white-browed	*Plocepasser mahali*
Starling, European	*Sturnus vulgaris*
Stonechat	*Saxicola torquatus*
Sunbird, orange-tufted	*Nectarinia osea*
Sunbird, splendid	*Nectarinia coccinigaster*
Swallow, bank	*Riparia riparia*
Swallow, barn	*Hirundo rustica*
Swallow, cliff	*Petrochelidon pyrrhonota*
Swallow, tree	*Tachycineta bicolor*
Tanager, summer	*Piranga rubra*
Tchagra, brown-headed	*Tchagra australis*
Thrasher, brown	*Toxostoma rufum*
Thrush, hermit	*Catharus guttatus*
Thrush, song	*Turdus philomelos*
Thrush, wood	*Hylocichla mustelina*
Tit, blue	*Cyanistes caeruleus*
Tit, coal	*Periparus ater*
Tit, great	*Parus major*
Tit, long-tailed	*Aegithalos caudatus*
Tit, willow	*Poecile montanus*
Tit. marsh	*Poecile palustris*
Titmouse, tufted	*Baeolophus bicolor*
Towhee, eastern	*Pipilo erythrophthalmus*
Treecreeper, Eurasian	*Certhia familiaris*
Treecreeper, short-toed	*Certhia brachydactyla*
Vireo, red-eyed	*Vireo olivaceus*

Vireo, white-eyed	*Vireo griseus*
Warbler, aquatic	*Acrocephalus paludicola*
Warbler, black-throated green	*Dendroica virens*
Warbler, blue-winged	*Vermivora pinus*
Warbler, Bonelli's	*Phylloscopus bonelli*
Warbler, chestnut-sided	*Dendroica pensylvanica*
Warbler, dusky	*Phylloscopus fuscatus*
Warbler, garden	*Sylvia borin*
Warbler, golden-winged	*Vermivora chrysoptera*
Warbler, great reed	*Acrocephalus arundinaceus*
Warbler, greenish	*Phylloscopus trochiloides*
Warbler, hooded	*Wilsonia citrina*
Warbler, Kentucky	*Oporornis formosus*
Warbler, marsh	*Acrocephalus palustris*
Warbler, reed	*Acrocephalus scirpaceus*
Warbler, Sardinian	*Sylvia melanocephala*
Warbler, sedge	*Acrocephalus schoenobaenus*
Warbler, Seychelles	*Acrocephalus sechellensis*
Warbler, willow	*Phylloscopus trochilus*
Warbler, wood	*Phylloscopus sibilatrix*
Warbler, yellow	*Dendroica petechia*
Whipbird, eastern	*Psophodes olivaceus*
Woodpecker, great-spotted	*Dendrocopos major*
Wren, banded	*Thryothorus pleurostictus*
Wren, bay	*Thryothorus nigracapillus*
Wren, Bewick's	*Thryomanes bewickii*
Wren, buff-breasted	*Thryothorus leucotis*
Wren, Carolina	*Thryothorus ludovicianus*
Wren, marsh	*Cistothorus palustris*
Wren, plain	*Thryothorus modestus*
Wren, plain-tailed	*Thryothorus euophrys*
Wren, rufous-and-white	*Thryothorus rufalbus*
Wren, sedge	*Cistothorus platensis*
Wren, stripe-backed	*Campylorhynchus nuchalis*
Wren, winter	*Troglodytes troglodytes*
Wrentit	*Chamaea fasciata*
Yellowhammer	*Emberiza citrinella*

REFERENCES

Adhikerana, A. S. 1992. The singing behaviour of coal tits (*Parus ater*). Ph.D. thesis, University of St Andrews.

Adhikerana, A. S. & Slater, P. J. B. 1993. Singing interactions in coal tits, *Parus ater*: An experimental approach. *Anim. Behav.*, 46, 1205–1211.

Adret, P. 1993. Operant conditioning, song learning and imprinting to taped song in the zebra finch. *Anim. Behav.*, 46, 149–159.

Adret-Hausberger, M. 1983. Variations dialectales des sifflements de l'Etourneau sansonnet (*Sturnus vulgaris*) sedentaire en Bretagne. *Z. Tierpsychol.*, 62, 55–71.

Adret-Hausberger, M. 1989. The species-repertoire of whistled songs in the European starling: species-specific characteristics and variability. *Bioacoustics*, 2, 137–162.

Adret-Hausberger, M. & Güttinger, H. R. 1984. Constancy of basic pattern in the songs of two populations of starlings (*Sturnus vulgaris*). A comparison of song variation between sedentary and migratory populations. *Z. Tierpsychol.*, 66, 309–327.

Adret-Hausberger, M. & Jenkins, P. F. 1988. Complex organisation of the warbling song in the European starling *Sturnus vulgaris*. *Behaviour*, 107, 138–156.

Agate, R. J., Grisham, W., Wade, J., Mann, S., Wingfield, J., Schanen, C., Palotie, A. & Arnold, A. P. 2003. Neural, not gonadal, origin of brain sex differences in a gynandromorphic finch. *Proc. Natl. Acad. Sci. USA*, 100, 4873–4878.

Airey, D. C., Buchanan, K. L., Catchpole, C. K., Szekely, T. & DeVoogd, T. J. 2000a. Song, sexual selection and a song control nucleus (HVC) in the brains of European sedge warblers. *J. Neurobiol.*, 44, 1–6.

Airey, D. C., Castillo-Juarez, H., Casella, G., Pollak, E. J. & DeVoogd, T. J. 2000b. Variation in the volume of zebra finch song control nuclei is heritable: developmental and evolutionary implications. *Proc. R. Soc. Lond.* B, 267, 2099–2104.

Airey, D. C. & DeVoogd, T. J. 2000. Greater song complexity is associated with augmented song system anatomy in zebra finches. *NeuroReport*, 11, 2339–2344.

Alatalo, R. V., Lundberg, A. & Glynn, C. 1990. Singing rate and female attraction in the pied flycatcher: an experiment. *Anim. Behav.*, 39, 601–603.

Alcock, J. 1989. *Animal Behaviour: An Evolutionary Approach*. Sunderland, Mass.: Sinauer.

Alvarez-Borda, B., Haripal, B. & Nottebohm, F. 2004. Timing of brain-derived neurotrophic factor exposure affects life expectancy of new neurons. *Proc. Natl. Acad. Sci. USA*, 101, 3957–3961.

Alvarez-Buylla, A., Kirn, J. R. & Nottebohm, F. 1990. Birth of projection neurons in adult avian brain may be related to perceptual or motor learning. *Science*, 249, 1444–1446.

Alvarez-Buylla, A. & Nottebohm, F. 1988. Migration of young neurons in adult avian brain. *Nature*, 335, 353–354.

Alvarez-Buylla, B. & Nottebohm, F. 2002. Gonads and singing play separate, additive roles in new neuron recruitment in adult canary brains. *J. Neurosci.*, 22, 8684–8690.

Alvarez-Buylla, A., Theelen, M. & Nottebohm, F. 1988. Birth of projection neurons in the higher vocal center of the canary forebrain before, during and after song learning. *Proc. Nat. Acad. Sci. USA*, 85, 8722–8726.

Amrhein, V. & Erne, N. 2006. Dawn singing reflects past territorial challenges in the winter wren. *Anim. Behav.*, 71, 1075–1080.

Amrhein, V., Kunc, H. P. & Naguib, M. 2004. Non-territorial nightingales prospect territories during the dawn chorus *Proc. R. Soc. Lond. B (Suppl.)*, 271, S167–S169.

Arcese, P., Stoddard, P. K. & Hiebert, S. M. 1988. The form and function of song in female song sparrows. *Condor*, 90, 44–50.

Anderson, M. E. & Conner, R. N. 1985. Cardinal song in three forest habitats in Eastern Texas. *Wilson Bull.*, 97, 436–449.

Anderson, R. C., Searcy, W. A. & Nowicki, S. 2005. Partial song matching in an eastern population of song sparrows, *Melospiza melodia*. *Anim. Behav.*, 69, 189–196.

Andersson, M. 1994. *Sexual Selection*. Princeton: Princeton University Press.

Andrew, R. J. 1962. Evolution of intelligence and vocal mimicking. *Science*, 137, 585–589.

Armstrong, E. A. 1973. *A Study of Bird Song*. New York: Dover.

Arnold, A. P. 1975. The effects of castration and androgen replacement on song, courtship and aggression in zebra finches *Poephila guttata*. *J. Exp. Zool.*, 191, 309–326.

Arnold, A. P. 1990. The passerine bird song system as a model in neuroendocrine research. *J. Exp. Zool. Suppl.*, 4, 22–30.

Aubin, T. & Jouventin, P. 1998. Cocktail-party effect in king penguin colonies. *Proc. R. Soc. Lond., B* 265, 1665–1673.

Avery, M. & Oring, L. W. 1977. Song dialects in the bobolink (*Dolichonyx oryzivorus*). *Condor*, 79, 113–118.

Aylor, D. 1972. Sound transmission through vegetation in relation to leaf area density, leaf width, and breadth of canopy. *J. Acoust. Soc. Am.*, 51, 411–414.

Badyaev, A. V. & Leaf, E. S. 1997. Habitat associations of song characteristics in *Phylloscopus* and *Hippolais* warblers. *Auk*, 114, 40–46.

Baker, M. C. 1974. Genetic structure of two populations of white-crowned sparrows with different song dialects. *Condor*, 76, 351–356.

Baker, M. C. 1975. Song dialects and genetic differences in white-crowned sparrows (*Zonotrichia leucophrys*). *Evolution*, 29, 226–241.

Baker, M. C. 1982. Vocal dialect recognition and population genetic consequences. *Am. Zool.*, 22, 561–569.

Baker, M. C. 1996. Depauperate meme pool of vocal signals in an island population of singing honeyeaters. *Anim. Behav.*, 51, 853–858.

Baker, M. C., Baker, E. M. & Baker, M. S. A. 2001. Island and island-like effects on vocal repertoire of singing honeyeaters. *Anim. Behav.*, 62, 767–774.

Baker, M. C., Bjerke, T., Lampe, H. & Espmark, Y. 1986. Sexual response of female great tits to variation in size of males song repertoire. *Am. Nat.*, 128, 491–8.

Baker, M. C. & Boylan, J. T. 1995. A catalog of song syllables of indigo and lazuli buntings. *Condor*, 97, 1028–40.

Baker, M. C. & Cunningham, M. A. 1985. The biology of bird-song dialects. *Behav. Brain Sci.*, 8, 85–133.

Baker, M. C. & Mewaldt, L. R. 1978. Song dialects as barriers to dispersal in white-crowned

sparrows (*Zonotrichia leucophrys nuttalli*). *Evolution*, 32, 712–22.

Baker, M. C., Spitler-Nabors, K. J. & Bradley, D. C. 1981. Early experience determines song dialect responsiveness of female sparrows. *Science*, 214, 819–21.

Baker, M. C., Spitler-Nabors, K. J. & Bradley, D. C. 1982. The response of female mountain white-crowned sparrows to songs from their natal dialect and an alien dialect. *Behav. Ecol. Sociobiol.*, 10, 175–179.

Baker, M. C., Spitler-Nabors, K. J., Thompson, A. D. & Cunningham, M. A. 1987. Reproductive behaviour of female white-crowned sparrows: effect of dialects and synthetic hybrid songs. *Anim. Behav.*, 35, 1766–1774.

Baker, M. C. & Thompson, D. B. 1985. Song dialects of white-crowned sparrows: historical processes inferred from patterns of geographic variation. *Condor*, 87, 127–41.

Balaban, E. 1988a. Cultural and genetic variation in swamp sparrows (*Melospiza georgiana*) I. Song variation, genetic variation, and their relationship. *Behaviour*, 105, 250–91.

Balaban, E. 1988b. Cultural and genetic variation in swamp sparrows (*Melospiza georgiana*) II. Behavioural salience of geographic song variants. *Behaviour*, 105, 292–322.

Ball, G. F. 1999. The neuroendocrine basis of seasonal changes in vocal behavior among songbirds. In *The Design of Animal Communication* (ed. M. D. Hauser & M. Konishi), pp. 213–53. Cambridge, MA: MIT Press.

Ball, G. F., Auger, C. J., Bernard, D. J. *et al.* 2004. Seasonal plasticity in the song control system. Multiple brain sites of steroid hormone action and the importance of the variation in song behavior. *Ann. NY. Acad. Sci.*, 1016, 1–25.

Ball, G. F. & MacDougall-Shackleton, S. A. 2001. Sex differences in songbirds 25 years later: what have we learned and where do we go? *Microsc. Res. Tech.*, 54, 327–34.

Ball, G. F., Riters, L. V. & Balthazart, J. 2002. Neuroendocrinology of song behavior and avian brain plasticity: multiple sites of sex steroid hormones. *Frontiers in Neuroendocrinology*, 23, 137–78.

Ballentine, B., Hyman, J. & Nowicki, S. 2004. Vocal performance influences female response to male bird song: an experimental test. *Behav. Ecol.*, 15, 163–8.

Baptista, L. F. 1975. Song dialects and demes in sedentary populations of the white-crowned sparrow (*Zonotrichia leucophrys nuttalli*). *Publ. Zool. Univ. Calif.*, 105, 1–52.

Baptista, L. F. 1977. Geographic variation in song and dialects of the Puget Sound white-crowned sparrow. *Condor*, 79, 356–70.

Baptista, L. F. 1978. Territorial, courtship and duet songs of the Cuban grassquit (*Tiaris canora*). *J. Ornithol.*, 119, 91–101.

Baptista, L. F. 1988. Imitations of white-crowned sparrow songs by a song sparrow. *Condor*, 90, 486–9.

Baptista, L. F. 1990. Dialectal variation in the raincall of the chaffinch (*Fringilla coelebs*). *Die Vogelwarte*, 35, 249–56.

Baptista, L. F. 1996. Nature and its nurturing in avian vocal development. In: *Ecology and Evolution of Acoustic Communication in Birds* (ed. D. E. Kroodsma & E. H. Miller), pp. 39–60. Ithaca, NY, and London: Comstock Publishing Associates.

Baptista, L. F. & Catchpole, C. K. 1989. Vocal mimicry and interspecific aggression in songbirds: Experiments using white-crowned sparrow imitation of song sparrow song. *Behaviour*, 109, 247–57.

Baptista, L. F. & Morton, M. L. 1981. Interspecific song acquisition by a white-crowned sparrow. *Auk*, 98, 383–5.

Baptista, L. F. & Morton, M. L. 1982. Song dialects and mate selection in montane white-crowned sparrows. *Auk*, 99, 537–47.

Baptista, L. F. & Morton, M. L. 1988. Song learning in montane white-crowned sparrows: from whom and when? *Anim. Behav.*, 36, 1753–64.

Baptista, L. F. & Petrinovich, L. 1984. Social interaction, sensitive phases and the song template hypothesis in the white-crowned sparrow. *Anim. Behav.*, 32, 172–81.

Baptista, L. F. & Petrinovich, L. 1986. Song development in the white-crowned sparrow: social factors and sex differences. *Anim. Behav.*, 34, 1359–71.

Baptista, L. F. & Schuchmann, K.-L. 1990. Song learning in the Anna hummingbird (*Calypte anna*). *Ethology*, 84, 15–26.

Baptista, L. F. & Trail, P. W. 1992. The role of song in the evolution of passerine diversity. *Syst. Biol.*, 41, 242–7.

Baptista, L. F., Trail, P. W., DeWolfe, B. B. & Morton, M. L. 1993. Singing and its functions in female white-crowned sparrows. *Anim. Behav.*, 46, 511–24.

Barrington, D. 1773. Experiments and observations on the singing of birds. *Phil. Trans. R. Soc. Lond.*, 63, 249–91.

Bartlett, P. & Slater, P. J. B. 1999. The effect of new recruits on the flock specific call of budgerigars (*Melopsittacus undulatus*). *Ethol. Ecol. Evol.*, 11, 139–47.

Baylis, J. R. 1982. Avian vocal mimicry: its function and evolution. In: *Acoustic Communication in Birds* (ed. D. E. Kroodsma & E. H. Miller), pp. 51–83. New York: Academic Press.

Becker, P. H. 1978. Der Einfluss des Lernens auf einfache und komplexe Gesangsstrophen der Sumpfmeise (*Parus palustris*). *J. Orn.*, 119, 388–411.

Becker, P. H. 1982. The coding of species-specific characteristics in bird sounds. In: *Acoustic Communication in Birds*, vol. 1 (ed. D. E. Kroodsma & E. H. Miller), pp. 213–52. New York: Academic Press.

Beebee, M. D. 2002. Song sharing by yellow warblers differs between two modes of singing: implications for song function. *Condor*, 104, 146–55.

Beebee, M. D. 2004. The functions of multiple singing modes: experimental tests in yellow warblers, *Dendroica petechia*. *Anim. Behav.*, 67, 1089–97.

Beecher, M. D. 1982. Development of parent-offspring recognition in birds. In: *Experiential Factors in Perceptual Development*. (ed. R. K. Aslin, J. R. Alberts & M. R. Petusa), pp. 45–66. New York: Academic Press.

Beecher, M. D. 1990. Evolution of parent-offspring recognition in swallows. In: *Contemporary Issues in Comparative Psychology* (ed. D. A. Dewsbury), pp. 360–380. Sundenand, MA: Sinauer.

Beecher, M. D., Beecher, I. M. & Hahn, S. 1981. Parent-offspring recognition in bank swallows *Riparia riparia*: Development and acoustic basis. *Anim. Behav.*, 29, 95–101.

Beecher, M. D. & Brenowitz, E. A. 2005. Functional aspects of song learning in songbirds. *Trends Ecol. Evol.*, 20, 143–9.

Beecher, M. D. & Campbell, S. E. 2005. The role of unshared songs in singing interactions between neighbouring song sparrows. *Anim. Behav.*, 70, 1297–304.

Beecher, M. D., Campbell, S. E. & Burt, J. M. 1994a. Song perception in the song sparrow: birds classify by song type but not by singer. *Anim. Behav.*, 47, 1343–51.

Beecher, M. D., Campbell, S. E., Burt, J. M., Hill, C. E. & Nordby, J. C. 2000a. Song-type matching between neighbouring song sparrows. *Anim. Behav.*, 59, 21–7.

Beecher, M. D., Campbell, S. E. & Nordby, J. C. 2000b. Territory tenure in song sparrows is related to song sharing with neighbours, but not to repertoire size. *Anim. Behav.*, 59, 29–37.

Beecher, M. D., Campbell, S. E. & Stoddard, P. K. 1994b. Correlation of song learning and territory establishment strategies in the song sparrow. *Proc. Natl. Acad. Sci. USA*, 91, 1450–54.

Beecher, M. D., Stoddard, P. K., Campbell, S. E. & Horning, C. L. 1996. Repertoire matching between neighbouring song sparrows. *Anim. Behav.*, 51, 917–23.

Beer, C. G. 1969. Laughing gull chicks: recognition of their parents voices. *Science*, 166, 1030–2.

Beletsky, L. D. 1982. Vocal behavior of the Northern oriole. *Wilson Bull.*, 94, 372–81.

Beletsky, L. D. 1983a. Aggressive and pair-bond maintenance songs of female red-winged blackbirds (Agelaius phoeniceus). *Z. Tierpsychol.*, 62, 47–54.

Beletsky, L. D. 1983b. Aggressive responses to 'self' songs by female red-winged blackbirds, *Agelaius phoeniceus*. *Can. J. Zool.*, 61, 462–5.

Beletsky, L. D. 1983c. Vocal mate recognition in male red-winged blackbirds, *Agelaius phoeniceus*. *Behaviour*, 84, 124–34.

Beletsky, L. D. 1983d. An investigation of vocal recognition by voice in female red-winged blackbirds. *Anim. Behav.*, 31, 355–62.

Beletsky, L. D. 1985. Intersexual song answering in red-winged blackbirds. *Can. J. Zool.*, 63, 735–7.

Beletsky, L. D. 1989. Communication and the cadence of bird song. *Am. Midl. Nat.*, 122, 298–306.

Beletsky, L. D. & Orians, G. H. 1985. Nest-associated vocalizations of female red-winged blackbirds, *Agelaius phoeniceus*. *Z. Tierpsychol.*, 69, 329–39.

Bell, B. D., Borowiec, M., McConkey, K. R. & Ranoszek, E. 1997. Settlement, breeding success and song repertoires of monogamous and polyg-

ynous sedge warblers (*Acrocephalus schoenobaenus*). *Die Vogelwarte* 39, 87–94.

Bensch, S. & Hasselquist, D. 1992. Evidence for female choice in a polygynous warbler. *Anim. Behav.*, 44, 301–11.

Berg, K. S., Brumfield, R. T. & Apanius, V. 2005. Phylogenetic and ecological determinants of the neotropical dawn chorus. *Proc. R. Soc. Lond.*, B 273, 999–1005.

Bergman, G. 1980. Die Veränderung der Gesangmelodie der Kohlmeise *Parus major* in Finnland und Schweden. *Ornis Fenn.*, 57, 97–111.

Bergmann, H.-H., Flottmann, E., Heitkamp, W., Stehn-Nix, K.-P. & Ubozak, F. 1988. Die Osnabrücker Dialektkarte von Regenrufen des Buchfinken *Fringilla coelebs* im Jahre 1987. *Vogelk. Ber. Nieders.*, 20, 89–96.

Bernard, D. J., Eens, M. & Ball, G. F. 1996. Age- and behavior-related variation in volumes of song control nuclei in male European starlings. *J. Neurobiol.*, 30, 329–39.

Bertram, B. 1970. The vocal behaviour of the Indian hill mynah. *Anim. Behav. Monogr.*, 3, 81–192.

Bezzel, E. 1988. Die Gesangzeiten des Buchfinken (*Fringilla coelebs*): eine Regionalstudie. *J. Orn.*, 129, 71–81.

Birkhead, T. R., Buchanan, K. L., DeVoogd, T. J., et al. 1997. Song, sperm quality and testes asymmetry in the sedge warbler. *Anim. Behav.*, 53, 965–71.

Birkhead, T. R. & Møller, A. P. 1992. *Sperm Competition in Birds*. London: Academic Press.

Bitterbaum, E. & Baptista, L. F. 1979. Geographical variation in songs of California house finches (*Carpodacus mexicanus*). *Auk*, 96, 462–74.

Bjerke, T. K. & Bjerke, T. H. 1981. Song dialects in the redwing, *Turdus iliacus*. *Ornis Scand.*, 12, 40–50.

Bjerke, T. K. 1980. Continuity and change in dialect structure of redwing *Turdus iliacus*. *Fauna norv. Ser. C Cinclus*, 3, 73–9.

Bjorklund, M. 1989. Microgeographic variation in the song of the scarlet rosefinch *Carpodacus erythrinus*. *Ornis Scand.*, 20, 255–64.

Bjorklund, M., Westman, B. & Allander, K. 1989. Song in Swedish great tits: intra- or intersexual communication. *Behaviour*, 111, 257–69.

Böhner, J. 1983. Song learning in the zebra finch (*Taeniopygia guttata*): selectivity in the choice of a tutor and accuracy of song copies. *Anim. Behav.*, 31, 231–7.

Bolhuis, J. J. & Eda-Fujiwara, H. 2003. Bird brains and songs: neural mechanisms of birdsong perception and memory. *Anim. Biol.*, 53, 129–45.

Bolhuis, J. J. & Gahr, M. 2006. Neural mechanisms of birdsong memory. *Nature Reviews Neurosci.*, 7, 347–57.

Borror, D. J. 1981. The songs and singing behavior of the red-eyed vireo. *Condor*, 83, 217–28.

Borror, D. J. 1987. Song in the white-eyed vireo. *Wilson Bull.*, 99, 377–97.

Boseret, G., Carere, C., Ball, G. F. & Balthazart, J. 2005. Social context affects testosterone-induced singing and the volume of song control nuclei in male canaries *Serinus canaria*. *J. Neurobiol.*, 66, 1044–60.

Bottjer, S. W., Miesner, E. A. & Arnold, A. P. 1984. Forebrain lesions disrupt development but not maintenance of song in passerine birds. *Science*, 224, 901–3.

Boughey, M. J. & Thompson, N. S. 1981. Song variety in the Brown Thrasher (*Toxostoma rufum*). *Z. Tierpsychol.*, 56, 47–58.

Brackenbury, J. H. 1980. Respiration and production of sounds by birds. *Biol. Rev.*, 55, 363–78.

Brackenbury, J. H. 1982. The structural basis of voice production and its relationship to sound characteristics. In: *Acoustic Communication in Birds*, vol. 1 (ed. D. E. Kroodsma & E. H. Miller), pp. 214–52. New York: Academic Press.

Brackenbury, J. H. 1989. Functions of the syrinx and the control of sound production. In: *Form and Function in Birds*, (ed. A. S. King & J. McLelland), pp. 193–220. San Diego, CA: Academic Press.

Bradley, R. A. 1981. Song variation in a population of white-eyed vireos (*Vireo griseus*). *Auk*, 98, 80–7.

Brainard, M. S. & Doupe, A. J. 2000. Auditory feedback in learning and maintenance of vocal behaviour. *Nature Reviews*, 1, 31–40.

Brainard, M. S. & Doupe, A. J. 2002. What songbirds teach us about learning. *Nature*, 417, 351–358.

Breitwisch, R. & Whitesides, G. H. 1987. Directionality of singing and non-singing behaviour of mated and unmated northern mockingbirds, *Mimus polyglottos*. *Anim. Behav.*, 35, 331–9.

Brémond, J.-C. 1968. Recherches sur la semantique et les elements vecteurs d'information dans les signaux acoustiques du rouge-gorge *Erithacus rubecula*. *La Terre et la Vie*, 2, 109–220.

Brémond, J.-C. 1976. Specific recognition in the song of Bonelli's warbler *Phylloscopus bonelli*. *Behaviour*, 58, 99–116.

Brémond, J.-C. 1978. Acoustic competition between the songs of the wren (*Troglodytes troglodytes*) and the songs of other species. *Behaviour*, 65, 89–98.

Brenowitz, E. A. 1982a. The active space of red-winged blackbird song. *J. comp. Physiol.*, 147, 511–22.

Brenowitz, E. A. 1982b. Aggressive response of red-winged blackbirds to mockingbird song imitation. *Auk*, 99, 584–6.

Brenowitz, E. A. 1991a. Evolution of the vocal control system in the avian brain. *Seminars in the Neurosciences*, 3, 399–407.

Brenowitz, E. A. 1991b. Altered perception of species-specific song by birds after lesions of a forebrain nucleus. *Science*, 251, 303–5.

Brenowitz, E. A. & Arnold, A. P. 1986. Interspecific comparisons of the size of neural song control

regions and song complexity in duetting birds: evolutionary implications. *J. Neurosci.*, 6, 2875–9.

Brenowitz, E. A., Nalls, B., Wingfield, J. C. & Kroodsma, D. E. 1991. Seasonal changes in avian song nuclei without seasonal changes in song repertoire. *J. Neurosci.*, 11, 1367–74.

Brindley, E. L. 1991. Response of European robins to playback of song: neighbour recognition and overlapping. *Anim. Behav.*, 41, 503–12.

Brockway, B. F. 1965. Stimulation of ovarian development and egg-laying by male courtship vocalizations in budgerigars *Melopsittacus undulatus*. *Anim. Behav.*, 13, 575–9.

Brooke, M.L. 1978. Sexual differences in the voice and individual vocal recognition in the manx shearwater *Puffinus puffinus*. *Anim. Behav.*, 26, 622–9.

Brooks, R. J. & Falls, J. B. 1975a. Individual recognition by song in white-throated sparrows. I. *Can. J. Zool.*, 53, 879–88.

Brooks, R. J. & Falls, J. B. 1975b. Individual recognition by song in white-throated sparrows. III. *Can. J. Zool.*, 53, 1749–61.

Brown, C. H. 1982. Ventroloquial and locatable vocalizations in birds. *Z. Tierpsychol.*, 59, 338–50.

Brown, E. D. & Farabaugh, S. M. 1991. Song sharing in a group-living songbird, the Australian magpie, *Gymnorhina tibicen*. Part III. Sex specificity and individual specificity of vocal parts in communal chorus and duet songs. *Behaviour*, 118, 244–74.

Brown, E. D., Farabaugh, S. M. & Veltman, C. J. 1988. Song sharing in a group-living songbird, the Australian magpie, *Gymnorhina tibicen* Part I. Vocal sharing within and among social groups. *Behaviour*, 104, 1–28.

Brown, R. N. & Lemon, R. E. 1979. Structure and evolution of song form in the wrens *Thryothorus sinaloa* and *T. felix*. *Behav. Ecol. Sociobiol.*, 5, 111–31.

Brown, T. J. & Handford, P. 2003. Why birds sing at dawn: the role of consistent sound transmission. *Ibis*, 145, 120–9.

Brumm, H. 2004. The impact of environmental noise on song amplitude in a terrirorial bird. *J. Anim. Ecol.*, 73, 434–40.

Brumm, H. 2006. Signalling through acoustic windows: nightingales avoid interspecific competition by short-term adjustment of song timing. *J. comp. Physiol. A*, 192, 1279–85.

Brumm, H. & Slabbekoorn, H. 2005. Acoustic communication in noise. *Adv. Study Behav.*, 35, 151–209.

Brumm, H. & Slater, P. J. B. 2006a. Ambient noise, motor fatigue and serial redundancy in chaffinch song. *Behav. Ecol. Sociobiol.*, 60, 475–81.

Brumm, H. & Slater, P. J. B. 2006b. Animals can vary signal amplitude with receiver distance: evidence from zebra finch song. *Anim. Behav.*, 71, 699–705.

Brumm, H. & Todt, D. 2002. Noise-dependent song amplitude regulation in a territorial songbird. *Anim. Behav.*, 63, 891–7.

Brumm, H. & Todt, D. 2003. Facing the rival: directional singing behaviour in nightingales. *Behaviour*, 140, 43–53.

Buchanan, K. L. & Catchpole, C. K. 1997. Female choice in the sedge warbler, *Acrocephalus schoenobaenus*: multiple cues from song and territory quality. *Proc. R. Soc. Lond.*, B 264, 521–6.

Buchanan, K. L. & Catchpole, C K. 2000a. Extra-pair paternity in the socially monogamous sedge warbler *(Acrocephalus schoenobaenus)* as revealed by multilocus DNA fingerprinting. *Ibis*, 142, 12–20.

Buchanan, K. L. & Catchpole, C. K. 2000b. Song as an indicator of male parental effort in the sedge warbler. *Proc. R. Soc. Lond.*, B 267, 321–6.

Buchanan, K. L., Catchpole, C. K., Lewis, J. W. & Lodge, A. 1999. Song as an indicator of parasitism in the sedge warbler. *Anim. Behav.*, 57, 307–14.

Buchanan, K. L., Leitner, S., Spencer, K. A., Goldsmith, A. R. & Catchpole, C. K. 2004. Developmental stress selectively affects the song control nucleus HVC in the zebra finch. *Proc. R. Soc. Lond,* B 271, 2381–2386.

Buchanan, K. L., Spencer, K. A., Goldsmith, A. R. & Catchpole, C. K. 2003. Song as an honest signal of past developmental stress in the European starling (*Sturnus vulgaris*). *Proc. R. Soc. Lond.,* B 270, 1149–56.

Burd, G. D. & Nottebohm, F. 1985. Ultrastructural characterization of synaptic terminals formed on newly generated neurons in a song control nucleus of the adult canary forebrain. *J. comp. Neurol.,* 240, 143–52.

Burnell, K. 1998. Cultural variation in savannah sparrow, *Passerculus sandwichensis,* songs: an analysis using the meme concept. *Anim. Behav.,* 56, 995–1003.

Burt, J. M., Bard, S. C., Campbell, S. E. & Beecher, M. D. 2002a. Alternative forms of song matching in song sparrows. *Anim. Behav.,* 63, 1143–51.

Burt, J. M., Campbell, S. E. & Beecher, M. D. 2002b. Song type matching as threat: a test using interactive playback. *Anim. Behav.,* 62, 1163–70.

Burt, J. M., Lent, K. L., Beecher, M. D. & Brenowitz, E.A. 2000. Lesions of the anterior forebrain song control pathway in female canaries affect song perception in an operant task. *J. Neurobiol.,* 42, 1–13.

Burt, J. M. & Vehrencamp, S. L. 2005. Dawn chorus as an interactive communication network. In: *Animal Communication Networks* (ed by P. McGregor), pp. 320–43. Cambridge: Cambridge University Press.

Byers, B. E. 1995. Song types, repertoires and song variability in a population of chestnut-sided warblers. *Condor,* 97, 390–401.

Byers, B. E. 1996a. Geographic variation of song form within and among chestnut-sided warbler populations. *Auk,* 113, 288–99.

Byers, B. E. 1996b. Messages encoded in the songs of chestnut-sided warblers. *Anim. Behav.,* 52, 691–705.

Calder, W.A. 1970. Respiration during song in the canary *(Serinus canaria). Comp. Biochem. Physiol.,* 32, 251–8.

Canady, R. A., Kroodsma, D. E. & Nottebohm, F. 1984. Population differences in complexity of a learned skill are correlated with the brain space involved. *Proc. Nat. Acad. Sci. USA,* 81, 6232–4.

Capp, M. S. 1992. Tests of the function of the song repertoire in bobolinks. *Condor,* 94, 468–79.

Capp, M. S. & Searcy, W. A. 1991. An experimental study of song type function in the bobolink (*Dolichonyx oryzivorus*). *Behav. Ecol. Sociobiol.,* 28, 179–86.

Catchpole, C. K. 1973. The functions of advertising song in the sedge warbler *Acrocephalus schoenobaenus* and the reed warbler *A. scirpaceus. Behaviour,* 46, 300–20.

Catchpole, C. K. 1976. Temporal and sequential organisation of song in the sedge warbler (*Acrocephalus schoenobaenus*). *Behaviour,* 59, 226–46.

Catchpole, C. K. 1978. Interspecific territorialism and competition in *Acrocephalus* warblers as revealed by playback experiments in areas of sympatry and allopatry. *Anim. Behav.,* 26, 1072–80.

Catchpole, C. K. 1979. *Vocal Communication in Birds.* London: Edward Arnold.

Catchpole, C. K. 1980. Sexual selection and the evolution of complex songs among warblers of the genus *Acrocephalus. Behaviour,* 74, 149–66.

Catchpole, C. K. 1982. The evolution of bird sounds in relation to mating and spacing behavior. In: *Acoustic Communication in Birds* (ed. D. Kroodsma & E. H. Miller), pp. 297–319. New York: Academic Press.

Catchpole, C. K. 1983. Variation in the song of the great reed warbler *Acrocephalus arundinaceus* in

relation to mate attraction and territorial defence. *Anim. Behav.*, 31, 1217–25.

Catchpole, C. K. 1986. Song repertoires and reproductive success in the great reed warbler *Acrocephalus arundinaceus*. *Behav. Ecol. Sociobiol.*, 19, 439–45.

Catchpole, C. K. 1987. Bird song, sexual selection and female choice. *Trends Ecol. Evol.*, 2, 94–7.

Catchpole, C. K. 1989. Pseudoreplicaton and external validity: some thoughts on the suggested redesign of playback experiments in avian bioacoustics. *Trends Ecol. Evol.*, 4, 286–7.

Catchpole, C. K. 1992. Integrating playback: a wider context. In: *Playback and Studies of Animal Communication* (ed. P. K. McGregor), pp. 35–46. New York: Plenum Press.

Catchpole, C. K. 2000. Sexual selection and the evolution of song and brain structure in *Acrocephalus* warblers. *Adv. Study Behav.*, 29, 45–97.

Catchpole, C. K. & Baptista, L. F. 1988. A test of the competition hypothesis of vocal mimicry, using song sparrow imitation of white-crowned sparrow song. *Behaviour*, 106, 119–28.

Catchpole, C. K., Dittami, J. & Leisler, B. 1984. Differential responses to male song repertoires in female songbirds implanted with oestradiol. *Nature*, 312, 563–4.

Catchpole, C. K. & Komdeur, J. 1993. The song of the Seychelles warbler *Acrocephalus sechellensis*, an island endemic. *Ibis*, 135, 190–5.

Catchpole, C. K. & Leisler, B. 1986. Interspecific territorialism in reed warblers: a local effect revealed by playback experiments. *Anim. Behav.*, 34, 299–330.

Catchpole, C. K. & Leisler, B. 1989. Variation in the song of the aquatic warbler *Acrocephalus paludicola* in response to playback of different song structures. *Behaviour*, 108, 125–38.

Catchpole, C. K., Leisler, B. & Dittami, J. 1986. Sexual differences in the responses of captive great reed warblers *Acrocephalus arundinaceus*, to variation in song structure and size. *Ethology*, 73, 69–77.

Catchpole, C. K. & McGregor, P. K. 1985. Sexual selection, song complexity and plumage dimorphism in European buntings of the genus *Emberiza*. *Anim. Behav.*, 33, 1378–80.

Catchpole, C. K. & Rowell, A. 1993. Song sharing and local dialects in a population of the European wren *Troglodytes troglodytes*. *Behaviour*, 125, 67–78.

Chappuis, C. 1971. Un example de l'influence du milieu sur les émissions vocales des oiseaux: L'évolution des chants en forêt équatoriale. *Terre et Vie*, 25, 183–202.

Chatfield, C. & Lemon, R. E. 1970. Analysing sequences of behavioural events. *J. theor. Biol*, 29, 427–45.

Chilton, G. & Lein, M. R. 1996. Long-term changes in songs and song dialect boundaries of Puget sound white-crowned sparrows. *Condor*, 98, 567–80.

Chilton, G., Lein, M. R. & Baptista, L. F. 1990. Mate choice by female white-crowned sparrows in a mixed dialect population. *Behav. Ecol. Sociobiol.*, 27, 223–7.

Chilton, G., Wiebe, M. O. & Handford, P. 2002. Large scale geographic variation in songs of Gambel's white-crowned sparrows. *Condor*, 104, 378–86.

Clark, A. J., Boersma, P. D. & Olmsted, D. M. 2006. Name that tune: call discrimination and individual recognition in Magellanic penguins. *Anim. Behav.*, 72, 1141–8.

Clayton, N. S. 1987. Song tutor choice in zebra finches. *Anim. Behav.*, 35, 714–21.

Clayton, N. S. 1988. Song discrimination learning in zebra finches. *Anim. Behav.*, 36, 1016–24.

Clayton, N. S. 1990a. Subspecies recognition and song learning in zebra finches. *Anim. Behav.*, 40, 1009–17.

Clayton, N. S. 1990b. Assortative mating in zebra finch subspecies, *Taeniopygia guttata guttata* and *T. g. castanotis*. *Phil. Trans. R. Soc. Lond.*, B 330, 351–70.

Cody, M. L. & Brown, J. H. 1969. Song asynchrony in neighbouring bird species. *Nature*, 222, 778–80.

Coles, R. B., Lewis, D. B., Hill, K. G., Hutchings, M. E. & Gower, D. M. 1980. Directional hearing in the Japanese quail (*Coturnix coturnix japonica*). II. Cochlear physiology. *J. exp. Biol.*, 86, 153–70.

Collins, S. 2004. Vocal fighting and flirting: the functions of birdsong. In: *Nature's Music: The Science of Birdsong* (ed. P. Marler & H. Slabbekoorn), pp. 42–82. San Diego, CA: Academic Press.

Cooney, R. & Cockburn, A. 1995. Territorial defence is the major function of female song in the superb fairy-wren, *Malurus cyaneus*. *Anim. Behav.* 49, 1635–47.

Cooper, B. G. & Goller, F. 2004. Multimodal signals: enhancement and constraint of song motor patterns by visual display. *Science*, 303, 544–6.

Cortopassi, K. A. & Bradbury, J. W. 2000. The comparison of harmonically rich sounds using spectrographic cross-correlation and principal coordinates analysis. *Bioacoustics*, 111, 89–127.

Cosens, S. E. & Falls, J. B. 1984a. A comparison of sound propagation and song frequency in temperate marsh and grassland habitats. *Behav. Ecol. Sociobiol.*, 15, 161–70.

Cosens, S. E. & Falls, J. B. 1984b. Structure and use of song in the yellow-headed blackbird (*Xanthocephalus xanthocephalus*). *Z. Tierpsychol.*, 66, 227–41.

Craig, W. 1943. The song of the wood pewee *Myiochanes virens* Linnaeus: A study of bird music. *N.Y. State Mus. Bull.*, 334, 1–186.

Cresswell, W. 1994. Song as a pursuit deterrent signal, and its occurrence relative to other anti-predator behaviours of skylarks (*Alauda arvensis*) on attack by merlins (*Falco columbarius*). *Behav. Ecol. Sociobiol.*, 34, 217–23.

Cunningham, M. A. & Baker, M. C. 1983. Vocal learning in white-crowned sparrows: sensitive phase and song dialects. *Behav. Ecol. Sociobiol.*, 13, 259–69.

Cuthill, I. C. & Hindmarsh, A. 1985. Increase in starling song activity with removal of mate. *Anim. Behav.*, 33, 326–8.

Cuthill, I. C. & Macdonald, W. A. 1990. Experimental manipulation of the dawn and dusk chorus in the blackbird *Turdus merula*. *Behav. Ecol. Sociobiol.*, 26, 209–16.

Cynx, J., Lewis, R., Tavel, B. & Tse, H. 1998. Amplitude regulation of vocalizations in noise by a songbird, *Taeniopygia guttata*. *Anim. Behav.*, 56, 107–13.

Cynx, J., Williams, H. & Nottebohm, F. 1992. Hemispheric differences in avian song discrimination. *Proc. Natl. Acad. Sci. USA*, 89, 1372–75.

Dabelsteen, T. 1992. Interactive playback: a finely tuned response. In: *Playback and Studies of Animal Communication* (ed. P. K. McGregor), pp. 97–110. New York: Plenum Press.

Dabelsteen, T., Larsen, O. N. & Pedersen, S. B. 1993. Habitat-induced degradation of sound signals: quantifying the effects of communication sounds and bird location of blur ratio, excess attenuation, and signal-to-noise ratio in blackbird song. *J. Acoust. Soc. Amer.*, 93, 2206–20.

Dabelsteen, T. & McGregor, P. K. 1996. Dynamic acoustic communication and interactive playback. In: *Ecology and Evolution of Acoustic Communication in Birds* (ed. D. E. Kroodsma & E. H. Miller), pp. 398–408. Ithaca, NY, and London: Comstock Publishing Associates.

Dabelsteen, T. & Pederson, S. 1990. Song and aggressive responses of blackbirds, *Turdus merula*: evidence from interactive playback experiments with territory owners. *Anim. Behav.* 40, 1158–68.

Dabelsteen, T. & Pedersen, S. B. 1991. A portable digital sound emitter for interactive playback of animal vocalisations. *Bioacoustics*, 3, 193–206.

Dabelsteen, T. & Pedersen, S. B. 1993. Song-based species discrimination and behaviour assessment by female blackbirds, *Turdus merula*. *Anim. Behav.*, 45, 759–71.

Dale, S., Amundsen, T., Lifjelt, J. T. & Slagsvold, T. 1990. Mate sampling behaviour of female pied flycatchers: evidence for active mate choice. *Behav. Ecol. Sociobiol.*, 27, 87–91.

Dale, S., Rinden, H. & Slagsvold, T. 1992. Competition for a mate restricts mate search of female pied flycatchers. *Behav. Ecol. Sociobiol.*, 30, 165–76.

Darwin, C. 1871. *The Descent of Man and Selection in Relation to Sex*. London: John Murray.

Date, E. M. & Lemon, R. E. 1993. Sound transmission: a basis for dialects in birdsong? *Behaviour*, 124, 291–312.

Dave, A. & Margoliash, D. 2000. Song replay during sleep and computational rules for sensorimotor vocal learning. *Science*, 290, 812–16.

Davies, N. B., Kilner, R. M. & Noble, D. G. 1998. Nestling cuckoos *Cuculus canorus* exploit hosts with begging calls that mimic a brood. *Proc. R. Soc. Lond*, B 265, 673–8.

Dawkins, M. S. & Guilford, T. 1991. The corruption of honest signalling. *Anim. Behav.*, 41, 865–74.

Dawson, S. M. & Jenkins, P. F. 1983. Chaffinch song repertoires and the Beau Geste hypothesis. *Behaviour*, 87, 256–69.

Deecke, V. B., Ford, J. K. B. & Spong, P. 1999. Quantifying complex patterns of bioacoustic variation: use of a neural network to compare killer whale (*Orcinus orca*) dialects. *J. Acoust. Soc. Amer.*, 105, 2499–507.

de Kort, S. R. & ten Cate, C. 2001. Response to interspecific vocalizations is affected by degree of relatedness in *Streptopelia* doves. *Anim. Behav.*, 61, 239–247.

del Negro, C., Gahr, M., Leboucher, G. & Kreutzer, M. 1998. The selectivity of sexual responses to song displays: effects of partial chemical lesion of HVC in female canaries. *Behav. Brain Res.*, 96, 151–9.

del Negro, C., Kreutzer, M. & Gahr, M. 2000. Sexual stimulating signals of canary (*Serinus canaria*) songs: evidence for a female-specific auditory representation in the HVC nucleus during the breeding season. *Behav. Neurosci.*, 114, 526–42.

Depraz, V., Leboucher, G. & Kreutzer, M. 2000. Early tutoring and adult reproductive behaviour in female domestic canary (*Serinus canaria*). *Anim. Cogn.*, 3, 45–51.

Derégnaucourt, S., Mitra, P. P., Fehér, O., Pytte, C. & Tchernichovski, O. 2005. How sleep affects the developmental learning of bird song. *Nature*, 433, 710–16.

DeVoogd, T. J. 1991. Endocrine modulation of the development and adult function of the avian song system. *Psychoneuroendocrinology*, 16, 41–66.

DeVoogd, T. J., Krebs, J. R., Healy, S. D. & Purvis, A. 1993. Relations between song repertoire size and the volume of brain nuclei related to song: comparative evolutionary analyses amongst oscine birds. *Proc. R. Soc. Lond.*, B 254, 75–82.

DeWolfe, B. B., Baptista, L. F. & Petrinovich, L. 1989. Song development and territory establishment in Nuttall's white-crowned sparrows. *Condor*, 91, 397–407.

DeWolfe, B. B., Kaska, D. D. & Peyton, L. J. 1974. Prominent variations in the songs of Gambel's white-crowned sparrows. *Bird Banding*, 45, 224–52.

Dilger, W. C. 1953. Duetting in the crimson-breasted barbet. *Condor*, 55, 220–1.

Dobkin, D. S. 1979. Functional and evolutionary relationships of vocal copying phenomena in birds. *Z. Tierpsychol.*, 50, 348–63.

Dobson, C. W. & Lemon, R. E. 1975. Re-examination of the monotony-threshold hypothesis in bird song. *Nature*, 257, 126–8.

Dobson, C. W. & Lemon, R. E. 1979. Markov sequences in songs of American thrushes. *Behaviour*, 68, 86–105.

Dooling, R. 2004. Audition: can birds hear everything they sing? In *Nature's Music: The Science of Birdsong* (ed. P. Marler & H. Slabbekoorn), pp. 209–28. San Diego, CA: Academic Press.

Doupe, A. J. & Konishi, M. 1991. Song-selective auditory circuits in the vocal control system of the zebra finch. *Proc. Nat. Acad. Sci. USA*, 88, 11339–43.

Doupe, A. J. & Kuhl, P. K. 1999. Birdsong and human speech: Common themes and mechanisms. *Annu. Rev. Neurosci.*, 22, 567–631.

Doutrelant, C., Aubin, T., Hitier, S. & Lambrechts, M. M. 1998. Two distinct song populations of blue tit *Parus caeruleus* in the French Mediterranean. *Bioacoustics*, 9, 1–16.

Doutrelant, C. & Lambrechts, M. M. 2001. Macrogeographic variation in song a test of competition and habitat effects in blue tits. *Ethology*, 107, 533–44.

Dowsett-Lemaire, F. 1979. The imitative range of the song of the marsh warbler *Acrocephalus palustris*, with special reference to imitations of African birds. *Ibis*, 121, 453–68.

Draganoiu, T. I., Nagle, L., Musseau, R. & Kreutzer, M. 2006. In a songbird, the black redstart, parents use acoustic cues to discriminate between their different fledglings. *Anim. Behav.*, 71, 1039–46.

Duffy, D. L. & Ball, G. F. 2002. Song predicts immunocompetence in male European starlings (*Sturnus vulgaris*). *Proc. R. Soc. Lond*, B 269, 847–52.

Dufty, A. M. 1985. Song sharing in the brown-headed cowbird (*Molothrus ater*). *Z. Tierpsychol.*, 69, 177–190.

Dussourd, N. & Ritchison, G. 2003. Singing behavior of male yellow-breasted chats: repertoires, rates, reproductive success, and a comparison with other wood-warblers. *Wilson Bull.*, 115, 52–7.

Eales, L. A. 1985. Song learning in zebra finches: Some effects of song model availability on what is learnt and when. *Anim. Behav.*, 33, 1293–300.

Eales, L. A. 1989. The influences of visual and vocal interaction on song learning in zebra finches. *Anim. Behav.*, 37, 507–8.

Eberhardt, L. S. 1994. Oxygen consumption during singing by male Carolina wrens (*Thryothorus ludovicianus*). *Auk*, 111, 124–30.

Eens, M. 1997. Understanding the complex song of the European starling: an integrated ethological approach. *Adv. Study Behav.*, 26, 355–434.

Eens, M., Pinxten, R. & Verheyen, R. F. 1989. Temporal and sequential organisation of song bouts in the starling. *Ardea*, 77, 75–86.

Eens, M., Pinxten, R. & Verheyen, R. F. 1990. On the function of singing and wing-waving in the European starling *Sturnus vulgaris* L. *Bird Study*, 37, 48–52.

Eens, M., Pinxten, R. & Verheyen, R. F. 1991a. Male song as a cue for mate choice in the European starling. *Behaviour*, 116, 210–38.

Eens, M., Pinxten, R. & Verheyen, R. F. 1991b. Organisation of song in the European starling: species specificity and individual differences. *Belg. J. Zool.*, 121, 257–78.

Eens, M., Pinxten, R. & Verheyen, R. F. 1992. Song learning in captive European starlings, *Sturnus vulgaris*. *Anim. Behav.*, 44, 1131–43.

Eens, M., Pinxten, R. & Verheyen, R. F. 1993. Function of the song and song repertoire in the European starling (*Sturnus vulgaris*): an aviary experiment. *Behaviour*, 125, 51–66.

Ellinger, N. & Hödl, W. 2003. Habitat acoustics of a neotropical lowland rainforest. *Bioacoustics*, 13, 297–321.

Emlen, S.T. 1972. An experimental analysis of the parameters of bird song eliciting species recognition. *Behaviour*, 41, 130–71.

Emlen, S. T., Rising, J. D. & Thompson, W. L. 1975. A behavioural and morphological study of sympatry in the indigo and lazuli buntings of the great plains. *Wilson Bull.*, 87, 145–302.

Eriksson, D. & Wallin, L. 1986. Male bird song attracts females - a field experiment. *Behav. Ecol. Sociobiol.*, 19, 297–9.

Espmark, Y. 1982. Microgeographical and individual variation in the morphology of song of the redwing *Turdus iliacus. Fauna norv. ser. C Cinclus*, 5, 73–83.

Espmark, Y. 1995. Individual and local variations in the song of the snow bunting *Plectrophenax nivalis* on Spitzbergen. *Bioacoustics*, 6, 117–34.

Espmark, Y. O. & Lampe, H. M. 1993. Variations in the song of the pied flycatcher within and between breeding seasons. *Bioacoustics*, 5, 33–65.

Espmark, Y. O., Lampe, H. M. & Bjerke, T. K. 1989. Song conformity and continuity in song dialects of redwings *Turdus iliacus* and some ecological correlates. *Ornis Scand.*, 20, 1–12.

Evans, E. J. & Bazley, E. N. 1956. The absorption of sound in air at audio frequencies. *Acustica*, 6, 238–45.

Falls, J. B. 1978. Bird song and territorial behavior. In: *Advances in the Study of Communication and Affect*, vol. 4: *Aggression, Dominance and Individual Spacing* (ed. L. Krames, P. Pliner & T. Alloway), pp. 61–89. New York: Plenum Press.

Falls, J. B. 1985. Song matching in Western Meadowlarks. *Can. J. Zool.*, 63, 2520–4.

Falls, J. B. 1987. Does song deter intrusion in white-throated sparrows *Zonotrichia alibicollis? Can. J. Zool.*, 66, 206–11.

Falls, J. B. 1992. Playback: a historical perspective. In: *Playback and Studies of Animal Communication*, (ed. P. K. McGregor), pp. 11–33. New York: Plenum.

Falls, J. B. & Brooks, R. J. 1975. Individual recognition of song in white-throated sparrows. II. *Can. J. Zool.*, 53, 1412–20.

Falls, J. B., Horn, A. G. & Dickinson, T. E. 1988. How western meadowlarks classify their song: evidence from song matching. *Anim. Behav.*, 36, 579–85.

Falls, J. B. & Krebs, J. R. 1975. Sequence of songs in repertoires of western meadowlarks (*Sturnella neglecta*). *Can. J. Zool.*, 53, 1165–78.

Farabaugh, S. M. 1982. The ecological and social significance of duetting. In: *Acoustic Communication in Birds* (ed. D. E. Kroodsma & E. H. Miller), pp. 85–124. New York: Academic Press.

Farabaugh, S. M., Brown, E. D. & Veltman, C. J. 1988. Song sharing in a group-living songbird, the Australian magpie. Part II. Vocal sharing between territorial neighbors, within and between geographic regions, and between sexes. *Behaviour*, 104, 105–25.

Farabaugh, S. M., Linzenbold, A. & Dooling, R. J. 1994. Vocal plasticity in budgerigars (*Melopsittacus undulatus*): evidence for social factors in the learning of contact calls. *J. comp. Psychol.*, 108, 81–92.

Felsenstein, J. 2004. *Inferring Phylogenies.* Sunderland, MA: Sinauer.

Ficken, R. W., Ficken, M. S. & Hailman, J. P. 1974. Temporal pattern shifts to avoid acoustic interference in singing birds. *Science*, 183, 762–3.

Ficken, R. W., Popp, J. W. & Matthiae, P. E. 1985. Avoidance of acoustic interference by ovenbirds. *Wilson Bull.*, 97, 569–71.

Fisher, R. A. 1954. Evolution and bird sociality. In: *Evolution as a Process* (ed. J. Huxley, A. Hardy & E. Ford), pp. 71–83. London: Allen & Unwin.

Forstmeier, W. & Balsby, J.S. 2002. Why mated dusky warblers sing so much: territory guarding and male quality announcement. *Behaviour*, 139, 89–111.

Forstmeier, W., Hasselquist, D., Bensch, S. & Leisler, B. 2006. Does song reflect age and viability? A comparison between two populations of the great

reed warbler *Acrocephalus arundinaceus*. *Behav. Ecol. Sociobiol.*, 59, 634–43.

Forstmeier, W. & Leisler, B. 2004. Repertoire size, sexual selection and offspring viability in the great reed warbler: changing patterns in space and time. *Behav. Ecol.*, 15, 555–63.

Fotheringham, J. R., Martin, P. R. & Ratcliffe, L. 1997. Song transmission and auditory perception of distance in wood warblers (Parulinae). *Anim. Behav.*, 53, 1271–85.

Fotheringham, J. R. & Ratcliffe, L. 1995. Song degradation and estimation of acoustic distance in black-capped chickadees (*Parus atricapillus*). *Can. J. Zool.*, 73, 858–68.

Franz, M. & Goller, F. 2002. Respiratory units of motor production and song imitation in the zebra finch. *J. Neurobiol.*, 51, 129–41.

Gaddis, P. K. 1983. Differential usage of song types by plain, bridled and tufted titmice. *Ornis Scand.*, 14, 16–23.

Gahr, M., Leitner, S., Fusani, L. & Rybak, F. 2002. What is the adaptive role of neurogenesis in adult birds? *Progr. Brain Res.*, 138, 233–54.

Gammon, D. E., Baker, M. C. & Tipton, J. R. 2005. Cultural divergence within novel song in the black-capped-chickadee (*Poecile atricapillus*). *Auk*, 122, 853–71.

Garamszegi, L. Z., Balsby, T. J. S., Bell, B. D., et al. 2005. Estimating the complexity of bird song by using the capture-recapture approaches from community ecology. *Behav. Ecol. Sociobiol.*, 57, 305–17.

Garamszegi, L. Z., Boulinier, T., Møller, A. P., et al. 2002. The estimation of size and composition of avian song repertoires. *Anim. Behav.*, 63, 623–30.

Garamszegi, L. Z. & Eens, M. 2004. Brain space for a learned task: strong intraspecific evidence for neural correlates of singing behavior in songbirds. *Brain Res. Rev.*, 44, 187–93.

Garamszegi, L. Z., Møller, A. P. & Erritzoe, J. 2003. The evolution of immune defense and song complexity in birds. *Evolution*, 57, 905–12.

Garamszegi, L. Z., Pavlova, D. Z., Eens, M. & Møller, A. P. 2007. The evolution of song in female birds in Europe. Behav. *Ecol.*, 18, 86–96.

Gaunt, A. S. & Nowicki, S. 1998. Sound production in birds: Acoustics and physiology revisited. In: *Animal Acoustic Communication: Sound Analysis and Research Methods* (ed. S. L. Hopp, M. J. Owren & C. S. Evans), pp. 291–322. Berlin: Springer-Verlag.

Gaunt, S. L. L., Baptista, L. F., Sanchez, J. E. & Hernandez, D. 1994. Song learning as evidenced from song sharing in two hummingbird species (*Colibri coruscans* and *C. thalassinus*). *Auk*, 111, 87–103.

Geberzahn, N. & Hultsch, H. 2003. Long-time storage of song types in birds: evidence from interactive playbacks. *Proc. R. Soc. Lond.*, B 270, 1085–90.

Geberzahn, N., Hultsch, H. & Todt, D. 2002. Latent song type memories are accessible through auditory stimulation in a hand-reared songbird. *Anim. Behav.*, 64, 783–90.

Gentner, T. Q. & Hulse, S. H. 2000. Female European starling preference and choice for variation in conspecific male song. *Anim. Behav.*, 59, 443–58.

Gibbs, H. L. 1990. Cultural evolution of male song types in Darwin's medium ground finches, *Geospiza fortis. Anim. Behav.*, 39, 253–63.

Gibson, R. M. 1989. Field playback of male display attracts females in a lek breeding sage grouse. *Behav. Ecol. Sociobiol.*, 24, 439–43.

Gibson, R. M. & Bradbury, J. W. 1985. Sexual selection in lekking sage grouse: phenotypic correlates of male mating success. *Behav. Ecol. Sociobiol.*, 18, 117–23.

Gil, D. & Gahr, M. 2002. The honesty of bird song: multiple constraints for multiple traits. *Trends Ecol. Evol.*, 17, 133–41.

Gil, D., Graves, J. A. & Slater, P. J. B. 1999. Seasonal patterns of singing in the willow warbler: evidence against the fertility announcement hypothesis. *Anim. Behav.*, 58, 995–1000.

Gil, D., Leboucher, G., Lacroix, A., Cue, R. & Kreutzer, M. 2004. Female canaries produce eggs with greater amounts of testosterone when exposed to preferred male song. *Horm. Behav.*, 45, 64–70.

Gill, F. B. & Murray, B. G. 1972. Song variation in sympatric blue-winged and golden-winged warblers. *Auk*, 89, 625–43.

Gill, S. A., Vonhof, M. J., Stutchbury, B. J. M., Morton, E. S. & Quinn, J. S. 2005. No evidence for acoustic mate-guarding in duetting buff-breasted wrens (*Thryothorus leucotis*). *Behav. Ecol. Sociobiol.*, 57, 557–65.

Gish, S. L. & Morton, E. S. 1981. Structural adaptations to local habitat acoustics in Carolina Wren songs. *Z. Tierpsychol.*, 56, 74–84.

Godard, R. 1991. Long-term memory of individual neighbours in a migratory songbird. *Nature*, 350, 228–9.

Godard, R. 1993. Tit for tat among neighboring hooded warblers, *Wilsonia citrina. Behav. Ecol. Sociobiol.*, 33, 45–50.

Goldman, S. A. & Nottebohm, F. 1983. Neuronal production, migration and differentiation in a vocal control nucleus of the adult canary brain. *Proc. Nat. Acad. Sci. USA*, 80, 2390–4.

Goller, F. & Larsen, O. N. 1997. A new mechanism of sound generation in songbirds. *Proc. Natl. Acad. Sci. USA*, 94, 14787–91.

Goller, F. & Larsen, O. N. 2002. New perspectives on mechanisms of sound generation in songbirds. *J. comp. Physiol*, A 188, 841–50.

Gonzalez, C. & Ornelas, J. F. 2005. Song structure and microgeographic song variation in wedge-tailed sabrewings (*Campylopterus curvipennis*) in Veracruz, Mexico. *Auk*, 122, 593–607.

Goodale, E. & Kotagama, S. W. 2006a. Context dependent vocal mimicry in a passerine bird. *Proc. R. Soc. Lond.*, B 273, 875–80.

Goodale, E. & Katagama, S. W. 2006b. Vocal mimicry by a passerine bird attracts other species involved in mixed-species flocks. *Anim. Behav.*, 72, 471–477.

Goodfellow, D. J. & Slater, P. J. B. 1986. A model of bird song dialects. *Anim. Behav.*, 34, 1579–80.

Goodfellow, D. J. & Slater, P. J. B. 1990. Can chaffinches change songs from year to year? *Bioacoustics*, 2, 249–51.

Goransson, G., Hogstedt, G., Karlsson, J., Kallander, H. & Ulfstrand, S. 1974. Sangensroll fur revirhallandet hos naktergal *Luscinia luscinia* – nagra experiment med playback-teknik. *Var Fagelvarld* 33, 201–9.

Grafe, T. U. & Bitz, J. H. 2004. Functions of duetting in the tropical boubou, *Laniarius aethiopicus*: territorial defence and mutual mate guarding. *Anim. Behav.*, 68, 193–201.

Grafe, T. U., Bitz, J. H. & Wink, M. 2004. Song repertoire and duetting behaviour of the tropical boubou, *Laniarius aethiopicus. Anim. Behav.*, 68, 181–91.

Grant, B. R. 1984. The significance of song variation in a population of Darwin's finches. *Behaviour*, 89, 90–116.

Grant, B. R. & Grant, P. R. 1979. Darwin's finches: population variation and sympatric speciation. *Proc. Nat. Acad. Sci. USA*, 76, 2359–63.

Grant, B. R. & Grant, P. R. 1996. Cultural inheritance of song and its role in the evolution of Darwin's finches. *Evolution*, 50, 2471–87.

Grant, P. R. 1986. *Ecology and evolution of Darwin's finches*. Princeton, NJ: Princeton University Press.

Greenewalt, C.H. 1968. *Bird Song: Acoustics and Physiology*. Washington, DC: Smithsonian Institution Press.

Greig-Smith, P. W. 1982a. Seasonal patterns of song production by male stonechats *Saxicola torquata*. *Ornis Scand.*, 13, 225–31.

Greig-Smith, P. W. 1982b. Song rates and parental care by individual male stonechats *Saxicola torquata*. *Anim. Behav.*, 30, 245–52.

Griessmann, B. & Naguib, M. 2002. Song sharing in neighboring and non-neighboring thrush nightingales (*Luscinia luscinia*) and its implications for communication. *Ethology*, 108, 377–87.

Grimes, L. G. 1974. Dialects and geographical variation in the song of the splendid sunbird *Nectarinia coccinigaster*. *Ibis*, 116, 314–329.

Groschupf, K. 1985. Changes in five-striped sparrow song in intra- and intersexual contexts. *Wilson Bull.*, 97, 102–5.

Groschupf, K. & Mills, G. S. 1982. Singing behavior of the five-striped sparrow. *Condor*, 84, 226–36.

Gurney, M. E. & Konishi, M. 1980. Hormone-induced sexual differentiation of brain and behaviour in zebra finches. *Science*, 208, 1380–3.

Gwinner, E. & Dorka, V. 1965. Beobachtungen an Zilpzalp-Fitis-Mischsängern. *Vogelwelt*, 86, 146–51.

Hafner, D. J. & Petersen, K. E. 1985. Song dialects and gene flow in the white-crowned sparrow, *Zonotrichia leucophrys nuttalli*. *Evolution*, 39, 687–94.

Hahnloser, R. H., Kozhevnikov, A. A. & Fee, M. S. 2002. An ultrasparse code underlies the generation of neural sequences in a songbird. *Nature*, 419, 65–70.

Hall, M. L. 2004. A review of hypotheses for the functions of avian duetting. *Behav. Ecol. Sociobiol.*, 55, 415–30.

Hall, M. L. 2006. Convergent vocal strategies of males and females are consistent with a cooperative function of duetting in Australian magpie-larks. *Behaviour*, 143, 425–49.

Hall, M. L., Illes, A. & Vehrencamp, S. L. 2006. Overlapping signals in banded wrens: long-term effects of prior experience on males and females. *Behav. Ecol.*, 17, 260–9.

Hall, M. L. & Magrath, R. D. 2000. Duetting and mate-guarding in Australian magpie-larks (*Grallina cyanoleuca*). *Behav. Ecol. Sociobiol.*, 47, 180–7.

Halliday, T. R. 1983. Information and communication. In: *Animal Behaviour, vol. 2: Communication*, (ed. T. R. Halliday & P. J. B. Slater), pp. 43–81. Oxford: Blackwell Scientific Publications.

Hamilton, W. D. & Zuk, M. 1982. Heritable true fitness and bright birds: a role for parasites? *Science*, 218, 384–7.

Handford, P. 1981. Vegetational correlates of variation in the song of *Zonotrichia capensis*. *Behav. Ecol. Sociobiol.*, 8, 203–6.

Handford, P. 1988. Trill rate dialects in the rufous-collared sparrow, *Zonotrichia capensis*, in north-western Argentina. *Can. J. Zool.*, 66, 2658–70.

Handford, P. & Lougheed, S. C. 1991. Variation in duration and frequency characters in the song of the rufous-collared sparrow, *Zonotrichia capensis*, with respect to habitat, trill dialects and body size. *Condor*, 93, 644–58.

Hansen, P. 1979. Vocal learning: its role in adapting sound structures to long-distance propagation and a hypothesis on its evolution. *Anim. Behav.*, 27, 1270–1.

Hansen, P. 1981. Coordinated singing in neighbouring yellowhammers (*Emberiza citrinella*). *Natura Jutlandica*, 19, 121–38.

Hansen, P. 1999. Long term stability of song elements in the yellowhammer *Emberiza citrinella*. *Bioacoustics*, 9, 281–95.

Harbison, H., Nelson, D. A. & Hahn, T. P. 1999. Long-term persistence of song dialects in the mountain white-crowned sparrow. *Condor*, 101, 133–48.

Harcus, J. L. 1977. The functions of mimicry in the vocal behaviour of the chorister robin. *Z. Tierpsychol.*, 44, 178–93.

Harris, C. M. 1966. Absorption of sound in air versus humidity and temperature. *J. Acoust. Soc. Am.*, 40, 148–59.

Hartley, R. S. 1990. Expiratory muscle activity during song production in the canary. *Respir. Physiol.*, 81, 177–88.

Hartshorne, C. 1956. The monotony threshold in singing birds. *Auk*, 83, 176–92.

Hartshorne, C. 1973. *Born to Sing*. Bloomington: Indiana University Press.

Harvey, P. H. & Pagel, M. D. 1991. *The Comparative Method in Evolutionary Biology*. Oxford: Oxford University Press.

Hasselquist, D., Bensch, S. & von Schantz, T. 1996. Correlation between male song repertoire, extra-pair paternity and offspring survival in the great reed warbler. *Nature*, 381, 229–32.

Hatchwell, B. J., Ross, D. J., Fowlie, M. K. & McGowan, A. 2001. Kin discrimination in cooperatively breeding long-tailed tits. *Proc. R. Soc. Lond*, B 268, 885–90.

Hausberger, M. 1993. How studies on vocal communication in birds contribute to a comparative approach of cognition. *Ethologia*, 3, 171–85.

Hausberger, M., Jenkins, P. F. & Keene, J. 1991. Species-specificity and mimicry in bird song: are they paradoxes? A reevaluation of song mimicry in the European starling. *Behaviour*, 117, 53–81.

Heinemann, D. 1981. Song dialects, migration, and population structure of Puget Sound white-crowned sparrows. *Auk*, 98, 512–21.

Helb, H.-W., Dowsett-Lemaire, F., Bergmann, H.-H. & Conrads, K. 1985. Mixed singing in European songbirds - a review. *Z. Tierpsychol.*, 69, 27–41.

Henwood, K. & Fabrick, A. 1979. A quantitative analysis of the dawn chorus: temporal selection for communicatory optimization. *Am. Nat.*, 114, 260–74.

Hernandez, A. M. & MacDougall-Shackleton, S. A. 2004. Effects of early song experience on song preferences and song control and auditory brain regions in female house finches (*Carpodacus mexicanus*). *J. Neurobiol.*, 59, 247–58.

Hessler, N. A. & Doupe, A. J. 1999. Singing-related neural activity in a dorsal forebrain basal ganglia circuit of adult zebra finches. *J. Neurosci.*, 19, 10461–81.

Heuwinkel, H. 1990. The effect of vegetation on the transmission of songs of selected European Passeriformes. *Acta Biol. Benrodis*, 2, 133–50.

Hiebert, S. M., Stoddard, P. K. & Arcese, P. 1989. Repertoire size, territory acquisition and reproductive success in the song sparrow. *Anim. Behav.*, 37, 266–73.

Hiett, J. C. & Catchpole, C. K. 1982. Song repertoires and seasonal song in the yellowhammer *Emberiza citrinella*. *Anim. Behav.*, 30, 568–74.

Highsmith, R. T. 1989. The singing behavior of golden-winged warblers. *Wilson Bull.*, 101, 36–50.

Hile, A. G., Plummer, T. K. & Striedter, G. F. 2000. Male vocal imitation produces call convergence during pair bonding in budgerigars, *Melopsittacus undulatus*. *Anim. Behav.*, 59, 1209–18.

Hile, A. G. & Striedter, G. F. 2000. Call convergence within groups of female budgerigars (*Melopsittacus undulatus*). *Ethology*, 106, 1105–14.

Hill, K. G., Lewis, D. B., Hutchings, M. E. & Coles, R. B. 1980. Directional hearing in the Japanese quail (*Coturnix coturnix japonica*). I. Acoustic properties of the auditory system. *J. exp. Biol.*, 86, 135–51.

Hinde, R. A. 1958. Alternative motor patterns in chaffinch song. *Anim. Behav.*, 6, 211–18.

Hinde, R. A. & Steele, E. 1976. The effect of male song on an oestrogen-dependent behaviour in the female canary *Serinus canarius*. *Horm. Behav.*, 7, 293–304.

Hindmarsh, A. M. 1984. Vocal mimicry in starlings. *Behaviour*, 90, 302–24.

Hindmarsh, A. M. 1986. The functional significance of vocal mimicry in song. *Behaviour*, 99, 87–100.

Hjorth, I. 1976. The divalent origin and adaptive radiation of grouse songs. *Ornis Scand.*, 7, 147–57.

Hoelzel, A. R. 1986. Song characteristics and response to playback of male and female robins *Erithacus rubecula*. *Ibis*, 128, 115–27.

Hoese, W. J., Podos, J., Boetticher, N. C. & Nowicki, S. 2000. Vocal tract function in birdsong production: experimental manipulation of beak movements. *J. exp. Biol.*, 203, 1845–55.

Holland, J., McGregor, P. K. & Rowe, C. L. 1996. Changes in microgeographic song variation in the corn bunting. *J. Avian Biol.*, 27, 47–55.

Hooker, T. & Hooker, B. I. 1969. Duetting. In: *Bird Vocalizations* (ed. R. A. Hinde), pp. 185–205. Cambridge: Cambridge University Press.

Horn, A. & Falls, J. B. 1988a. Structure of Western meadowlark (*Sturnella neglecta*) song repertoires. *Can. J. Zool.*, 66, 284–8.

Horn, A. G. & Falls, J. B. 1988b. Repertoires and countersinging in western meadowlarks. *Ethology*, 77, 337–43.

Horn, A. G. & Falls, J. B. 1988c. Responses of western meadowlarks, *Sturnella neglecta*, to song repetition and contrast. *Anim. Behav.*, 36, 291–3.

Hough, G. E., Nelson, D. A. & Volman, S. F. 2000. Re-expression of songs deleted during vocal development in white-crowned sparrows, *Zonotrichia leucophrys*. *Anim. Behav.*, 60, 279–87.

Houston, A. I. & McNamara, J. M. 1987. Singing to attract a mate: a stochastic dynamic game. *J. theor. Biol.*, 129, 57–68.

Houtman, A. M. 1992. Female zebra finches choose extra-pair copulations with genetically attractive males. *Proc. R. Soc. Lond*, B 249, 3–6.

Houx, B. B. & ten Cate, C. 1999. Song learning from playback in zebra finches: is there an effect of operant contingency? *Anim. Behav.* 57, 837–45.

Houx, B. B., ten Cate, C. & Feuth, E. 2000. Variations in zebra finch song copying: an examination of the relationship with tutor song quality and pupil behaviour. *Behaviour*, 137, 1377–89.

Howard, E. 1920. *Territory in Bird Life*. London: Murray.

Howard, R. D. 1974. The influence of sexual selection and interspecific competition on mocking bird song (*Mimus polyglottos*). *Evolution*, 28, 428–38.

Huber, S. K. & Podos, J. 2006. Beak morphology and song production covary in a population of Darwin's finches. *Biol. J. Linn. Soc.*, 88, 489–98.

Hughes, M., Nowicki, S., Searcy, W. A. & Peters, S. 1998. Song-type sharing in song sparrows: implications for repertoire function and song learning. *Behav. Ecol. Sociobiol.*, 42, 437–46.

Hulse, S. H. 2002. Auditory scene analysis in animal communication. *Adv. Study Behav.*, 31, 163–200.

Hulse, S. H., MacDougall-Shackleton, S. A. & Wisniewski, A. B. 1997. Auditory scene analysis by songbirds: stream segregation of bird song by European starlings (*Sturnus vulgaris*). *J. comp. Psychol.*, 111, 3–13.

Hultsch, H. 1991. Early experience can modify singing styles: evidence from experiments with nightingales, *Luscinia megarhynchos*. *Anim. Behav.*, 42, 883–9.

Hultsch, H. & Kopp, M.-L. 1989. Early auditory learning and song improvisation in nightingales, *Luscinia megarhynchos*. *Anim. Behav.*, 37, 510–512.

Hultsch, H. & Todt, D. 1982. Temporal performance roles during vocal interactions in nightingales (*Luscinia megarhynchos*). *Behav. Ecol. Sociobiol.*, 11, 253–60.

Hultsch, H. & Todt, D. 1989a. Context memorization in the song learning of birds. *Naturwissenschaften*, 76, 584–6.

Hultsch, H. & Todt, D. 1989b. Memorization and reproduction of songs in nightingales: evidence

for package formation. *J. comp. Physiol, A.* 165, 197–203.

Hultsch, H. & Todt, D. 1992. The serial order effect in the song acquisition of birds: relevance of exposure frequency to song models. *Anim. Behav.*, 44, 590–2.

Hunter, M. L. 1989. Himalayan birds face uphill while singing. *Auk*, 106, 728–9.

Hunter, M. L., Kacelnik, A., Roberts, J. & Vuillermoz, M. 1986. Directionality of avian vocalizations: a laboratory study. *Condor*, 88, 371–5.

Hunter, M. L. & Krebs, J. R. 1979. Geographical variation in the song of the great tit (*Parus major*) in relation to ecological factors. *J. Anim. Ecol.*, 48, 759–85.

Hutchinson, J. M. C., McNamara, J. M. & Cuthill, I. C. 1993. Song, sexual selection, starvation and strategic handicaps. *Anim. Behav.*, 45, 1153–77.

Hyman, J. & Hughes, M. 2006. Territory owners discriminate between aggressive and nonaggressive neighbours. *Anim. Behav.*, 71, 209–15.

Immelmann, K. 1963. Drought adaptations in Australian desert birds. *Proc. XIII Int. Orn Congr.*, 2, 649–57.

Immelmann, K. 1969. Song development in the zebra finch and other estrildid finches. In: *Bird Vocalizations* (ed. R. A. Hinde), pp. 61–74. Cambridge: Cambridge University Press.

Ince, S. A. & Slater, P. J. B. 1985. Versatility and continuity in the songs of thrushes *Turdus* spp. *Ibis*, 127, 355–64.

Ince, S. A., Slater, P. J. B. & Weismann, C. 1980. Changes with time in the songs of a population of chaffinches. *Condor*, 82, 285–90.

Irwin, D. E. 2000. Song variation in an avian ring species. *Evolution*, 54, 998–1010.

Irwin, D. E., Bensch, S. & Price, T. D. 2001. Speciation in a ring. *Nature*, 409, 333–7.

Irwin, D. E. & Irwin, J. H. 2004. Speciation in a ring: the role of song. In *Nature's Music: The Science of Birdsong* (ed. P. Marler & H. Slabbekoorn), p. 204. San Diego, CA: Academic Press.

Irwin, R. E. 1990. Directional selection cannot explain variation in song repertoire size in the New World blackbirds (Icterinae). *Ethology*, 85, 212–24.

Isler, M. L., Isler, P. R. & Brumfield, R. T. 2005. Clinal variation in vocalisations of an antbird (Thamnophilidae) and implications for defining species limits. *Auk*, 122, 433–44.

Janik, V. M. 1999. Pitfalls in the categorization of behaviour: a comparison of dolphin whistle classification methods. *Anim. Behav.*, 57, 133–43.

Janik, V. M. & Slater, P. J. B. 2000. The different roles of social learning in vocal communication. *Anim. Behav.*, 60, 1–11.

Jarvi, T., Radesater, T. & Jakobsson, S. 1980. The song of the willow warbler *Phylloscopus trochilus* with special reference to singing behaviour in agonistic situations. *Ornis Scand.*, 11, 236–42.

Jarvis, E. D. 2004. Brains and birdsong. In *Nature's Music: The Science of Birdsong* (ed. P. Marler & H. Slabbekoorn), pp. 229–275. San Diego, CA: Academic Press.

Jarvis, E. D. & Nottebohm, F. 1997. Motor-driven gene expression. *Proc. Natl. Acad. Sci. USA*, 94, 4097–102.

Jarvis, E. D., Smith, V. A., Wada, K., *et al.* 2002. A framework for integrating the songbird brain. *J. comp. Physiol*, A 188, 961–80.

Jenkins, P. F. 1978. Cultural transmission of song patterns and dialect development in a free-living bird population. *Anim. Behav.*, 26, 50–78.

Jenkins, P. F. & Baker, A. J. 1984. Mechanisms of song differentiation in introduced populations of chaffinches *Fringilla coelebs* in New Zealand. *Ibis*, 126, 510–24.

Jilka, A. & Leisler, B. 1974. Die Einpassung dreier Rohrsängerarten (*Acrocephalus schoenabaenus*, *A. scirpaceus*, *A. arundinaceus*) in ihre Lebensräume

in bezug auf das Frequenzspektrum ihrer Reviergesänge. *J. Orn.*, 115, 192–212.

Jin, H. & Clayton, D.F. 1997. Localized changes in immediate early gene regulation during sensory and motor learning in zebra finches. *Neuron*, 19, 1049–59.

Johnson, L. S. 1987. Pattern of song type use for territorial defence in the plain titmouse *Parus inornatus*. *Ornis Scand.*, 18, 24–32.

Johnson, L. S. & Searcy, W. A. 1996. Female attraction to male song in house wrens (*Troglodytes aedon*). *Behaviour*, 133, 357–66.

Jones, A. E. & Slater, P. J. B. 1993. Do young male zebra finches prefer to learn songs that are familiar to females with which they are housed? *Anim. Behav.*, 46, 616–17.

Jones, A. E. & Slater, P. J. B. 1996. The role of aggression in song tutor choice in the zebra finch: cause or effect? *Behaviour*, 133, 103–15.

Jones, A. E., ten Cate, C. & Bijleveld, C. C. J. H. 2001. The interobserver reliability of scoring sonagrams by eye: a study on methods, illustrated on zebra finch songs. *Anim. Behav.*, 62, 791–801.

Jones, A. E., ten Cate, C. & Slater, P. J. B. 1996. Early experience and plasticity of song in adult male zebra finches (*Taeniopygia guttata*). *J. comp. Psychol.*, 110, 354–69.

Jouventin, P. & Aubin, T. 2002 Acoustic systems are adapted to breeding ecologies: individual recognition in nesting penguins. *Anim. Behav.*, 64, 747–57.

Jouventin, P., Aubin, T. & Lengagne, T. 1999 Finding a parent in a king penguin colony: the acoustic system of individual recognition. *Anim. Behav.*, 57, 1175–83.

Jouventin, P., Guillotin, M. & Cornet, A. 1979. Le chant du manchot empereur et sa signification adaptive. *Behaviour*, 70, 231–50.

Kacelnik, A. & Krebs, J. R. 1983. The dawn chorus in the great tit *(Parus major)*: proximate and ultimate causes. *Behaviour*, 83, 287–309.

Kelsey, M. G. 1989. A comparison of the song and territorial behaviour of a long-distance migrant, the marsh warbler *Acrocephalus palustris*, in summer and winter. *Ibis*, 131, 403–414.

Kempenaers, B., Verheyen, G. R. & Dhondt, A. A. 1997. Extrapair paternity in the blue tit *Parus caeruleus*: female choice, male characteristics and offspring quality. *Behav. Ecol.*, 8, 481–92.

Kempenaers, B., Verheyen, G. R., van den Broeck, M., Burke, T., van Broeckhoven, C. & Dhondt, A.A. 1992. Extra-pair paternity results from female preference for high quality males in the blue tit. *Nature*, 357, 494–96.

Khayutin, S. N. 1985. Sensory factors in the behavioral ontogeny of altricial birds. *Adv. Study Behav.*, 15, 105–52.

Kilner, R. M., Noble, D. G. & Davies, N.B. 1999. Signals of need in a parent-offspring communication and their exploitation by the common cuckoo. *Nature*, 397, 667–72.

King, A. P. & West, M. J. 1977. Species identification in the North American cowbird: Appropriate responses to abnormal song. *Science*, 195, 1002–4.

King, A. P. & West, M. J. 1983. Epigenesis of cowbird song – a joint endeavour of males and females. *Nature*, 305, 704–6.

King, J. R. 1972. Variation in the song of the rufous collared sparrow, *Zonotrichia capensis*, in northwestern Argentina. *Z. Tierpsychol.*, 30, 344–73.

Kipper, S., Mundry, R., Hultsch, H. & Todt, D. 2004. Long-term persistence of performance rules in nightingales (*Luscinia megarhynchos*): a longitudinal field study on repertoire size and composition. *Behaviour*, 141, 371–90.

Kirn, J. R., Alvarez-Buylla, A. & Nottebohm, F. 1991. Production and survival of projection neurons in a forebrain vocal centre of adult male canaries. *J. Neurosci.*, 11, 1756–62.

Kirn, J. R., Clower, R. P., Kroodsma, D. E. & DeVoogd, T. J. 1989. Song-related brain regions

in the red-winged blackbird are affected by sex and season but not repertoire size. *J. Neurobiol.*, 20, 139–63.

Kirn, J. R. & DeVoogd, T. J. 1989. The genesis and death of vocal control neurons during sexual differentiation in the zebra finch. *J. Neurosci.*, 9, 3176–87.

Kleeman, P. M. & Gilardi, J. D. 2005. Geographical variation of St Lucia parrot flight vocalizations. *Condor*, 107, 62–8.

Klump, G. M., Kretzschmar, E. & Curio, E. 1986. The hearing of an avian predator and its avian prey. *Behav. Ecol. Sociobiol.*, 18, 317–23.

Klump, G. M. & Shalter, M. D. 1984. Acoustic behaviour of birds and mammals in the predator context. *Z. Tierpsychol.*, 66, 189–226.

Konishi, M. 1963. The role of auditory feedback in the vocal behavior of the domestic fowl. *Z. Tierpsychol.*, 20, 349–67.

Konishi, M. 1964. Effects of deafening on song development in two species of juncos. *Condor*, 66, 85–102.

Konishi, M. 1965a. Effects of deafening in song development in American robins and black-headed grosbeaks. *Z. Tierpsychol.*, 22, 584–99.

Konishi, M. 1965b. The role of auditory feedback in the control of vocalisation in the white-crowned sparrow. *Z. Tierpsychol.*, 22, 770–8.

Konishi, M. 1970. Comparative neurophysiological studies of hearing and vocalization in songbirds. *Z. vergl. Physiol.*, 66, 257–72.

Konishi, M. & Nottebohm, F. 1969. Experimental studies in the ontogeny of avian vocalizations. In: *Bird Vocalizations* (ed. R. A. Hinde), pp. 29–48. Cambridge: Cambridge University Press.

Kopuchian, C., Lijtmaer, D. A., Tubaro, P. L. & Handford, P. 2004. Temporal stability and change in a microgeographical pattern of song variation in the rufous-collared sparrow. *Anim. Behav.*, 68, 551–9.

Kramer, H. G. & Lemon, R. E. 1983. Dynamics of territorial singing between neighboring song sparrows. *Behaviour*, 85, 198–223.

Krams, I. 2001. Perch selection by singing chaffinches: a better view of surroundings and the risk of predation. *Behav. Ecol.*, 12, 295–300.

Krebs, J. R. 1976. Habituation and song repertoires in the great tit. *Behav. Ecol. Sociobiol.*, 1, 215–27.

Krebs, J. R. 1977a. The significance of song repertoires: the Beau Geste hypothesis. *Anim. Behav.*, 25, 475–8.

Krebs, J. R. 1977b. Song and territory in the great tit. In: *Evolutionary Ecology*, (ed. B. Stonehouse & C. Perrins), pp. 47–62. New York: Macmillan.

Krebs, J. R., Ashcroft, R. & van Orsdol, K. 1981a. Song matching in the great tit *Parus major* L. *Anim. Behav.*, 29, 918–23.

Krebs, J. R., Ashcroft, R. & Webber, M. 1978. Song repertoires and territory defence in the great tit. *Nature*, 271, 539–42.

Krebs, J. R., Avery, M. & Cowie, R.J. 1981b. Effect of removal of mate on the singing behaviour of great tits. *Anim. Behav.*, 29, 635–7.

Krebs, J. R. Davies, N. B. 1993. *An Introduction to Behavioural Ecology*. Oxford: Blackwell Scientific Publications.

Krebs, J. R. & Dawkins, R. 1984. Animal signals: mind-reading and manipulation. In: *Behavioural Ecology: An Evolutionary Approach* (ed. J. R. Krebs & N. B. Davies), pp. 380–402. Oxford: Blackwell Scientific Publications.

Krebs, J. R. & Kroodsma, D. E. 1980. Repertoires and geographical variation in bird song. *Adv. Study Behav.*, 11, 143–77.

Kriner, E. & Schwabl, H. 1991. Control of winter song and territorial aggression of female robins (*Erithacus rubecula*) by testosterone. *Ethology*, 87, 37–44.

Kroodsma, D.E. 1976. Reproductive development in a female songbird: differential stimulation by quality of male song. *Science*, 192, 574–5.

Kroodsma, D.E. 1977. Correlates of song organization among North European wrens. *Am. Nat.*, 111, 995–1008.

Kroodsma, D. E. 1978a. Aspects of learning in the ontogeny of birdsong. In: *The Development of Behavior: Comparative and Evolutionary Aspects* (ed. G. M. Burghardt & M. Bekoff), pp. 215–30. New York: Garland.

Kroodsma, D. E. 1978b. Continuity and versatility in bird song: support for the monotony-threshold hypothesis. *Nature*, 274, 681–3.

Kroodsma, D. E. 1982. Song repertoires: Problems in their definition and use. In: *Acoustic Communi-cation In Birds* (ed. D. E. Kroodsma & E. H. Miller), pp. 125–46. New York: Academic Press.

Kroodsma, D. E. 1984. Songs of the alder flycatcher (*Empidonax alnorum*) and willow flycatcher (*Empidonax traillii*) are innate. *Auk*, 101 13–24.

Kroodsma, D. E. 1985. Development and use of two song forms by the Eastern phoebe. *Wilson Bull.*, 97, 21–9.

Kroodsma, D. E. 1986. Design of song playback experiments. *Auk*, 103, 640–2.

Kroodsma, D. E. 1988. Song types and their use: developmental flexibility of the male blue-winged warbler. *Ethology*, 79, 235–47.

Kroodsma, D. E. 1989a. Male eastern phoebes (*Sayornis phoebe*; Tyrannidae, Passeriformes) fail to imitate songs. *J. comp. Psychol.*, 103, 227–32.

Kroodsma, D. E. 1989b. Suggested experimental designs for song playbacks. *Anim. Behav.*, 37, 600–9.

Kroodsma, D. E. 1990a. Patterns in songbird singing behaviour: Hartshorne vindicated. *Anim. Behav.*, 39, 994–6.

Kroodsma, D. E. 1990b. Using appropriate experi-mental designs for intended hypotheses in 'song'

playbacks, with examples for testing effects of song repertoire sizes. *Anim. Behav.*, 40, 1138–50.

Kroodsma, D. 2004. The diversity and plasticity of birdsong. In: *Nature's Music: The Science of Birdsong* (ed. P. Marler & H. Slabbekoorn), pp. 111–34. San Diego, CA: Academic Press.

Kroodsma, D. E. 2005. *The Singing Life of Birds*. New York: Houghton Mifflin.

Kroodsma, D. E. & Baylis, J. R. 1982. A world survey of evidence for vocal learning in birds. In: *Acoustic Communication in Birds* (ed. D. E. Kroodsma & E. H. Miller), pp. 311–37. New York: Academic Press.

Kroodsma, D. E., Bereson, R. C., Byers, B. E. & Minear, E. 1989. Use of song types by the chestnut-sided warbler: evidence for both intra- and inter-sexual functions. *Can. J. Zool.*, 67, 447–56.

Kroodsma, D. E. & Byers, B. E. 1991. The func-tion(s) of bird song. *Amer. Zool.*, 31, 318–28.

Kroodsma, D. E., Byers, B. E., Goodale, E., Johnson, S. & Liu, W.-C. 2001. Pseudoreplica-tion in playback experiments, revisited a decade later. *Anim. Behav.*, 61, 1029–33.

Kroodsma, D. E. & Canady, R.A. 1985. Differences in repertoire size, singing behaviour and associ-ated neuroanatomy among marsh wren popula-tions have a genetic basis. *Auk*, 102, 439–46.

Kroodsma, D. E., Houlihan, P. W., Fallon, P. A. & Wells, J. A. 1997. Song development in grey cat-birds. *Anim. Behav.*, 54, 457–64.

Kroodsma, D. E. & Konishi, M. 1991. A suboscine bird (eastern phoebe, *Sayornis phoebe*) develops normal song without auditory feedback. *Anim. Behav.*, 42, 477–87.

Kroodsma, D. E., Liu, W.-C., Goodwin, E. & Bedell, P. A. 1999a. The ecology of song improv-isation as illustrated by North American sedge wrens. *Auk*, 116, 373–86.

Kroodsma, D. E. & Parker, L. D. 1977. Vocal vir-tuosity in the brown thrasher. *Auk*, 94, 783–5.

Kroodsma, D. E. & Pickert, R. 1980. Environmentally dependent sensitive periods for avian vocal learning. *Nature*, 288, 477–9.

Kroodsma, D. E., Sanchez, J., Stemple, D. W., Goodwin, E., Silva, M. L. d. & Vielliard, J. M. E. 1999b. Sedentary lifestyle of neotropical sedge wrens promotes song imitation. *Anim. Behav.*, 57, 855–63.

Kroodsma, D. E. & Verner, J. 1987. Use of song repertoires among marsh wren populations. *Auk*, 104, 63–72.

Kroodsma, D. E., Woods, R. W. & Goodwin, E. A. 2002. Falkland island sedge wrens (*Cistothorus platensis*) imitate rather than improvise large song repertoires. *Auk*, 119, 523–8.

Kunc, H. P., Amrhein, V. & Naguib, M. 2005. Seasonal variation in dawn song characteristics in the common nightingale. *Anim. Behav.*, 70, 1265–71.

Kunc, H. P., Amrhein, V. & Naguib, M. 2006. Vocal interactions in nightingales, *Luscinia megarhynchos*: more aggressive males have higher pairing success. *Anim. Behav.*, 72, 25–30.

Lachlan, R. F. & Servedio, M. R. 2004. Song learning accelerates allopatric speciation. *Evolution*, 58, 2049–63.

Lachlan, R. F. & Slater, P. J. B. 1999. The maintenance of vocal learning by gene-culture interaction: the cultural trap hypothesis. *Proc. R. Soc. Lond.*, B 266, 701–6.

Lachlan, R. F. & Slater, P. J. B. 2003. Song learning in chaffinches: how accurate, and from where? *Anim. Behav.*, 65, 957–69.

Lack, D. 1946. *The Life of the Robin*. London: Witherby.

Laiolo, P. & Tella, J. L. 2005. Habitat fragmentation affects culture transmission: patterns of song-matching in Dupont's lark. *J. Appl. Ecol.*, 42, 1183–93.

Lambrechts, M. 1988. Great tit song output is determined by both motivation and by constraints in singing ability: a reply to Weary et al. *Anim. Behav.*, 36, 1244–6.

Lambrechts, M. & Dhondt, A. 1986. Male quality, reproduction and survival in the great tit *Parus major*. *Behav. Ecol. Sociobiol.*, 19, 57–63.

Lambrechts, M. & Dhondt, A. A. 1987. Differences in singing performance between male great tits. *Ardea*, 75, 43–52.

Lambrechts, M. & Dhondt, A. A. 1988. The anti-exhaustion hypothesis: a new hypothesis to explain song performance and song switching in the great tit. *Anim. Behav.*, 36, 327–34.

Lampe, H. M. 1991. The response of male redwings *Turdus iliacus* to playback of conspecific songs with or without the terminating twitter. *Ornis Scand.*, 22, 137–42.

Lampe, H. M. & Espmark, Y. O. 1987. Singing activity and song pattern of the redwing *Turdus iliacus* during the breeding season. *Ornis Scand.*, 18, 179–85.

Lang, A. L. & Barlow, J. C. 1997. Cultural evolution in the Eurasian tree sparrow: divergence between introduced and ancestral populations. *Condor*, 99, 413–23.

Langmore, N. E. 1998. Functions of duet and solo songs of female birds. *Trends Ecol. Evol.*, 13, 136–40.

Langmore, N. E. & Davies, N. B. 1997. Female dunnocks use vocalizations to compete for males. *Anim. Behav.*, 53, 881–90.

Langmore, N. E., Davies, N. B., Hatchwell, B. J. & Hartley, I. R. 1996. Female song attracts males in the alpine accentors *Prunella collaris*. *Proc. R. Soc. Lond*, B 263, 141–6.

Larom, D., Garstang, M., Payne, K., Raspet, R. & Lindeque, M. 1997. The influence of surface atmospheric conditions on the range and area reached by animal vocalizations. *J. exp. Biol.*, 200, 421–31.

Larsen, O. N. 2004. Does the environment constrain avian sound localization? *An. Acad. Bras. Ciênc.*, 76, 267–73.

Larsen, O. N. & Dabelsteen, T. 1990. Directionality of blackbird vocalization. Implications for vocal communication and its further study. *Ornis Scand.*, 21, 37–45.

Larsen, O. N. & Goller, F. 1999. Role of syringeal vibrations in bird vocalizations. *Proc. R. Soc. Lond*, B 266, 1609–15.

Larsen, O. N. & Goller, F. 2002. Direct observation of syringeal muscle function in songbirds and a parrot. *J. exp. Biol.*, 205, 25–35.

Latruffe, C., McGregor, P. K., Tavares, J. P. & Mota, P. G. 2000. Microgeographic variation in corn bunting (*Miliaria calandra*) song: quantitative and discrimination aspects. *Behaviour*, 137, 1241–55.

Lauay, C., Gerlach, N. M., Adkins-Regan, E. & DeVoogd, T. J. 2004. Female zebra finches require early song exposure to prefer high-quality song as adults. *Anim. Behav.*, 68, 1249–55.

Leader, N., Wright, J. & Yom-Tov, Y. 2000. Microgeographic song dialects in the orange-tufted sunbird (*Nectarinia osea*). *Behaviour*, 137, 1613–27.

Leader, N., Wright, J. & Yom-Tov, Y. 2005. Acoustic properties of two urban song dialects in the orange-tufted sunbird (*Nectarinia osea*). *Auk*, 122, 231–45.

Leger, D. W. & Mountjoy, D. J. 2003. Geographic variation in song of the bright-rumped attila (Tyrannidae: *Attila spadiceus*): implications for species status. *Auk*, 120, 69–74.

Lehtonen, L. 1983. The changing song patterns of the great tit *Parus major*. *Ornis Fenn.*, 60, 16–21.

Lein, M. R. 1972. Territorial and courtship songs of birds. *Nature*, 237, 48–9.

Lein, M. R. 1978. Song variation in a population of chestnut-sided warblers (*Dendroica pensylvanica*): Its nature and suggested significance. *Can. J. Zool.*, 56, 1266–83.

Leitão, A., van Dooren, T. J. M. & Riebel, K. 2004. Temporal variation in chaffinch *Fringilla coelebs* song: interrelations between the trill and flourish. *J. Avian Biol.*, 35, 199–203.

Leitner, S. & Catchpole, C. K. 2002. Female canaries that respond and discriminate more between male songs of different quality have a larger song control nucleus (HVC) in the brain. *J. Neurobiol.*, 52, 294–301.

Leitner, S. & Catchpole, C. K. 2004. Syllable repertoire and the size of the song control system in captive canaries (*Serinus canaria*). *J. Neurobiol.*, 60, 21–7.

Leitner, S., Marshall, R. C., Leisler, B. & Catchpole, C. K. 2006. Male song quality, egg size and offspring sex in captive canaries (*Serinus canaria*). *Ethology*, 112, 554–63.

Leitner, S., Nicholson, J., Leisler, B., DeVoogd, T. J. & Catchpole, C. K. 2002. Song and the song control pathway in the brain can develop independently of exposure to song in the sedge warbler. *Proc. R. Soc. Lond.*, B 269, 2519–24.

Leitner, S., Voigt, C., Garcia-Segura, L., Van't Hof, T. & Gahr, M. 2001. Seasonal activation of song motor memories in wild canaries is not reflected in neuroanatomical changes of forebrain song areas. *Horm. Behav.*, 40, 160–8.

Leitner, S., Voigt, C., Metzdorf, R. & Catchpole, C. K. 2005. Immediate early gene (ZENK, Arc) expression in the auditory forebrain of female canaries varies in response to male song quality. *J. Neurobiol.*, 64, 275–84.

Lemon, R. E. 1968. The relation between organization and function of song in cardinals. *Behaviour*, 32, 158–78.

Lemon, R. E. 1974. Song dialects, song matching and species recognition by cardinals *Richmondena cardinalis*. *Ibis*, 116, 545–8.

Lemon, R. E. 1975. How birds develop song dialects. *Condor*, 77, 385–406.

Lemon, R. E. & Chatfield, C. 1971. Organization of song in cardinals. *Anim. Behav.*, 19, 1–17.

Lemon, R. E., Cotter, R., MacNally, R. C. & Monette, S. 1985. Song repertoires and song sharing by American redstarts. *Condor*, 87, 457–70.

Lemon, R. E., Dobson, C. W. & Clifton, P. G. 1993. Songs of American redstarts (*Setophaga ruticilla*): sequencing rules and their relationships to repertoire size. *Ethology*, 93, 198–210.

Lemon, R. E., Falls, J. B., Dickinson, T., Perreault, S. & Tittler, R. 2000. Song clustering by meadowlarks: is it related to repertoire size? *Behaviour*, 137, 75–92.

Lemon, R. E., Fieldes, M. A. & Struger, J. 1981a. Testing the monotony threshold of bird song. *Z. Tierpsychol.*, 56, 359–79.

Lemon, R. E., Monette, S. & Roff, D. 1987. Song repertoires of American warblers (Parulinae): Honest advertising or assessment? *Ethology*, 74, 265–84.

Lemon, R. E., Perreault, S. & Weary, D. M. 1994. Dual strategies of song development in American redstarts (*Setophaga ruticilla*). *Anim. Behav.*, 47, 317–29.

Lemon, R. E., Struger, J., Lechowicz, M. J. & Norman, R. F. 1981b. Song features and singing heights of American warblers: maximization or optimization of distance? *J. Acoust. Soc. Amer.*, 69, 1169–76.

Leonard, M. L. & Horn, A. G. 2001. Begging calls and parental feeding decisions in tree swallows, *Tachycineta bicolor. Behav. Ecol. Sociobiol.*, 49, 170–5.

Leonard, M. L., Horn, A. G., Brown, C. R. & Fernandez, N. J. 1997. Parent-offspring recognition in tree swallows, *Tachycineta bicolor. Anim. Behav.*, 54, 1107–16.

Leonardo, A. & Konishi, M. 1999. Decrystallization of adult birdsong by perturbation of auditory feedback. *Nature*, 399, 466–70.

Leppelsack, H. J. & Vogt, M. 1976. Responses of auditory neurons in the forebrain of a songbird to stimulation with species-specific sounds. *J. comp. Physiol.*, 107, 263–74.

Levin, R. N. 1996a. Song behaviour and reproductive strategies in a duetting wren, *Thryothorus nigricapillus*. I. Removal experiments. *Anim. Behav.*, 52, 1093–106.

Levin, R. N. 1996b. Song behaviour and reproductive strategies in a duetting wren, *Thryothorus nigricapillus*. II. Playback experiments. *Anim. Behav.*, 52, 1107–17.

Lewis, B. & Coles, R. 1980. Sound localization in birds. *Trends Neurosci.*, 3, 102–5.

Li, X. C., Jarvis, E. D., Alvarez-Borda, B., Lim, D. A. & Nottebohm, F. 2000. A relationship between behavior, neurotrophin expression and new neuron survival. *Proc. Natl. Acad. Sci. USA*, 97, 8584–9.

Lind, H., Dabelsteen, T. & McGregor, P. K. 1996. Female great tits can identify mates by song. *Anim. Behav.*, 52, 667–71.

Lindell, C. 1998. Limited geographical variation in the vocalizations of a neotropical funariid, *Synallaxis albescens. Wilson Bull.*, 110, 368–74.

Liu, W.-C. 2004. The effects of neighbours and females on dawn and daytime singing behaviours by male chipping sparrows. *Anim. Behav.*, 68, 39–44.

Logan, C. A. 1983. Reproductively dependent song cyclicity in mated male mockingbirds *Mimus polyglottos. Auk*, 100, 404–13.

Logan, C. A., Hyatt, L. E. & Gregorcyk, L. 1990. Song playback initiates nest building during clutch overlap in mockingbirds, *Mimus polyglottos. Anim. Behav.*, 39, 943–53.

Logue, D. M. 2006. The duet code of the female back-bellied wren. *Condor*, 108, 326–35.

Logue, D. M. & Gammon, D. E. 2004. Duet song and sex roles during territory defence in a tropical

bird, the black-bellied wren, *Thryothorus fasciato-ventris*. *Anim. Behav.*, 68, 721–31.

Lougheed, S. C. & Handford, P. 1992. Vocal dialects and the structure of geographic variation in morphological and allozymic characters in the rufous-collared sparrow, *Zonotrichia capensis*. *Evolution*, 46, 1443–56.

Lougheed, S. C., Handford, P. & Baker, A. J. 1993. Mitochondrial DNA hyperdiversity, vocal dialects and subspecies in the rufous-collared sparrow (*Zonotrichia capensis*). *Condor*, 95, 889–95.

Lougheed, S. C., Lougheed, A. J., Rae, M. & Handford, P. 1989. Analysis of a dialect boundary in chaco vegetation in the rufous-collared sparrow. *Condor*, 91, 1002–05.

Lovell, S. F. & Lein, M. R. 2004. Neighbor-stranger discrimination by song in a suboscine bird, the alder flycatcher, *Empidonax alnorum*. *Behav. Ecol.*, 15, 799–804.

Lovell, S. F. & Lein, M. R. 2005 Individual recognition of neighbors by song in a suboscine bird, the alder flycatcher *Empidonax alnorum*. *Behav. Ecol. Sociobiol.*, 57, 623–30.

Luschi, P. 1993. Improvisation of new notes during singing by male Sardinian warblers. *Bioacoustics*, 4, 235–44.

Lynch, A. & Baker, A. J. 1986. Congruence of morphometric and cultural evolution in Atlantic island chaffinch populations. *Can. J. Zool.*, 64, 1576–80.

Lynch, A. & Baker, A. J. 1993. A population memetics approach to cultural evolution in chaffinch song: meme diversity within populations. *Am. Nat.*, 141, 597–620.

Lynch, A. & Baker, A. J. 1994. A population memetics approach to cultural evolution in chaffinch song: differentiation among populations. *Evolution*, 48, 351–9.

Lynch, A., Plunkett, G. M., Baker, A. J. & Jenkins, P. F. 1989. A model of cultural evolution of chaffinch song derived with the meme concept. *Am. Nat.*, 133, 634–53.

MacDonald, I. F., Kempster, B., Zanette, L. & MacDougall-Shackleton, S. A. 2006. Early nutritional stress impairs development of a song control brain region in both male and female juvenile song sparrows. *Proc. R. Soc. Lond.*, B 273, 2559–64.

MacDougall-Shackleton, E. & Harbison, H. 1998. Singing behavior of lekking green hermits. *Condor*, 100, 149–52.

MacDougall-Shackleton, S. A., Hulse, S. H. & Ball, G. F. 1998. Neural correlates of singing behaviour in male zebra finches *Taeniopygia guttata*. *J. Neurobiol.*, 36, 421–30.

MacDougall-Shackleton, E. & MacDougall-Shackleton, S. A. 2001. Cultural and genetic evolution in mountain white-crowned sparrows: song dialects are associated with population structure. *Evolution*, 55, 2568–75.

Mace, R. H. 1986. Importance of female behaviour in the dawn chorus. *Anim. Behav.*, 34, 621–2.

Mace, R. H. 1987a. The dawn chorus in the great tit *Parus major* is related to female fertility. *Nature*, 330, 745–6.

Mace, R. H. 1987b. Why do birds sing at dawn. *Ardea*, 75, 123–32.

Mace, R. H. 1989. The relationship between daily routines of singing and foraging: an experiment on captive great tits *Parus major*. *Ibis*, 131, 415–20.

MacNally, R. C. & Lemon, R. E. 1985. Repeat and serial singing modes in American redstarts (*Setophaga ruticilla*): a test of functional hypotheses. *Z. Tierpsychol.*, 69, 191–202.

Madden, J. R. & Davies, N. B. 2006. a host-race difference in begging calls of nestling cuckoos *Cuculus canorus* develops through experience and increases host provisioning. *Proc. R. Soc. Lond.*, B 273, 2343–51.

Malacarne, G., Cucco, M. & Camanni, S. 1991. Coordinated visual displays and vocal duetting in different ecological situations among Western Palearctic non-passerine birds. *Ethol. Ecol. Evol.*, 3, 207–19.

Mann, N. I., Dingess, K. A. & Slater, P. J. B. 2005. Antiphonal four-part synchronized chorusing in a Neotropical wren. *Biol. Lett.*, 2, 1–4.

Mann, N. I., Marshall-Ball, L. & Slater, P. J. B. 2003. The complex song duet of the plain wren. *Condor*, 105, 672–82.

Mann, N. I. & Slater, P. J. B. 1994. What causes young male zebra finches, *Taeniopygia guttata*, to choose their father as song tutor? *Anim. Behav.*, 47, 671–7.

Mann, N. I., Slater, P. J. B., Eales, L. A. & Richards, C. 1991. The influence of visual stimuli on song tutor choice in the zebra finch, *Taeniopygia guttata*. *Anim. Behav.*, 42, 285–93.

Margoliash, D. 1983. Acoustic parameters underlying the responses of song-specific neurons in the white-crowned sparrow. *J. Neurosci.*, 3, 1039–57.

Margoliash, D. 1986. Preference for autogenous song by auditory neurons in a song system nucleus of the white-crowned sparrow. *J. Neurosci.*, 6, 1643–61.

Margoliash, D. & Fortune, E. S. 1992. Temporal and harmonic combination-sensitive neurons in the zebra finch's HVC. *J. Neurosci.*, 12, 4309–26.

Marler, P. 1952. Variations in the song of the chaffinch, *Fringilla coelebs*. *Ibis*, 94, 458–72.

Marler, P. 1955. Characteristics of some animal calls. *Nature*, 176, 6–8.

Marler, P. 1960. Bird songs and mate selection. In: *Animal Sounds and Communication* (ed. W. E. Lanyon & W. N. Tavolga), pp. 348–67. Washington, D.C.: American Institute of Biological Sciences.

Marler, P. 1969. Tonal quality in bird sounds. In: *Bird Vocalizations* (ed. R. A. Hinde), pp. 5–18. Cambridge: Cambridge University Press.

Marler, P. 1970. A comparative approach to vocal learning: song development in white-crowned sparrows. *J. comp. physiol. Psychol.*, 71 (Suppl.), 1–25.

Marler, P. 1981. Birdsong: the acquisition of a learned motor skill. *Trends Neurosci.*, 3, 88–94.

Marler, P. 1987. Sensitive periods and the roles of specific and general sensory stimulation in birdsong learning. In: *Imprinting and Cortical Plasticity* (ed. J. P. Rauschecker & P. Marler), pp. 99–135. New York: Wiley.

Marler, P. 1990. Song learning: the interface between behaviour and neuroethology. *Phil. Trans. R. Soc. Lond.*, B, 329, 109–14.

Marler, P. 1991. Song-learning behavior: the interface with neuroethology. *Trends Neurosci.*, 14, 199–206.

Marler, P. 2004. Bird calls: a cornucopia for communication. In: *Nature's Music: The Science of Birdsong* (ed. P. Marler & H. Slabbekoorn), pp. 135–80. San Diego, CA: Academic Press.

Marler, P. & Nelson, D. 1992. Neuroselection and song learning in birds: species universals in culturally transmitted behavior. *Seminars in the Neurosciences*, 4, 415–23.

Marler, P. & Peters, S. 1977. Selective vocal learning in a sparrow. *Science*, 198, 519–21.

Marler, P. & Peters, S. 1981. Sparrows learn adult song and more from memory. *Science*, 213, 780–2.

Marler, P. & Peters, S. 1982a. Developmental overproduction and selective attrition: new processes in the epigenesis of bird song. *Dev. Psychobiol.*, 15, 369–78.

Marler, P. & Peters, S. 1982b. Structural change in song ontogeny in the swamp sparrow. *Auk*, 99, 446–58.

Marler, P. & Peters, S. 1987. A sensitive period for song acquisition in the song sparrow, *Melospiza melodia*: a case of age-limited learning. *Ethology*, 76, 89–100.

Marler, P. & Peters, S. 1988a. Sensitive periods for song acquisition from tape recordings and live tutors in the swamp sparrow, *Melospiza georgiana*. *Ethology*, 77, 76–84.

Marler, P. & Peters, S. 1988b. The role of song phonology and syntax in vocal learning preferences in the song sparrow, *Melospiza melodia*. *Ethology*, 77, 125–49.

Marler, P., Peters, S., Ball, G. F., Dufty, A. M. & Wingfield, J. C. 1988. The role of sex steroids in the acquisition and production of bird song. *Nature*, 336, 770–2.

Marler, P. & Pickert, R. 1984. Species-universal microstructure in a learned birdsong: the swamp sparrow (*Melospiza georgiana*). *Anim. Behav.*, 32, 673–89.

Marler, P. & Sherman, V. 1983. Song structure without auditory feedback: emendations of the auditory template hypothesis. *J. Neurosci.*, 3, 517–31.

Marler, P. & Tamura, M. 1964. Song "dialects" in three populations of white-crowned sparrows. *Science*, 146, 1483–6.

Marshall, R. C., Buchanan, K. L. & Catchpole, C. K. 2003. Sexual selection and individual genetic diversity in a songbird. *Proc. R. Soc. Lond.*, B 270 (Suppl) S248–S250.

Marshall, R. C., Buchanan, K. L. & Catchpole, C. K. 2007. Song and female choice for extra-pair copulations in the sedge warbler *Acrocephalus schoenobaenus*. *Anim. Behav.*, 73, 629–35.

Marshall, R. C., Leisler, B., Catchpole, C. K. & Schwabl, H. 2005. Male song quality affects circulating but not yolk steroid concentrations in female canaries *(Serinus canaria)*. *J. exp. Biol.*, 208, 4593–8.

Marshall-Ball, L., Mann, N. I. & Slater, P. J. B. 2006. Multiple functions to duet singing: hidden conflict and apparent cooperation. *Anim. Behav.*, 71, 823–31.

Marshall-Ball, L. & Slater, P. J. B. 2004. Duet singing and repertoire use in threat signaling of the individual and the pair. *Proc. R. Soc. Lond.*, B 271, S440–S443.

Marten, K. & Marler, P. 1977. Sound transmission and its significance for animal vocalization. I. Temperate habitats. *Behav. Ecol. Sociobiol.*, 2, 271–90.

Marten, K., Quine, D. & Marler, P. 1977. Sound transmission and its significance for animal vocalization. II. Tropical forest habitats. *Behav. Ecol. Sociobiol.*, 2, 291–302.

Martens, J. & Geduldig, G. 1990. *Acoustic adaptations of birds living close to Himalayan torrents*. In: *Proc. Int. 100 Meet. Deutschen Ornithologen-Gesellschaft*, pp. 123–131. Bonn.

Martens, J. & Kessler, P. 2000. Territorial song and song neighbourhoods in the scarlet rosefinch *Carpodacus erythrinus*. *J. Avian Biol.*, 31, 399–411.

Martens, M. J. M. 1980. Foliage as a low pass filter: experiments with model forests in an anechoic chamber. *J. Acoust. Soc. Am.*, 67, 66–72.

Martin, D. J. 1990. Songs of the fox sparrow. III. Ordering of song. *Wilson Bull.*, 102, 655–71.

Mathevon, N., Aubin, T. & Dabelsteen, T. 1996. Song degradation during propagation: importance of song post for the wren *Troglodytes troglodytes*. *Ethology*, 102, 397–412.

Matyjasiak, P. 2004 Birds associate species-specific acoustic and visual cues: recognition of heterospecific rivals by male blackcaps. *Behav. Ecol.*, 16, 467–71.

Maynard Smith, J. & Harper, D. 2003. *Animal Signals*. Oxford: Oxford University Press.

McCasland, J. S. 1987. Neuronal control of bird song production. *J. Neurosci.*, 7, 23–39.

McCracken, K. G. & Sheldon, F. H. 1997. Avian vocalizations and phylogenetic signal. *Proc. Natl. Acad. Sci. USA*, 94, 3833–6.

McDonald, M. V. 1989. Function of song in Scott's seaside sparrow. *Anim. Behav.*, 38, 468–85.

McGregor, P. K. 1980. Song dialects in the corn bunting (*Emberiza calandra*). *Z. Tierpsychol.*, 54, 285–97.

McGregor, P. K. 1983. The response of corn buntings to playback of dialects. *Z. Tierpsychol.*, 62, 256–60.

McGregor, P. K. 1986. Song types in the corn bunting *Emberiza calandra*: matching and discrimination. *J. Orn.*, 127, 37–42.

McGregor, P. K. 1988. Song length and 'male quality' in the chiffchaff. *Anim. Behav.*, 36, 606–8.

McGregor, P. K. 1989. Bird song and kin recognition: potential, constraints and evidence. *Ethol. Ecol. Evol.*, 1, 123–7.

McGregor, P. K. 1991. The singer and the song: on the receiving end of bird song. *Biol. Rev.*, 66, 57–81.

McGregor, P. K. (ed.) 1992a. *Playback and Studies of Animal Communication*. New York: Plenum Press.

McGregor, P. K. 1992b. Quantifying responses to playback: one, many, or composite multivariate measures. In: *Playback and Studies of Animal Communication* (ed. P. K. McGregor), pp. 79–96. New York: Plenum Press.

McGregor, P. K. 1993. Signalling in territorial systems: a context for individual identification, ranging and eavesdropping. *Phil. Trans. R. Soc. Lond.*, B 340, 237–44.

McGregor, P. K. (ed) 2005. *Animal Communication Networks*. Cambridge: Cambridge University Press.

McGregor, P. K. & Avery, M. I. 1986. The unsung songs of great tits *Parus major*: learning neighbours' songs for discrimination. *Behav. Ecol. Sociobiol.*, 18, 311–16.

McGregor, P. K., Catchpole, C. K., Dabelsteen, T. *et al.* 1992a. Design of playback experiments: the Thornbridge Hall NATO ARW consensus. In: *Playback and Studies of Animal Communication* (ed. P. K. McGregor), pp. 1–9. New York: Plenum Press.

McGregor, P. K. & Dabelsteen, T. 1996. Communication networks. In: *Ecology and Evolution of Acoustic Communication in Birds* (ed. D. E. Kroodsma & E. H. Miller), pp. 409–25. Ithaca, NY, and London: Comstock Publishing Associates.

McGregor, P. K., Dabelsteen, T., Shepherd, M. & Pedersen, S. B. 1992b. The signal value of matched singing in great tits: evidence from interactive playback experiments. *Anim. Behav.*, 43, 987–98.

McGregor, P. K. & Falls, J. B. 1984. The response of western meadowlarks (*Sturnella neglecta*) to the playback of undegraded and degraded songs. *Can. J. Zool.*, 62, 2125–28.

McGregor, P. K. & Krebs, J. R. 1982a. Song types in a population of great tits (*Parus major*): their distribution, abundance and acquisition by individuals. *Behaviour*, 79, 126–52.

McGregor, P. K. & Krebs, J. R. 1982b. Mating and song types in the great tit. *Nature*, 297, 60–1.

McGregor, P. K. & Krebs, J. R. 1984a. Song learning and deceptive mimicry. *Anim. Behav.*, 32, 280–7.

McGregor, P. K. & Krebs, J. R. 1984b. Sound degradation as a distance cue in great tit (*Parus major*) song. *Behav. Ecol. Sociobiol.*, 16, 49–56.

McGregor, P. K., Krebs, J. R. & Ratcliffe, L. M. 1983. The reaction of great tits (*Parus major*) to playback of degraded and undegraded songs: the effect of familiarity with the stimulus song type. *Auk*, 100, 898–906.

McGregor, P. K. & Thompson, D. B. A. 1988. Constancy and change in local dialects of the corn bunting. *Ornis Scand.*, 19, 153–9.

McGregor, P. K., Walford, V. R. & Harper, D. G. C. 1988. Song inheritance and mating in a songbird with local dialects. *Bioacoustics*, 1, 107–29.

McGregor, P. K., Krebs, J. R. & Perrins, C. M. 1981. Song repertoires and lifetime reproductive success in the great tit *Parus major*. *Am. Nat.*, 118, 149–59.

McNamara, J. M., Mace, R. H. & Houston, A.I. 1987. Optimal daily routines of singing and foraging in a bird singing to attract a mate. *Behav. Ecol. Sociobiol.*, 20, 399–405.

Medvin, M. B. & Beecher, M. D. 1986. Parent-offspring recognition in the barn swallow. *Anim. Behav.*, 34, 1627–39.

Mello, C. V., Nottebohm, F. & Clayton, D. 1995. repeated exposure to one song leads to a rapid and persistent decline in an immediate early gene's response to that song in the zebra finch telencephalon. *J. Neurosci.*, 15, 6919–25.

Mello, C. V., Vicario, D. S. & Clayton, D. F. 1992. Song presentation induces gene expression in the songbird's forebrain. *Proc. Natl. Acad. Sci. USA*, 89, 6818–22.

Mennill, D. J. 2006. Aggressive responses of male and female rufous-and-white wrens to stereo duet playback. *Anim. Behav.*, 71, 219–26.

Mennill, D. J., Boag, P. T. & Ratcliffe, L. M. 2003. The reproductive choices of eavesdropping female black-capped chickadees, *Poecile atricapillus*. *Naturwissenschaften*, 90, 577–82.

Mennill, D. J., Ratcliffe, L. M. & Boag, P. T. 2002. Female eavesdropping on male song contests in songbirds. *Science*, 296, 873.

Mennill, D. J. & Rogers, A. C. 2006. Whip it good! Geographic consistency in male songs and variability in female songs of the duetting eastern whipbird *Psophodes olicaceus*. *J. Avian Biol.*, 37, 93–100.

Michelsen, A. 1983. Biophysical basis of sound communication. In: *Bioacoustics: A Comparative Approach* (ed. B. Lewis), pp. 3–38. London: Academic Press.

Miller, D. B. 1979a. Long-term recognition of father's song by female zebra finches. *Nature*, 280, 89–91.

Miller, D. B. 1979b. The acoustic basis of mate recognition by female zebra finches (*Taeniogpygia guttata*). *Anim. Behav.* 27, 376–80.

Miller, E. H. 1982. Character and variance shift in acoustic signals of birds. In: *Acoustic Communication in Birds*, Vol. 1 (ed. D. E. Kroodsma & E. H. Miller), pp. 253–95. New York: Academic Press.

Milligan, M. M. & Verner, J. 1971. Interpopulational song discrimination in the white-crowned sparrow. *Condor*, 73, 208–13.

Millington, S. J. & Price, T. D. 1985. Song inheritance and mating patterns in Darwin's finches. *Auk*, 102, 342–6.

Møller, A. P. 1986. On song post selection and the timing of song in the corn bunting (*Miliaria calandra*). *Ökol. Vögel*, 8, 57–66.

Møller, A. P. 1988. Spatial and temporal distribution of song in the yellowhammer *Emberiza cintrinella*. *Ethology*, 78, 321–31.

Møller, A. P. 1991. Why mated songbirds sing so much: mate guarding and male announcement of mate fertility status. *Am. Nat.*, 138, 994–1014.

Møller, A. P., Henry, P. Y. & Erritzoe, J. 2000. The evolution of song repertoires and immune defence in birds. *Proc. R. Soc. Lond.*, B 267, 165–9.

Møller, A. P., Saino, N., Taramino, G., Galeotti, P. & Ferrario, S. 1998. Paternity and multiple signalling: effects of a secondary sexual character and song on paternity in the barn swallow. *Am. Nat.* 151, 236–42.

Molles, L. E. & Vehrencamp, S. L. 1999. Repertoire size, repertoire overlap, and singing modes in the banded wren (*Thryothorus pleurostictus*). *Auk*, 116, 677–89.

Molles, L. E. & Vehrencamp, S. L. 2001 Neighbour recognition by resident males in the banded wren, *Thryothorus pleurostictus*, a tropical song bird with high song type sharing. *Anim. Behav.* 61, 119–27.

Molles, L. E. & Waas, J. R. 2006. Are two heads better than one? Responses of the duetting kokako to one- and two-speaker playback. *Anim. Behav.*, 71, 131–8.

Morrison, M. L. & Hardy, J. W. 1983. Vocalizations of the black-throated grey warbler. *Wilson Bull.*, 95, 640–3.

Morse, D. H. 1966. The context of songs in the yellow warbler. *Wilson Bull.*, 78, 444–55.

Morse, D. H. 1970. Territorial and courtship songs of birds. *Nature*, 226, 659–61.

Morse, D. H. 1989. Song patterns of warblers at dawn and dusk. *Wilson Bull.*, 101, 26–35.

Morton, E. S. 1975. Ecological sources of selection on avian sounds. *Am. Nat.*, 109, 17–34.

Morton, E. S. 1986. Predictions from the ranging hypothesis for the evolution of long distance signals in birds. *Behaviour*, 99, 65–86.

Morton, E. S. 1987. The effects of distance and isolation on song-type sharing in the Carolina wren. *Wilson Bull.*, 99, 601–10.

Morton, E. S. 1996. Why songbirds learn songs: an arms race over ranging? *Poultry & Avian Biology Reviews*, 7, 65–71.

Morton, E. S. 1998a. Degradation and signal ranging in birds: memory matters. *Behav. Ecol. Sociobiol.*, 42, 135–7.

Morton, E. S. 1998b. Reply to Naguib and Wiley. *Behav. Ecol. Sociobiol.*, 42, 147–8.

Morton, E. S. & Derrickson, K. C. 1996. Song ranging by the dusky antbird, *Cercomacra tyrannina* – ranging without song learning. *Behav. Ecol. Sociobiol.*, 39, 195–201.

Morton, E. S., Gish, S. L. & van der Voort, M. 1986. On the learning of degraded and undegraded songs in the Carolina wren. *Anim. Behav.*, 34, 815–820.

Mountjoy, D. J. & Lemon, R. E. 1991. Song as an attractant for male and female European starlings, and the influence of song complexity on their response. *Behav. Ecol. Sociobiol.*, 28, 97–100.

Mountjoy, D. J. & Lemon, R. E. 1995. Extended song learning in wild European starlings. *Anim. Behav.*, 49, 357–66.

Mountjoy, D. J. & Lemon, R. E. 1996. Female choice for complex song in the European starling: a field experiment. *Behav. Ecol. Sociobiol.*, 38, 65–71.

Mulder, R. A., Bishop, H., Cooper, M., Dennis, S., Koetsveld, M., Marshall, J., Saunders, B. L. & Langmore, N. E. 2003. Alternate functions for duet and solo songs in magpie-larks, *Grallina cyanoleuca*. *Aust. J. Zool.*, 51, 25–30.

Mundinger, P. C. 1975. Song dialects and colonization in the house finch, *Carpodacus mexicanus*, on the east coast. *Condor*, 77, 407–22.

Mundinger, P. C. 1982. Microgeographic and macrogeographic variation in the acquired vocalizations of birds. In: *Acoustic Communication in Birds* (ed. D. E. Kroodsma & E. H. Miller), pp. 147–208. New York: Academic Press.

Mundry, R., Hau, B. & Böhner, J. 1994. Individual- und strophenspezifischer Einsatz des "kit" im Gesang des Buchfinken (*Fringilla coelebs*). *J. Orn.*, 135, 223–31.

Naguib, M. 1996. Ranging by song in Carolina wrens *Thyothorus ludovicianus*: effects of environmental acoustics and strength of song degradation. *Behaviour*, 133, 541–59.

Naguib, M. 1997. Ranging of songs in Carolina wrens: effects of familiarity of the song type on use of different cues. *Behav. Ecol. Sociobiol.*, 40, 385–93.

Naguib, M. 1998. Perception of degradation in acoustic signals and its implications for ranging. *Behav. Ecol. Sociobiol.*, 42, 139–42.

Naguib, M. 1999. Effects of song overlapping and alternating on nocturnally singing nightingales. *Anim. Behav.*, 58, 1061–7.

Naguib, M., Altenkamp, R. & Griessmann, B. 2001. Nightingales in space: song and extra-territorial forays of radio tagged song birds. *J. Orn.*, 142, 306–12.

Naguib, M. & Kipper, S. 2006. Effects of different levels of song overlapping on singing behaviour in male territorial nightingales (*Luscinia megarhynchos*). *Behav. Ecol. Sociobiol.*, 59, 419–26.

Naguib, M. & Todt, D. 1997. Effects of dyadic vocal interactions on other conspecific receivers in nightingales. *Anim. Behav.*, 54, 1535–43.

Naugler, C. T. & Ratcliffe, L. 1994. Character release in bird song: a test of the acoustic competition hypothesis using American tree sparrows *Spizella arborea*. *J. Avian Biol.*, 25, 142–8.

Naugler, C. T. & Smith, P. C. 1991. Song similarity in an isolated population of fox sparrows (*Passerella iliaca*). *Condor*, 93, 1001–3.

Nelson, D. A. 1988. Feature weighting in species song recognition by the field sparrow *Spizella pusilla*. *Behaviour*, 106, 158–82.

Nelson, D. A. 1989. The importance of invariant and distinctive features in species recognition of bird song. *Condor*, 91, 120–30.

Nelson, D. A. 1992. Song overproduction and selective attrition lead to song sharing in the field sparrow (*Spizella pusilla*). *Behav. Ecol. Sociobiol.*, 30, 415–424.

Nelson, D. A. 1997. Social interaction and sensitive phases for song learning: a critical review. In: *Social Influences on Vocal Development* (ed. C. T. Snowdon & M. Hausberger), pp. 7–22. Cambridge: Cambridge University Press.

Nelson, D. A. 1998. External validity and experimental design: the sensitive phase for song learning. *Anim. Behav.*, 56, 487–91.

Nelson, D. A., Khanna, H. & Marler, P. 2001. Learning by instruction or selection: Implications for patterns of geographic variation in bird song. *Behaviour*, 138, 1137–60.

Nelson, D. A. & Marler, P. 1990. The perception of bird song and an ecological concept of signal space. In: *Comparative Perception*, Vol. 2. (ed. W. C. Stebbins & M. A. Berkley), pp. 443–78. New York: John Wiley.

Nelson, D. A. & Marler, P. 1993. Innate recognition of song in white-crowned sparrows: a role in selective vocal learning? *Anim. Behav.*, 46, 806–8.

Nelson, D. A., Marler, P. & Morton, M. L. 1996. Overproduction in song development: an evolutionary correlate with migration. *Anim. Behav.*, 51, 1127–40.

Nelson, D. A., Marler, P. & Palleroni, A. 1995. A comparative approach to vocal learning: intraspecific variation in the learning process. *Anim. Behav.*, 50, 83–97.

Nelson, D. A., Marler, P., Soha, J. A. & Fullerton, A. L. 1997. The timing of song memorization differs in males and females: a new assay for avian vocal learning. *Anim. Behav.*, 54, 587–97.

Nicholls, J. A. & Goldizen, A. W. 2006. Habitat type and density influence vocal signal design in satin bowerbirds. *J. Anim. Ecol.*, 75. (In press.)

Nicholson, J., Buchanan, K. L., Marshall, R. C. & Catchpole, C. K. 2007. Song sharing and repertoire size in the sedge warbler *Acrocephalus schoenobaenus*: changes within and between years. *Anim. Behav.* (In press.)

Nicolai, J. 1959. Familientradition in der Gesangsentwicklung des Gimpels (*Pyrrhula pyrrhula* L.). *J. Orn.*, 100, 39–46.

Nolan, P. M. & Hill, G. E. 2004. Female choice for song characteristics in the house finch. *Anim. Behav.* 67, 403–10.

Nordby, J. C., Campbell, S. E. & Beecher, M. D. 1999. Ecological correlates of song learning in song sparrows. *Behav. Ecol.*, 10, 287–97.

Nordby, L. C., Campbell, S. E. & Beecher, M. D. 2001. Late song learning in song sparrows. *Anim. Behav.*, 61, 835–46.

Nordby, J. C., Campbell, S. E. & Beecher, M. D. 2002. Adult song sparrows do not alter their song repertoires. *Ethology*, 108, 39–50.

Nordeen, K. W. & Nordeen, E. J. 1988. Projection neurons within a vocal motor pathway are born during song learning in zebra finches. *Nature*, 334, 149–51.

Nordeen, K. W. & Nordeen, E. J. 1992. Auditory feedback is necessary for the maintenance of stereotyped song in adult zebra finches. *Behav. Neur. Biol.*, 57, 58–66.

Nottebohm, F. 1968. Auditory experience and song development in the chaffinch *Fringilla coelebs*. *Ibis*, 110, 549–68.

Nottebohm, F. 1969a. The "critical period" for song learning in birds. *Ibis*, 111, 386–7.

Nottebohm, F. 1969b. The song of the chingolo (*Zonotrichia capensis*) in Argentina: description and evaluation of a system of dialects. *Condor*, 71, 299–315.

Nottebohm, F. 1971. Neural lateralization of vocal control in a passerine bird. 1 Song. *J. exp. Zool.*, 177, 229–62.

Nottebohm, F. 1972. The origins of vocal learning. *Am. Nat.*, 106, 116–40.

Nottebohm, F. 1975. Continental patterns of song variability in *Zonotrichia capensis*: some possible ecological correlates. *Am. Nat.*, 109, 605–24.

Nottebohm, F. 1991. Reassessing the mechanisms and origins of vocal learning in birds. *Trends Neurosci.*, 14, 206–11.

Nottebohm, F. 2000. The anatomy and timing of vocal learning in birds. In *The Design of Animal Communication* (ed. M. D. Hauser & M. Konishi), pp. 63–110. Cambridge, MA: MIT Press.

Nottebohm, F. 2002. Why are some neurons replaced in the adult brain? *J. Neurosci.*, 22, 624–28.

Nottebohm, F. 2005. The neural basis of birdsong. *PLoS Biol.*, 3(5), e164.

Nottebohm, F., Alvarez-Buylla, A., Cynx, J., *et al.* 1990. Song learning in birds: the relation between perception and production. *Phil. Trans. R. Soc. Lond.*, B 329, 115–24.

Nottebohm, F. & Arnold, A.P. 1976. Sexual dimorphism in vocal control areas of the songbird brain. *Science*, 194, 211–13.

Nottebohm, F., Kasparian, S. & Pandazis, C. 1981. Brain space for a learned task. *Brain Research*, 213, 99–109.

Nottebohm, F. & Nottebohm, M. E. 1971. Vocalizations and breeding behaviour of surgically deafened ring doves. *Anim. Behav.*, 19, 313–27.

Nottebohm, F. & Nottebohm, M.E. 1976. Left hypoglossal dominance in the control of canary and white-crowned sparrow song. *J. comp. Physiol.*, 108, 171–92.

Nottebohm, F. & Nottebohm, M. E. 1978. Relationship between song repertoire and age in the canary *Serinus canaria*. *Z. Tierpsychol.*, 46, 298–305.

Nottebohm, F., Nottebohm, M. E. & Crane, L. 1986. Developmental and seasonal changes in canary song and their relation to changes in the anatomy of song control nuclei. *Behav. Neural Biol.*, 46, 445–71.

Nottebohm, F., Nottebohm, M. E., Crane, L. A. & Wingfield, J. C. 1987. Seasonal changes in gonadal hormone levels of adult male canaries and their relation to song. *Behav. Neural Biol.*, 47, 197–211.

Nottebohm, F., Stokes, T. M. & Leonard, C. M. 1976. Central control of song in the canary *Serinus canarius*. *J. comp. Neurol.*, 165, 457–86.

Nowicki, S. 1987. Vocal tract resonances in oscine bird sound production: evidence from bird songs in a helium atmosphere. *Nature*, 325, 53–5.

Nowicki, S. & Marler, P. 1988. How do birds sing? *Music Perception*, 5, 391–426.

Nowicki, S., Marler, P., Maynard, A. & Peters, S. 1992. Is the tonal quality of bird song learned? Evidence from song sparrows. *Ethology*, 90, 225–35.

Nowicki, S., Peters, S. & Podos, J. 1998a. Song learning, early nutrition and sexual selection in songbirds. *Amer. Zool.*, 38, 179–90.

Nowicki, S., Peters, S., Searcy, W. A. & Clayton, C. 1999. The development of within-song type variation in song sparrows. *Anim. Behav.*, 57, 1257–64.

Nowicki, S. & Searcy, W. A. 2005. Song and mate choice in birds: how the development of behavior helps us understand function. *Auk*, 122, 1–14.

Nowicki, S., Searcy, W. A. & Hughes, M. 1998b. The territory defense function of song in song sparrows: a test with the speaker occupation design. *Behaviour*, 135, 615–28.

Nowicki, S., Searcy, W. A. & Peters, S. 2002. Brain development, song learning and mate choice in birds: a review and experimental test of the "nutritional stress hypothesis". *J. comp. Physiol. A*, 188, 1003–14.

Nuechterlein, G. L. 1981. Variations and multiple functions of the advertising display of western grebes. *Behaviour*, 76, 289–317.

Nürnberger, F., Siebold, D. & Bergmann, H.-H. 1989. Annual changes of learned behaviour – variation of song pattern in free-living chaffinches, *Fringilla coelebs*, during the breeding season. *Bioacoustics*, 1, 273–86.

Oberweger, K. & Goller, F. 2001. The metabolic cost of birdsong production. *J. exp. Biol.*, 204, 3379–85.

Okanoya, K. & Dooling, R. J. 1988. Hearing in the swamp sparrow *Melospiza georgiana*, and the song sparrow *Melospiza melodia*. *Anim. Behav.*, 36, 726–32.

Ornelas, J. F., Gonzalez, C. & Uribe, J. 2002. Complex vocalizations and aerial displays of the amethyst-throated hummingbird (*Lampornis amethystinus*). *Auk*, 119, 1141–9.

Otter, K., Chruszcz, B. & Ratcliffe, L. 1997. Honest advertisement and song output during the dawn chorus of black-capped chickadees. *Behav. Ecol.*, 8, 167–73.

Otter, K., McGregor, P. K., Terry, A. M. R., *et al.* 1999. Do female great tits (*Parus major*) assess males by eavesdropping? A field study using interactive song playback. *Proc. R. Soc. Lond.*, B 266, 1305–9.

Otter, K., Stewart, I. R., McGregor, P. K. *et al.* 2001. Extra-pair paternity among great tits *Parus major* following manipulation of male signals. *J. Avian Biol.*, 32, 338–44.

Part, T. 1991. Is dawn singing related to paternity insurance? The case of the collared flycatcher. *Anim. Behav.*, 41, 451–6.

Paton, J. A. & Nottebohm, F. 1984. Neurons generated in the adult brain are recruited into functional circuits. *Science*, 225, 1046–8.

Patricelli, G. L., Dantzker, M. S. & Bradbury, J. W. 2007. Differences in acoustic directionality among vocalizations of the male red-winged blackbird (*Agelaius pheoniceus*) are related to function in communication. *Behav. Ecol. Sociobiol.* 69, 1099–110.

Payne, R. B. 1978a. Local dialects in the wingflaps of flappet larks *Mirafra rufocinnamomea*. *Ibis*, 120, 204–7.

Payne, R. B. 1978b. Microgeographic variation in songs of splendid sunbirds *Nectarinia coccinigaster*; population phenetics, habitats and song dialects. *Behaviour*, 65, 282–308.

Payne, R. B. 1981a. Population structure and social behavior: models for testing the ecological significance of song dialects in birds. In: *Natural Selection and Social Behavior: Recent Research and New Theory* (ed. R. D. Alexander & D. W. Tinkle), pp. 108–20. New York: Chiron Press.

Payne, R. B. 1981b. Song learning and social interaction in indigo buntings. *Anim. Behav.*, 29, 688–97.

Payne, R. B. 1982. Ecological consequences of song matching: breeding success and intraspecific song mimicry in indigo buntings. *Ecology*, 63, 401–11.

Payne, R. B. 1983. The social context of song mimicry: song-matching dialects in indigo buntings (*Passerina cyanea*). *Anim. Behav.*, 31, 788–805.

Payne, R. B. 1985. Behavioral continuity and change in local song populations of village indigobirds *Vidua chalybeata*. *Z. Tierpsychol.*, 70, 1–44.

Payne, R. B. 1986. Bird songs and avian systematics. *Curr. Ornithol.*, 87–126.

Payne, R. B. 1987. Song dialects and neighborhood habitats in the indigobirds *Vidua chalybeata* and *V. purpurascens* at Lochinvar National Park, Zambia. *J. Field Ornithol.*, 58, 152–70.

Payne, R. B. & Groschupf, K. D. 1984. Sexual selection and interspecific competition: a field experiment on territorial behavior of nonparental finches (*Vidua* spp.). *Auk*, 101, 140–5.

Payne, R. B. & Payne, K. 1977. Social organization and mating success in local song populations of village indigobirds *Vidua chalybeata*. *Z. Tierpsychol.*, 45, 133–73.

Payne, R. B. & Payne, L. L. 1993. Song copying and cultural transmission in indigo buntings. *Anim. Behav.*, 46, 1045–65.

Payne, R. B., Payne, L. L. & Doehlert, S. M. 1988a. Biological and cultural success of song memes in indigo buntings. *Ecology*, 69, 104–17.

Payne, R. B., Payne, L, & Rowley, I. 1988b. Kin and social relationships in splendid fairy wrens:

recognition by song in a cooperative bird. *Anim. Behav.*, 36, 1341–51.

Payne, R. B., Payne, L. L. & Woods, J. L. 1998. Song learning in brood parasitic indigobirds *Vidua chalybeata*: song mimicry of the host species. *Anim. Behav.*, 55, 1537–53.

Payne, R. B., Payne, L. L., Woods, J. L. & Sorenson, M. D. 2000. Imprinting and the origin of parasite-host species associations in brood parasitic indigobirds, *Vidua chalybeata. Anim. Behav.*, 59, 69–81.

Payne, R. B., Thompson, W. L., Fiala, K. L. & Sweany, L. L. 1981. Local song traditions in indigo buntings: cultural transmission of behavior patterns across generations. *Behaviour*, 77, 199–221.

Peake, T. M. 2005. Eavesdropping in communication networks. In *Animal Communication Networks* (ed. P. K. McGregor), pp. 13–37. Cambridge: Cambridge University Press.

Peake, T. M., Otter, K. A., Terry, A. M. R. & McGregor, P. K. 2000. Screech: an interactive playback program for PCs. *Bioacoustics*, 11, 69–75.

Peake, T. M., Terry, A. M., McGregor, P. K. & Dabelsteen, T. 2001. Male great tits eavesdrop on simulated male-to-male vocal interactions. *Proc. R. Soc. Lond.*, B. 268, 1183–7.

Peek, F.W. 1972. An experimental study of the territorial function of vocal and visual display in the male red-winged blackbird. *Anim. Behav.*, 20, 112–18.

Pepperberg, I. M. 1981. Functional vocalizations by an African grey parrot. *Z. Tierpsychol.*, 55, 139–60.

Pepperberg, I. M. 1985. Social modeling theory: a possible framework for understanding avian vocal learning. *Auk*, 102, 854–64.

Pepperberg, I. M. 1990. Learning to communicate: the effects of social interaction. In: *Perspectives in*

Ethology, Volume 9 (ed. P. Bateson & P. H. Klopfer), pp. 119–64. New York: Plenum.

Pepperberg, I. M. 1994. Vocal learning in gray parrots (*Psittacus erithacus*): effects of social interaction, reference, and context. *Auk*, 111, 300–13.

Peters, S., Marler, P. & Nowicki, S. 1992. Song sparrows learn from limited exposure to song models. *Condor*, 94, 1016–19.

Peters, S., Searcy, W. A., Beecher, M. D. & Nowicki, S. 2000. Geographic variation in the organization of song sparrow repertoires. *Auk*, 117, 936–42.

Petrinovich, L. 1985. Factors influencing song development in the white-crowned sparrow (*Zonotrichia leucophrys*). *J. comp. Psychol.*, 99, 15–29.

Petrinovich, L. & Baptista, L. F. 1987. Song development in the white-crowned sparrow: modification of learned song. *Anim. Behav.*, 35, 961–74.

Petrinovich, L. & Patterson, T. L. 1981. The responses of white-crowned sparrows to songs of different dialects and subspecies. *Z. Tierpsychol.*, 57, 1–14.

Petrinovich, L., Patterson, T. & Baptista, L. 1981. Song dialects as barriers to dispersal: a re-evaluation. *Evolution*, 35, 180–88.

Pinxten, R. & Eens, M. 1998. Male starlings sing most in the late morning, following egg-laying: a strategy to protect their paternity? *Behaviour*, 135, 1197–211.

Planck, R. J., McLaren, G. & Konishi, M. 1975. Temporal pattern shifts in singing birds: a critique. *Science*, 190, 292.

Podos, J. 1996. Motor constraints on vocal development in a songbird. *Anim. Behav.*, 51, 1061–70.

Podos, J. 1997. A performance constraint on the evolution of trilled vocalizations in a songbird family (Passeriformes: Emberizidae). *Evolution*, 5, 537–51.

Podos, J. 2001. Correlated evolution of morphology and vocal signature structure in Darwin's finches. *Nature*, 409, 185–8.

Podos, J. & Nowicki, S. 2004. Performance limits on bird song. In *Nature's Music: The Science of Birdsong* (ed. P. Marler & H. Slabbekoorn), pp. 322–46. San Diego, CA: Academic Press.

Poesel, A. & Dabelsteen, T. 2005. Territorial responses of male blue tits to simulated dynamic intrusions: effects of song overlap and intruder location. *Anim. Behav.*, 70, 1419–27.

Poesel, A., Foerster, K. & Kempenaers, B. 2001. The dawn song of the blue tit *Parus caeruleus* and its role in sexual selection. *Ethology*, 107, 521–31.

Poesel, A., Kunc, H., Foerster, K., Johnsen, A. & Kempenaers, B. 2006. Early birds are sexy: male age, dawn song and extrapair paternity in blue tits, *Cyanistes* (formerly *Parus*) *caeruleus*. *Anim. Behav.*, 72, 531–8.

Popp, J. W. & Ficken, R.W. 1987. Effects of nonspecific singing on the song of the ovenbird. *Bird Behav.*, 7, 22–6.

Popp, J. W., Ficken, R. W. & Reinartz, J.A. 1985. Short-term temporal avoidance of interspecific acoustic interference among forest birds. *Auk*, 102, 744–8.

Potash, L. M. 1972. Noise-induced changes in calls of the Japanese quail. *Psychon. Sci.*, 26, 252–4.

Prescott, D. R.C. 1987. Territorial responses to song playback in allopatric and sympatric populations of alder *Empidonax alnorum* and willow *E. traillii* flycatchers. *Wilson Bull.*, 99, 611–19.

Price, J.J. 1999. Recognition of family-specific calls in stripe-backed wrens. *Anim. Behav.*, 57, 483–92.

Price, J. J. & Lanyon, S. M. 2002. Reconstructing the evolution of complex bird song in the oropendolas. *Evolution*, 56, 1514–30.

Price, J. J. & Lanyon, S. M. 2004. Patterns of song evolution and sexual selection in the oropendolas and caciques. *Behav. Ecol.*, 15, 485–97.

Putland, D. A., Nicholls, J. A., Noad, M. J. & Goldizen, A. W. 2006. Imitating the neighbours: vocal dialect matching in a mimic-model system. *Biol. Lett.*, 2, 367–70.

Pytte, C. L., Rusch, K. M. & Ficken, M. S. 2003. Regulation of vocal amplitude by the blue-throated hummingbird, *Lampornis clemenciae*. *Anim. Behav.*, 66, 703–10.

Radesater, T. & Jakobsson, S. 1989. Song rate correlation of replacement territorial willow warblers *Phylloscopus trochilus*. *Ornis. Scand.*, 20, 71–3.

Radesater, T., Jakobsson, S., Andbjer, N., Bylin, A. & Nystrom, K. 1987. Song rate and pair formation in the willow warbler *Phylloscopus trochilus*. *Anim. Behav.*, 35, 1645–51.

Rasmussen, R. & Dabelsteen, T. 2002. Song repertoires and repertoire sharing in a local group of blackbirds. *Bioacoustics*, 13, 63–76.

Ratcliffe, L. & Weisman, R. G. 1986. Song sequence discrimination in the black-capped chickadee *Parus atricapillus*. *J. comp. Psychol.*, 100, 361–367.

Read, A. F. & Weary, D.M. 1990. Sexual selection and the evolution of bird song: A test of the Hamilton-Zuk hypothesis. *Behav. Ecol. Sociobiol.*, 26, 47–56.

Read, A. F. & Weary, D. M. 1992. The evolution of bird song: comparative analyses. *Phil. Trans. Roy. Soc. Lond.*, B 338, 165–87.

Rechten, C. 1978. Interspecific mimicry in bird song: does the Beau Geste hypothesis apply? *Anim. Behav.*, 26, 305.

Reed, T.M. 1982. Interspecific territoriality in the chaffinch and the great tit on islands and the mainland of Scotland. Anim. *Behav.*, 30, 171–81.

Reid, J. M., Arcese, P., Cassidy, A. L. E. V. *et al.* 2004. Song repertoire size predicts initial mating success in male song sparrows, *Melospiza melodia*. *Anim. Behav.*, 68, 1055–63.

Reid, J. M., Arcese, P., Cassidy, A. L. E. V. *et al.* 2005. Hamilton and Zuk meet heterozygosity?

Song repertoire size indicates inbreeding and immunity in song sparrows (*Melospiza melodia*). *Proc. R. Soc. Lond.*, B 272, 481–7.

Reid, M. L. 1987. Costliness and reliability in the singing vigour of Ipswich sparrows. *Anim. Behav.*, 35, 1735–43.

Rich, T. 1981. Microgeographic variation in the song of the sage sparrow. *Condor*, 83, 113–19.

Richards, D. G. 1981a. Alerting and message components in songs of rufous-sided towhees. *Behaviour*, 76, 223–49.

Richards, D. G. 1981b. Estimation of distance of singing conspecifics by the Carolina wren. *Auk*, 98, 127–33.

Richards, D. G. & Wiley, R. H. 1980. Reverberations and amplitude fluctuations in the propagation of sound in a forest: implications for animal communication. *Am. Nat.*, 115, 381–99.

Riebel, K. 2000. Early exposure leads to repeatable preferences for male song in female zebra finches. *Proc. R. Soc. Lond.*, B 267, 2553–8.

Riebel, K. 2003a. Developmental influences on auditory perception in female zebra finches – is there a sensitive phase for song preference learning? *Anim. Biol.*, 53, 73–87.

Riebel, K. 2003b. The 'mute' sex revisited: vocal production and perception learning in female songbirds. *Adv. Study Behav.*, 33, 49–86.

Riebel, K. & Slater, P. J. B. 1999a. Do male chaffinches *Fringilla coelebs* copy song sequencing and bout length from their tutors. *Ibis*, 141, 680–83.

Riebel, K. & Slater, P. J. B. 1999b. Song type switching in the chaffinch, *Fringilla coelebs*: timing or counting. *Anim. Behav.*, 57, 655–61.

Riebel, K. & Slater, P. J. B. 2003. Temporal variation in male chaffinch song depends on the singer and the song type. *Behaviour*, 140, 269–88.

Riebel, K., Smallegange, I. M., Terpstra, N. J. & Bolhuis, J. J. 2002. Sexual equality in zebra finch song preference: evidence for a dissociation

between song recognition and production learning. *Proc. R. Soc. Lond.*, B 269, 729–33.

Ritchison, G. 1981. Variation in the songs of vesper sparrows. *Am. Midl. Nat.*, 106, 392–8.

Ritchison, G. 1983a. The function of singing in female black-headed grosbeaks *(Pheucticus melanocephalus)*: family group maintenance. *Auk*, 100, 105–16.

Ritchison, G. 1983b. Possible 'deceptive' use of song by female black-headed grosbeaks. *Condor*, 85, 250–1.

Ritchison, G. 1985. Variation in the songs of female black-headed grosbeaks. *Wilson Bull.*, 97, 47–56.

Rivers, J. W. & Kroodsma, D. E. 2000. Singing behavior of the hermit thrush. *J. Field Orn.*, 71, 467–71.

Robinson, A. 1949. The biological significance of bird song in Australia. *Emu*, 48, 291–315.

Robinson, F. N. 1974. The function of vocal mimicry in some avian displays. *Emu*, 74, 9–10.

Robinson, F. N. 1975. Vocal mimicry and the evolution of bird song. *Emu*, 75, 23–7.

Robinson, F. N. 1991. Phatic communication in bird song. *Emu*, 91, 61–3.

Rogers, A. C. 2005. Male and female song structure and singing behaviour in the duetting eastern whipbird. *Aust. J. Zool.*, 53, 157–66.

Rogers, A. C., Ferguson, J. E., Harrington, H. M., McDowell, S., Miller, A. & Panagos, J. 2004. Use of stereo duet playback to investigate traditional duet playback methods and mechanisms of cooperative territorial defence in magpie-larks. *Behaviour*, 141, 741–53.

Rogers, A. C., Mulder, R. A. & Langmore, N. E. 2006. Duet duels: sex differences in song matching in duetting eastern whipbirds. *Anim. Behav.*, 72, 53–61.

Rogers, D. J. 2003. Geographic song variation within and between populations and subspecies of the rufous bristlebird *(Dasyornis broadbenti)*. *Aust. J. Zool.*, 51, 1–14.

Rost, R. 1987. *Entstehung, Fortbestand und funktionelle Bedeutung von Gesangsdialekten bei der Sumpfmeise Parus palustris: ein Test von Modellen.* Konstanz: Hartung-Gorre Verlag.

Rost, R. 1990. Hormones and behaviour: a joint examination of studies on seasonal variation in song production and plasma levels of testosterone in the great tit *Parus major*. *J. Orn.*, 131, 403–11.

Rost, R. 1992. Hormones and behaviour: a comparison of studies on seasonal changes in song production and testosterone plasma levels in the willow tit *Parus montanus*. *Ornis Fenn.*, 69, 1–6.

Rothstein, S. I. & Fleischer, R. C. 1987. Vocal dialects and their possible relation to honest status signalling in the brown-headed cowbird. *Condor*, 89, 1–23.

Ruppell, W. 1933. Physiologie und Akustik der Vogelstimme. *J. Orn.*, 81, 433–542.

Ryan, M. J. & Brenowitz, E. A. 1985. The role of body size, phylogeny, and ambient noise in the evolution of bird song. *Am. Nat.*, 126, 87–100.

Ryan, M. J., Fox, J. H., Wilczynski, W. & Rand, A. S. 1990. Sexual selection for sensory exploitation in the frog *Physalaemus pustulosus*. *Nature*, 343, 66–7.

Saunders, J. C. & Henry, W. J. 1989. The peripheral auditory system in birds: structural and functional contributions to auditory perception. In: *The Comparative Psychology of Audition: Perceiving Complex Sounds* (ed. R. J. Dooling & S. H. Hulse), pp. 35–6. New Jersey: Erlbaum.

Saunders, J. C., Duncan, R. K., Doan, D. E. & Werner, Y.L. 2000. The middle ear of reptiles and birds. In: *Comparative Hearing: Birds and Reptiles* (ed. A. N. Popper), pp. 13–69. New York: Springer-Verlag.

Schleidt, W. M. 1964. Über die Spontaneität von Erbkoordinationen. *Z. Tierpsychol.*, 21, 235–6.

Schroeder, D. J. & Wiley, R. H. 1983a. Communication with repertoires of song themes in tufted titmice. *Anim. Behav.*, 31, 1128–38.

Schroeder, D. J. & Wiley, R. H. 1983b. Communication with shared song themes in tufted titmice. *Auk*, 100, 414–24.

Schwabl, H. & Kriner, E. 1991. Territorial aggression and song of male European robins *(Erithacus rubecula)* in autumn and spring: effects of antiandrogen treatment. *Horm. Behav.*, 25, 180–94.

Searcy, W.A. 1984. Song repertoire size and female preferences in song sparrows. *Behav. Ecol. Sociobiol.*, 14, 281–6.

Searcy, W. A. 1988. Dual intersexual and intrasexual functions of song in red-winged blackbirds. *Proc. XIX Int. Congr. Ornithol.*, 1, 1373–81.

Searcy, W. A. 1989. Pseudoreplication, external validity and the design of playback experiments. *Anim. Behav.*, 38, 715–17.

Searcy, W. A. 1990. Species recognition of song by female red-winged blackbirds. *Anim. Behav.*, 40, 1119–27.

Searcy, W. A. 1992a. Song repertoire and mate choice in birds. *Amer. Zool.*, 32, 71–80.

Searcy, W. A. 1992b. Measuring responses of female birds to male song. In: *Playback and Studies of Animal Communication* (ed. P. K. McGregor), pp. 175–89. New York: Plenum Press.

Searcy, W. A. & Andersson, M. 1986. Sexual selection and the evolution of song. *Ann. Rev. Ecol. Syst.*, 17, 507–33.

Searcy, W. A. & Brenowitz, E. A. 1988. Sexual differences in species recognition of avian song. *Nature*, 332, 152–4.

Searcy, W. A., Coffman, S. & Raikow, D. F. 1994. Habituation, recovery and the similarity of song types within repertoires in red-winged blackbirds (*Agelaius phoeniceus*) (Aves, Emberizidae). *Ethology*, 98, 38–49.

Searcy, W. A. & Marler, P. 1981. A test for responsiveness to song structure and programming in female sparrows. *Science*, 213, 926–8.

Searcy, W. A. & Marler, P. 1984. Interspecific differences in the response of female birds to song repertoires. *Z. Tierpsychol.*, 66, 128–42.

Searcy, W. A., McArthur, P. D. & Yasukawa, K. 1985. Song repertoire size and male quality in song sparrows. *Condor*, 87, 222–8.

Searcy, W. A & Nowicki, S. 2005. *The Evolution of Animal Communication: Reliability and Deception in Signaling Systems*. Princeton, NJ: Princeton University Press.

Searcy, W. A., Nowicki, S. & Hughes, M. 1997. The response of male and female song sparrows to geographic variation in song. *Condor*, 99, 651–7.

Searcy, W. A., Nowicki, S., Hughes, M. & Peters, S. 2002. Geographic song discrimination in relation to dispersal distances in song sparrows. *Am. Nat.*, 159, 221–30.

Searcy, W. A., Nowicki, S. & Peters, S. 2003. Phonology and geographic song discrimination in song sparrows. *Ethology*, 109, 23–35.

Searcy, W. A. & Yasukawa, K. 1990. Use of the song repertoire in intersexual and intrasexual contexts by male red-winged blackbirds. *Behav. Ecol. Sociobiol.*, 27, 123–8.

Searcy, W. A. & Yasukawa, K. 1996. Song and female choice. In *Ecology and Evolution of Acoustic Communication in Birds* (ed. D. E. Kroodsma & E. H. Miller), pp. 454–73. Ithaca, NY, and London: Comstock Publishing Associates.

Seddon, N. 2002. The structure, context and possible functions of solos, duets and choruses in the subdesert mesite (*Monias benschi*). *Behaviour*, 139, 645–76.

Seddon, N. 2005. Ecological adaptation and species recognition drives vocal evolution in neotropical suboscine birds. *Evolution*, 59, 200–15.

Seddon, N., Butchart, S. H. M. & Odling-Smee, L. 2002. Duetting in the subdesert mesite *Monias benschi*: evidence for acoustic mate defence? *Behav. Ecol. Sociobiol.*, 52, 7–16.

Seddon, N. & Tobias, J. A. 2003. Communal singing in the cooperatively breeding subdesert mesite *Monias benschi*: evidence of numerical assessment? *J. Avian. Biol.*, 34, 72–80.

Seddon, N. & Tobias, J. A. 2006. Duets defend mates in a suboscine passerine, the warbling antbird (*Hypocnemis cantator*). *Behav. Ecol.*, 17, 73–83.

Seibt, U., Wickler, W., Kleindienst, H.-U. & Sonnenschein, E. 2002. Structure, geography and origin of dialects in the traditive song of the forest weaver *Ploceus bicolor sclateri* in Natal, S. Africa. *Behaviour*, 139, 1237–65.

Shutler, D. & Weatherhead, P. 1990. Targets of sexual selection: song and plumage of wood warblers. *Evolution*, 44, 1967–77.

Shy, E. 1983. The relation of geographical variation in song to habitat characteristics and body size in North American tanagers (Thraupinae: *Piranga*). *Behav. Ecol. Sociobiol.*, 12, 71–6.

Shy, E. & Morton, E. S. 1986. Adaptation of amplitude structure of songs to propagation in field habitat in song sparrows. *Z. Tierpsychol.*, 72, 177–84.

Simpson, B. S. 1984. Tests of habituation to song repertoires by Carolina wrens. *Auk*, 101, 244–54.

Simpson, H. B. & Vicario, D. S. 1990. Brain pathways for learned and unlearned vocalizations differ in zebra finches. *J. Neurosci.*, 10, 1541–56.

Slabbekoorn, H. 2004. Singing in the wild: the ecology of birdsong. In: *Nature's Music: The Science of Birdsong* (ed. P.Marler & H. Slabbekoorn), pp. 181–208. San Diego, CA: Academic Press.

Slabbekoorn, H. & den Boer-Visser, A. 2006. Cities change the songs of birds. *Curr. Biol.* 16, 2326–31.

Slabbekoorn, H., Ellers, J. & Smith, T. B. 2002. Birdsong and sound transmission: the benefits of reverberations. *Condor*, 104, 564–73.

Slabbekoorn, H. & Peet, M. 2003. Birds sing at a higher pitch in urban noise. *Nature*, 424, 267.

Slabbekoorn, H. & Smith, T. B. 2002a. Bird song, ecology and speciation. *Phil. Trans. R. Soc. Lond,* B 357, 493–503.

Slabbekoorn, H. & Smith, T. B. 2002b. Habitat dependent song divergence in the little greenbul: an analysis of environmental selection pressures on acoustic signals. *Evolution*, 56, 1849–58.

Slagsvold, T. 1977. Bird song activity in relation to breeding cycle, spring weather and environmental phenology. *Ornis Scand.*, 8, 197–222.

Slagsvold, T., Dale, S. & Saetre, G.-P. 1994. Dawn singing in the great tit (*Parus major*): mate attraction, mate guarding, or territorial defence? *Behaviour* 131, 115–38.

Slater, P. J. B. 1978. Beau Geste has problems. *Anim. Behav.*, 26, 304–5.

Slater, P. J. B. 1981. Chaffinch song repertoires: observations, experiments and a discussion of their significance. *Z. Tierpsychol.*, 56, 1–24.

Slater, P. J. B. 1983a. Bird song learning: theme and variations. In: *Perspectives in Ornithology*. (ed. A. H. Brush & G. A. Clark), pp. 475–99. Cambridge: Cambridge University Press.

Slater, P. J. B. 1983b. Sequences of song in chaffinches. *Anim. Behav.*, 31, 272–81.

Slater, P. J. B. 1983c. The study of communication. In: *Animal Behaviour*. Vol. 2: Communication (ed. T. R. Halliday & P. J. B. Slater), pp. 9–42. Oxford: Blackwell Scientific Publications.

Slater, P. J. B. 1983d. The development of individual behaviour. In: *Animal Behaviour. Vol. 3: Genes, Development and Learning* (ed. T. R. Halliday & P. J. B. Slater), pp. 82–113. Oxford: Blackwell Scientific Publications.

Slater, P. J. B. 1986. The cultural transmission of bird song. *Trends Ecol. Evol.*, 1, 94–7.

Slater, P. J. B. 1989. Bird song learning: causes and consequences. *Ethol. Ecol. Evol.*, 1, 19–46.

Slater, P. J. B., Clements, F. A. & Goodfellow, D. J. 1984. Local and regional variations in chaffinch song and the question of dialects. *Behaviour*, 88, 76–97.

Slater, P. J. B., Eales, L. A. & Clayton, N. S. 1988. Song learning in zebra finches: progress and prospects. *Adv. Study Behav.*, 18, 1–34.

Slater, P. J. B., Gil, D., Barlow, C. R. & Graves, J. A. 2002. Male led duets in the moho (*Hypergerus atriceps*) and yellow-crowned gonolek (*Lanarius barbarus*): mate guarding by females. *Ostrich*, 73, 49–51.

Slater, P. J. B. & Ince, S. A. 1979. Cultural evolution in chaffinch song. *Behaviour*, 71, 146–66.

Slater, P. J. B. & Ince, S. A. 1982. Song development in chaffinches: what is learnt and when? *Ibis*, 124, 21–6.

Slater, P. J. B., Ince, S. A. & Colgan, P. W. 1980. Chaffinch song types: their frequencies in the population and distribution between the repertoires of different individuals. *Behaviour*, 75, 207–18.

Slater, P. J. B. & Lachlan, R. F. 2003. Is innovation in bird song adaptive? In: *Animal Innovation* (ed. K. N. Laland & S. Reader), pp. 117–35. Oxford: Oxford University Press.

Slater, P. J. B., Lachlan, R. F. & Riebel, K. 2000. The significance of learning in signal development: The curious case of the chaffinch. In: *Animal Signals. Signalling and Signal Design in Animal Communication* (ed. Y. Espmark, T. Amundsen & G. Rosenqvist), pp. 341–52. Trondheim, Norway: Tapir Publishers.

Slater, P. J. B. & Mann, N. I. 1990. Do male zebra finches learn their fathers' songs? *Trends Ecol. Evol.*, 5, 415–17.

Slater, P. J. B. & Mann, N. I. 1991. Early experience and song learning in zebra finches (Taeniopygia guttata). In: *Acta XX Congressus Internationalis Ornithologici*, pp. 1074–80. Christchurch, New Zealand: Ornithological Congress Trust Board.

Slater, P. J. B. & Mann, N. I. 2004. Why do the females of many bird species sing in the tropics? *J. Avian Biol.*, 35, 289–94.

Smith, D. G. 1976. An experimental analysis of the function of red-winged blackbird song. *Behaviour*, 56, 136–56.

Smith, D. G. 1979. Male singing ability and territory integrity in red-winged blackbirds *Agelaius phoeniceus*. *Behaviour*, 68, 193–206.

Smith, D. G. & Norman, D. O. 1979. 'Leader-follower' singing in red-winged blackbirds. *Condor*, 81, 83–4.

Smith, W. J. 1977. *The Behavior of Communicating*. Cambridge, MA: Harvard University Press.

Soha, J. A. & Marler, P. 2000. A species-specific acoustic cue for selective song learning in the white-crowned sparrow. *Anim. Behav.*, 60, 297–306.

Soha, J. A., Nelson, D. A. & Parker, P. G. 2004. Genetic analysis of song dialect populations in Puget sound white-crowned sparrows. *Behav. Ecol.*, 15, 636–46.

Sohrabji, F., Nordeen, E. J. & Nordeen, K. W. 1990. Selective impairment of song learning following lesions of a forebrain nucleus in the juvenile zebra finch. *Behav. Neural Biol.*, 53, 51–63.

Sonnenschein, E. & Reyer, H.-U. 1983. Mate-guarding and other functions of antiphonal duets in the slate-coloured boubou (*Laniarius funebris*). *Z. Tierpsychol.*, 63, 112–40.

Sorjonen, J. 1986a. Mixed singing and interspecific territoriality - consequences of secondary contact of two ecologically and morphologically similar nightingale species in Europe. *Ornis Scand.*, 17, 53–67.

Sorjonen, J. 1986b. Singing strategies in Northern European passerines. *Univ. of Joensu Publications in Sciences*, 9, 1–21.

Sorjonen, J. 1987. Temporal and spatial differences in traditions and repertoires in the song of the thrush nightingale (*Luscinia luscinia*). *Behaviour*, 102, 196–212.

Sorjonen, J. & Merilä, J. 2000. Response of male bluethroats *Luscinia svecica* to song playback: evidence of territorial function of song and song flights. *Ornis Fennica*, 77, 43–7.

Spector, D. A., McKim, L. K. & Kroodsma, D. E. 1989. Yellow warblers are able to learn songs and situations in which to use them. *Anim. Behav.*, 38, 723–5.

Speirs, E. A. & Davis, L. S. 1991. Discrimination by adelie penguins *Pygoscelis adeliae*, between the loud mutual calls of mates, neighbours and strangers. *Anim. Behav.*, 41, 937–44.

Spencer, K. A., Buchanan, K. A., Goldsmith, A. R. & Catchpole, C. K. 2003. Song as an honest indicator of developmental history in the zebra finch (*Taeniopygia guttata*). *Horm. Behav.*, 44, 132–9.

Spencer, K. A., Buchanan, K. A., Leitner, S., Goldsmith, A. R. & Catchpole, C. K. 2005a. Parasites affect song complexity and neural development in a songbird. *Proc. R. Soc. Lond.*, B 272, 2037–43.

Spencer, K. A., Wimprenny, J. H., Buchanan, K. L. *et al.* 2005b. Developmental stress affects the attractiveness of male song and female choice in the zebra finch (*Taeniopygia guttata*). *Behav. Ecol. Sociobiol.*, 58, 423–8.

Staicer, C. A. 1989. Characteristics, use, and significance of two singing behaviors in Grace's warbler. *Auk*, 106, 49–63.

Staicer, C. A., Spector, D. A. & Horn, A. G. 1996. The dawn chorus and other diel patterns in acoustic signalling. In: *Ecology and Evolution of Acoustic Communication in Birds* (ed. D. E. Kroodsma & E.

H. Miller), pp. 426–53. Ithaca, NY, and London: Comstock Publishing Associates.

Stoddard, P. K. 1996. Vocal recognition of neighbours by territorial passerines. In: *Ecology and Evolution of Acoustic Communication in Birds* (ed. D. E. Kroodsma & E. H. Miller), pp. 356–74. Ithaca, NY: Cornell University Press.

Stoddard, P. K. & Beecher, M. D. 1983. Parent recognition of offspring in the cliff swallow. *Auk*, 100, 795–9.

Stoddard, P. K., Beecher, M. D., Campbell, S. E. & Horning, C. L. 1992. Song-type matching in the song sparrow. *Can. J. Zool.*, 70, 1440–4.

Stoddard, P. K., Beecher, M. D., Horning, C. L. & Campbell, S. E. 1990. Strong neighbour-stranger discrimination in song sparrows. *Condor*, 92, 1051–6.

Suthers, R.A. 1990. Contributions to bird song from the left and right sides of the intact syrinx. *Nature*, 347, 473–7.

Suthers, R. A. 1999. The motor basis of vocal performance in songbirds. In *The Design of Animal Communication* (ed. M. D. Hauser & M. Konishi), pp. 37–62. Cambridge, MA: MIT Press.

Suthers, R. A. 2004. How birds sing and why it matters. In *Nature's Music: The Science of Birdsong* (ed. P. Marler & H. Slabbekoorn), pp. 276–99. San Diego, CA: Academic Press.

Suthers, R. A., Goller, F. & Pytte, C. 1999. The neuromuscular control of bird song. *Phil. Trans. R. Soc. Lond.*, B 354, 927–39.

Suthers, R. A., Vallet, E., Tanvez, A. & Kreutzer, M. 2004. Bilateral song production in domestic canaries. *J. Neurobiol.*, 60, 381–93.

Szekely, T., Catchpole, C. K., DeVoogd, A., Marchl, Z & DeVoogd, T. J. 1996. Evolutionary changes in a song control area of the brain (HVC) are associated with evolutionary changes in song repertoire among European warblers (Sylviidae). *Proc. R. Soc. Lond.*, B 263, 607–10.

Tanvez, A., Beguin, N., Chastel, O., Lacroix, A. & Leboucher, G. 2004. Sexually attractive phrases increase yolk androgens deposition in canaries *Serinus canaria*. *Gen. comp. Endocrinol.*, 138, 113–20.

Tchernichovski, O., Mitra, P. P., Lints, T. & Nottebohm, F. 2001. Dynamics of the vocal imitation process: how a zebra finch learns its song. *Science*, 291, 2564–9.

Tchernichovski, O., Nottebohm, F., Ho, C. E., Pesaran, B. & Mitra, P. P. 2000. A procedure for automated measurement of song similarity. *Anim. Behav.*, 59, 1167–76.

Temrin, H. 1986. Singing behaviour in relation to polyterritorial polygyny in the wood warbler (*Phylloscopus sibilatrix*). *Anim. Behav.*, 34, 146–52.

ten Cate, C., Slabbekoorn, H. & Ballintijn, M. 2002. Birdsong and male-male competition: causes and consequences of vocal variability in the collared dove (*Streptopelia decaocto*). *Adv. Study Behav.*, 31, 31–75.

Thielcke, G. 1969. Geographic variation in bird vocalizations. In: *Bird Vocalizations* (ed. R. A. Hinde), pp. 310–39. Cambridge: Cambridge University Press.

Thielcke, G. 1973. On the origin of divergence of learned signals (songs) in isolated populations. *Ibis*, 115, 511–16.

Thielcke, G. 1986. Constant proportions of mixed singers in treecreeper populations (*Certhia familiaris*). *Z. Tierpsychol.*, 72, 154–64.

Thielcke, G. 1987. Langjährige Dialektkonstanz beim Gartenbaumläufer (*Certhia brachydactyla*). *J. Orn.*, 128, 171–80.

Thielcke, G. 1988. Buchfinken (*Fringilla coelebs*) eliminieren erlernte Gesänge von Baumpiepern (*Anthus trivialis*). *Die Vogelwarte*, 34, 319–36.

Thielcke, G. & Krome, M. 1989. Experimente über sensible Phasen und Gesangsvariabilität beim Buchfinken (*Fringilla coelebs*). *J. Orn.*, 130, 435–53.

Thielcke, G. & Krome, M. 1991. Chaffinches *Fringilla coelebs* do not learn song during autumn and early winter. *Bioacoustics*, 3, 207–12.

Thielcke-Poltz, H. & Thielcke, G. 1960. Akustisches Lernen verschieden alter schallisolierter Amseln (*Turdus merula* L.) und die Entwicklung erlernter Motive ohne und mit künstlichen Einfluss von Testosteron. *Z. Tierpsychol.*, 17, 211–44.

Thomas, R. J. 1999a. Two tests of a stochastic dynamic programming model of daily singing routines in birds. *Anim. Behav.* 57, 277–84.

Thomas, R. J. 1999b. The effect of variability in the food supply on the daily singing routines of European robins: a test of a stochastic dynamic programming model. *Anim. Behav.* 57, 365–9.

Thomas, R. J. & Cuthill, I. C. 2002. Body mass regulation and the daily singing routines of European robins. *Anim. Behav.* 63, 285–95.

Thomas, R. J., Szekely, T., Cuthill, I. C., Harper, D. G. C., Newson, S. E., Frayling, T. D. & Wallis, P. D. 2002. Eye size in birds and the timing of song at dawn. *Proc. R. Soc. Lond*, B 269, 831–7.

Thorpe, W. H. 1954. The process of song learning in the chaffinch as studied by means of the sound spectrograph. *Nature*, 173, 465.

Thorpe, W. H. 1958. The learning of song patterns by birds, with especial reference to the song of the chaffinch *Fringilla coelebs*. *Ibis*, 100, 535–70.

Thorpe, W. H. 1961. *Bird Song*. Cambridge: Cambridge University Press.

Thorpe, W. H. 1972. Duetting and antiphonal song in birds, its extent and significance. *Behaviour Suppl.*, 18, 1–197.

Todt, D. 1975a. Short term inhibition of outputs occurring in the vocal behaviour of blackbirds (*Turdus m. merula* L.). *J. comp. Physiol.*, 98, 289–306.

Todt, D. 1975b. Spontaneous recombinations of vocal patterns in parrots. *Naturwissenschaften*, 62, 399.

Todt, D. 1981. On functions of vocal matching: effect of counter-replies on song post choice and singing. *Z. Tierpsychol.*, 57, 73–93.

Todt, D. & Geberzahn, N. 2003. Age-dependent effects of song exposure: song crystallization sets a boundary between fast and delayed vocal imitation. *Anim. Behav.*, 65, 971–9.

Todt, D. & Hultsch, H. 1982. Impairment of vocal signal exchange in the monogamous duet-singer *Cossypha heuglini* (Turdidae): Effects on pair bond maintenance. *Z. Tierpsychol.*, 60, 265–74.

Todt, D. & Hultsch, H. 1998. How songbirds deal with large amounts of serial information: retrieval rules suggest a hierarchical song memory. *Biol. Cybern.*, 79, 487–500.

Todt, D., Hultsch, H. & Duvall, F. P. 1982. Behavioural significance and social function of vocal and non-vocal displays in the monogamous duet-singer *Cossypha heuglini* H. *Zool. Beitr.*, 27, 421–48.

Todt, D. & Naguib, M. 2000. Vocal interactions in birds: the use of song as a model in communication. *Adv. Study Behav.*, 29, 247–96.

Tomback, D. F. & Baker, M. C. 1984. Assortative mating by white-crowned sparrows at song dialect boundaries. *Anim. Behav.*, 32, 465–9.

Tomback, D. F., Thompson, D. B. & Baker, M. C. 1983. Dialect discrimination by white-crowned sparrows: reactions to near and distant dialects. *Auk*, 100, 452–60.

Trainer, J. M. 1989. Cultural evolution in song dialects of yellow-rumped caciques in Panama. *Ethology*, 80, 190–204.

Trainer, J. M., McDonald, D. B. & Learn, W. A. 2002. The development of coordinated singing in cooperatively displaying long-tailed manakins. *Behav. Ecol.*, 13, 65–9.

Trainer, J. M. & Parsons, R. J. 2001. Uniformity of long-tailed manakin songs from three localities in Costa Rica. *Wilson Bull.*, 113, 431–4.

Trivers, R. L. 1972. Parental investment and sexual selection. In: *Sexual Selection and the Descent of Man* (ed. B. Campbell), pp. 137–79. London: Heinemann.

Tschanz, B. 1968. Trottellummen. *Z. Tierpsychol. Suppl.* 4.

Tsipoura, N. & Morton, E. S. 1988. Song-type distribution in a population of Kentucky warblers. *Wilson Bull.*, 100, 9–16.

Tubaro, P. L. & Mahler, B. 1998. Acoustic frequencies and body mass in New World doves. *Condor*, 100, 54–61.

Tubaro, P. L. & Segura, E. T. 1994. Dialect differences in the song of *Zonotrichia capensis* in the southern pampas: a test of the acoustic adaptation hypothesis. *Condor*, 96, 1084–8.

Tubaro, P. L., Segura, E. T. & Handford, P. 1993. Geographic variation in the song of the rufous-collared sparrow in Eastern Argentina. *Condor*, 95, 588–95.

Turner, W. C. & Barber, C. A. 2004. Male song sparrows *Melospiza melodia* do not announce their female's fertility. *J. Avian Biol.*, 35, 483–6.

Vallet, E., Beme, I., & Kreutzer, M. 1998. Two-notes syllables in the canary songs elicit high levels of sexual display. *Anim. Behav.*, 55, 291–7.

Vallet, E. & Kreutzer, M. 1995. Female canaries are sexually responsive to special song phrases. *Anim. Behav.*, 49, 1603–10.

van Buskirk, J. 1997. Independent evolution of song structure and note structure in American wood warblers. *Proc. R. Soc. Lond.*, B 264, 755–61.

Veerman, P. A. 1994. Batesian acoustic mimicry by the regent honeyeater *Xanthomyza phrygia*. *Austral. Bird Watcher*, 15, 250–9.

Voigt, C., Leitner, S. & Gahr, M. 2006. Repertoire and structure of duet and solo songs in

cooperatively breeding white-browed sparrow weavers. *Behaviour*, 143, 159–82.

Volman, S. F. & Khanna, H. 1995. Convergence of untutored song in group-reared zebra finches. *J. comp. Psychol.*, 109, 211–21.

von Haartman, L. & von Numers, M. 1992. Rain call dialects of the chaffinch *Fringilla coelebs* in the archipelago of SW Finland. *Ornis Fennica*, 69, 65–71.

von Helversen, D. 1980. Structure and function of antiphonal duets. In: *Acta XVII Congr. Int. Orn. Berlin.*

Waas, J. R. 1988. Song pitch-habitat relationships in white-throated sparrows: cracks in acoustic windows? *Can. J. Zool.*, 66, 2578–81.

Wallin, L. 1986. Divergent character displacement in the song of two allospecies: the pied flycatcher *Ficedula hypoleuca*, and the collared flycatcher *Ficedula albicollis*. *Ibis*, 128, 251–9.

Wallschläger, D. 1980. Correlation of song frequency and body weight in passerine birds. *Experientia*, 36, 412.

Wanker, R., Sugama, Y. & Prinage, S. 2005. Vocal labelling of family members in spectacled parrotlets (*Forpus conspicillatus*). *Anim. Behav.*, 70, 111–18.

Ward, B., Nordeen, E. & Nordeen, K. 2001. Anatomical and ontogenetic factors producing variation in HVC neuron number in zebra finches. *Brain Res.*, 904, 318–26.

Ward, S. & Slater, P. J. B. 2005a. Heat transfer and energetic cost of singing in canaries *Serinus canaria*. *J. comp. Physiol.*, 191, 953–64.

Ward, S. & Slater, P. J. B. 2005b. Raised thermoregulatory costs at exposed song posts increase the energetic cost of singing for willow warblers, *Phylloscopus trochilus*. *J. Avian Biol.*, 36, 280–6.

Ward, S., Lampe, H. M. & Slater, P. J. B. 2004. Singing is not energetically demanding for pied flycatchers *Ficedula hypoleuca*. *Behav. Ecol.*, 15, 477–84.

Ward, S., Speakman, J. R. & Slater, P. J. B. 2003. The energy cost of song in the canary, *Serinus canaria*. *Anim. Behav.*, 66, 893–902.

Wasserman, F. E. 1977. Intraspecific acoustical interference in the white-throated sparrow (*Zonotrichia albicollis*). *Anim. Behav.*, 25, 949–52.

Wasserman, F. E. 1979. The relationship between habitat and song in the white-throated sparrow *Zonotrichia albicollis*. *Condor*, 81, 424–6.

Wasserman, F. E. & Cigliano, J. A. 1991. Song output and stimulation of the female in white-throated sparrows. *Behav. Ecol. Sociobiol.*, 29, 55–9.

Weary, D. M. 1992. Bird song and operant experiments: a new tool to investigate song perception. In: *Playback and Studies of Animal Communication* (ed. P. K. McGregor), pp. 201–10. New York: Plenum Press.

Weary, D. M., Lambrechts, M. M. & Krebs, J. R. 1991. Does singing exhaust male great tits? *Anim. Behav.*, 41, 540–2.

Weary, D. M. & Lemon, R. E. 1988. Evidence against the continuity-versatility relationship in birdsong. *Anim. Behav.*, 36, 1379–83.

Weary, D. M. & Lemon, R. E. 1990. Kroodsma refuted. *Anim. Behav.*, 39, 996–8.

Weary, D. M., Lemon, R. E. & Perreault, S. 1992. Song repertoires do not hinder neighbor-stranger discrimination. *Behav. Ecol. Sociobiol.*, 31, 441–7.

Weary, D. M., Lemon, R. E. & Perreault, S. 1994a. Different responses to different song types in American redstarts. *Auk*, 111, 730–4.

Weary, D. M. & Krebs, J. R. 1992. Great tits classify songs by individual voice characteristics. *Anim. Behav.*, 43, 283–7.

Weary, D. M., Krebs, J. R., Eddyshaw, R., McGregor, P. K. & Horn, A. 1988. Decline in song output by

great tits: exhaustion or motivation? *Anim. Behav.*, 36, 1242–4.

Weary, D. M. & Mountjoy, D. J. 1992. On designs for testing the effect of song repertoire size. *Anim. Behav.*, 44, 577–9.

Weary, D. M., Norris, K. J. & Falls, J. B. 1990. Song features birds use to identify individuals. *Auk*, 107, 623–5.

Weedon, J. S. & Falls, J. B. 1959. Differential response of male ovenbirds to recorded songs of neighbouring and more distant individuals. *Auk*, 76, 343–51.

Weisman, R. G. & Ratcliffe, L. 1989. Absolute and relative pitch processing in black-capped chickadees *Parus atricapillus*. *Anim. Behav.*, 38, 685–92.

Weisman, R. G., Ratcliffe, L., Johnsrude, I. S. & Hurly, T. A. 1990. Absolute and relative pitch production in the song of the black-capped chickadee *Parus atricapillus*. *Condor*, 92, 118–24.

West, M. J. & King, A. P. 1988. Female visual displays affect the development of male song in the cowbird. *Nature*, 334, 244–6.

West, M. J. & King, A. P. 1990. Mozart's starling. *Amer. Sci.*, 78, 106–14.

West, M. J., Stroud, A. N. & King, A. P. 1983. Mimicry of the human voice by European starlings: the role of social interaction. *Wilson Bull.*, 95, 635–40.

Westcott, D. 1992. Inter and intrasexual selection: the role of song in a lek mating system. *Anim. Behav.*, 44, 695–703.

Westneat, M. W., Long, J. H., Hoese, W. & Nowicki, S. 1993. Kinematics of bird song: functional correlation of cranial movements and acoustic features in sparrows. *J. exp. Biol.*, 182, 147–71.

Whaling, C. S., Solis, M. M., Doupe, A. J., Soha, J. A. & Marler, P. 1997. Acoustic and neural bases for innate recognition of song. *Proc. Natl. Acad. Sci. USA*, 94, 12694–8.

White, G. 1789. *The Natural History and Antiquities of Selborne*. London: Chiswick Press.

White, S. J. 1971. Selective responses by the gannet *Sula bassana* to played-back calls. *Anim. Behav.*, 19, 125–31.

Whitney, C. L. 1981. Patterns of singing in the varied thrush: II. A model of control. *Z. Tierpsychol.*, 57, 141–62.

Whitney, C. L. 1990. Similarity of songs within songbird repertoires: analytic problems. *Anim. Behav.*, 39, 806–9.

Whitney, C. L. & Miller, J. 1983. Song matching in the wood thrush (*Hylocichla mustellina*): a function of song dissimilarity. *Anim. Behav.*, 31, 457–61.

Wickler, W. 1976. Duetting songs in birds: biological significance of stationary and non-stationary processes. *J. theor. Biol*, 61, 493–7.

Wickler, W. 1980. Vocal duetting and the pair bond: I. Coyness and partner commitment, a hypothesis. *Z. Tierpsychol.*, 52, 201–9.

Wiens, J. A. 1982. Song pattern variation in the sage sparrow (*Amphispiza belli*): dialects or epiphenomena. *Auk*, 99, 208–29.

Wilczynski, W., Ryan, M. J. & Brenowitz, E. A. 1989. The display of the blue-black grassquit: the acoustic advantage of getting high. *Ethology*, 80, 218–22.

Wild, J. M., Goller, F. & Suthers, R. A. 1998. Inspiratory muscle activity during birdsong. *J. Neurobiol.*, 36, 441–53.

Wild, J. M., Williams, M. N. & Suthers, R. A. 2000. Neural pathways for bilateral vocal control in songbirds. *J. comp. Neurol.*, 423, 413–26.

Wildenthal, J. L. 1965. Structure in primary song of the mockingbird (*Mimus polyglottos*). *Auk*, 82, 161–89.

Wiley, R. H. 1971. Song groups in a singing assembly of little hermits. *Condor*, 73, 28–35.

Wiley, R. H. 1991. Associations of song properties with habitats for territorial oscine birds of eastern North America. *Am. Nat.*, 138, 973–93.

Wiley, R. H. 1998. Ranging reconsidered. *Behav. Ecol. Sociobiol.*, 42, 143–6.

Wiley, R. H. 2005. Individuality in songs of Acadian flycatchers and recognition of neighbours. *Anim. Behav.*, 70, 237–47.

Wiley, R. H. & Godard, R. 1996. Ranging of conspecific songs by Kentucky warblers and its implications for interactions of territorial males. *Behaviour*, 133, 81–102.

Wiley, R. H., Hatchwell, B. J. & Davies, N. B. 1991. Recognition of individual males' songs by female dunnocks: a mechanism increasing the number of copulatory partners and reproductive success. *Ethology*, 88, 145–53.

Wiley, R. H. & Richards, D. G. 1978. Physical constraints on acoustic communication in the atmosphere: implications for the evolution of animal vocalizations. *Behav. Ecol. Sociobiol.*, 3, 69–94.

Wiley, R. H. & Richards, D. G. 1982. Adaptations for acoustic communication in birds: sound transmission and signal detection. In: *Acoustic Communication in Birds* (ed. D. E. Kroodsma & E. H. Miller), pp. 131–81. New York: Academic Press.

Williams, H. 1989. Multiple representations and auditory motor interactions in the avian song system. *Ann. N.Y. Acad. Sci.*, 563, 148–64.

Williams, H., Crane, L. A., Hale, T. K., Esposito, M. A. & Nottebohm, F. 1992. Right-sided dominance for song control in the zebra finch. *J. Neurobiol.*, 23, 1006–20.

Williams, J. M. & Slater, P. J. B. 1990. Modelling bird song dialects: the influence of repertoire size and numbers of neighbours. *J. theor. Biol.*, 145, 487–96.

Williams, J. M. & Slater, P. J. B. 1991. Simulation studies of song learning in birds. In: *Simulation of Adaptive Behavior* (ed. J.-A. Meyer & S. Wilson), pp. 281–7. Cambridge, MA: MIT Press.

Williams, J. M. & Slater, P. J. B. 1993. Does chaffinch *Fringilla coelebs* song vary with the habitat in which it is sung? *Ibis*, 135, 202–8.

Williams, L. & MacRoberts, M. H. 1977. Individual variation in songs of dark-eyed juncos. *Condor*, 79, 106–12.

Wilson, P. L., Towner, M. C. & Vehrencamp, S. L. 2000. Survival and song-type sharing in a sedentary subspecies of the song sparrow. *Condor*, 102, 355–63.

Wisniewski, A. B. & Hulse, S. H. 1997. Auditory scene analysis in European starlings (*Sturnus vulgaris*): discrimination of starling song elements, their segregation from conspecific songs, and evidence for conspecific song categorization. *J. comp. Psychol.*, 111, 337–50.

Wittenberger, J. F. 1983. A contextual analysis of two song variants in the bobolink. *Condor*, 85, 172–84.

Wolffgramm, J. 1979. Vergleich der hochvariablen Gesangsvorträge benachbarter Amseln. *Verh. Dtsch. Zool. Ges.*, 1979, 237.

Wright, J. & Cuthill, I. 1992. Monogamy in the European starling. *Behaviour*, 120, 262–85.

Wright, T. F. 1996. Regional dialects in the contact call of a parrot. *Proc. R. Soc. Lond.*, B 263, 867–72.

Wright, T. F. & Wilkinson, G. S. 2001. Population genetic structure and vocal dialects in an Amazon parrot. *Proc. R. Soc. Lond.*, B 268, 1–8.

Yamaguchi, A. 2001. Sex differences in vocal learning in birds. *Nature*, 411, 257–8.

Yasukawa, K. 1981. Song repertoires in the red-winged blackbird *Agelaius phoeniceus*: a test of the Beau Geste hypothesis. *Anim. Behav.*, 29, 114–25.

Yasukawa, K., Blank, J. L. & Patterson, C. B. 1980. Song repertoires and sexual selection in the red-winged blackbird. *Behav. Ecol. Sociobiol.*, 7, 233–8.

Yasukawa, K., Boley, R. A. & Simon, S. 1987. Seasonal change in the vocal behaviour of female red-winged blackbirds, *Agelaius phoeniceus. Anim. Behav.*, 35, 1416–23.

Yasukawa, K. & Searcy, W. 1985. Song repertoires and density assessment in red-winged blackbirds: further tests of the Beau Geste hypothesis. *Behav. Ecol. Sociobiol.*, 16, 171–5.

Ydenberg, R. C., Giraldeau, L. A. & Falls, J. B. 1988. Neighbours, strangers and the asymmetric war of attrition. *Anim. Behav.*, 36, 343–7.

Yu, A. C. & Margoliash, D. 1996. Temporal hierarchical control of singing in birds. *Science*, 273, 1871–5.

Zahavi, A. 1979. Ritualization and the evolution of movement signals. *Behaviour*, 72, 77–81.

Zahavi, A. 1987. The theory of signal selection and some of its implications. In: *International Symposium of Biological Evolution* (ed. V. P. Delfino), pp. 305–27. Bari: Adriatica Editrice.

Zann, R. 1990. Song and call learning in wild zebra finches in south-east Australia. *Anim. Behav.*, 40, 811–28.

Zann, R. 1993. Structure, sequence and evolution of song elements in wild Australian zebra finches. *Auk*, 110, 702–15.

Zink, R. M. & Barrowclough, G. F. 1984. Allozymes and song dialects: a reassessment. *Evolution*, 38, 444–8.

Index

Printed in the United States
By Bookmasters